NUREG-1804
Revision 2

Yucca Mountain Review Plan

Final Report

U.S. Nuclear Regulatory Commission
Office of Nuclear Material Safety and Safeguards
Washington, DC 20555-0001

AVAILABILITY OF REFERENCE MATERIALS IN NRC PUBLICATIONS

NRC Reference Material

As of November 1999, you may electronically access NUREG-series publications and other NRC records at NRC's Public Electronic Reading Room at http://www.nrc.gov/reading-rm.html.
Publicly released records include, to name a few, NUREG-series publications; *Federal Register* notices; applicant, licensee, and vendor documents and correspondence; NRC correspondence and internal memoranda; bulletins and information notices; inspection and investigative reports; licensee event reports; and Commission papers and their attachments.

NRC publications in the NUREG series, NRC regulations, and *Title 10, Energy*, in the Code of *Federal Regulations* may also be purchased from one of these two sources.

1. The Superintendent of Documents
   U.S. Government Printing Office
   Mail Stop SSOP
   Washington, DC 20402–0001
   Internet: bookstore.gpo.gov
   Telephone: 202-512-1800
   Fax: 202-512-2250
2. The National Technical Information Service
   Springfield, VA 22161–0002
   www.ntis.gov
   1–800–553–6847 or, locally, 703–605–6000

A single copy of each NRC draft report for comment is available free, to the extent of supply, upon written request as follows:

Address: Office of the Chief Information Officer,
Reproduction and Distribution
Services Section
U.S. Nuclear Regulatory Commission
Washington, DC 20555-0001
E-mail: DISTRIBUTION@nrc.gov
Facsimile: 301–415–2289

Some publications in the NUREG series that are posted at NRC's Web site address http://www.nrc.gov/reading-rm/doc-collections/nuregs are updated periodically and may differ from the last printed version. Although references to material found on a Web site bear the date the material was accessed, the material available on the date cited may subsequently be removed from the site.

Non-NRC Reference Material

Documents available from public and special technical libraries include all open literature items, such as books, journal articles, and transactions, *Federal Register* notices, Federal and State legislation, and congressional reports. Such documents as theses, dissertations, foreign reports and translations, and non-NRC conference proceedings may be purchased from their sponsoring organization.

Copies of industry codes and standards used in a substantive manner in the NRC regulatory process are maintained at—

The NRC Technical Library
Two White Flint North
11545 Rockville Pike
Rockville, MD 20852–2738

These standards are available in the library for reference use by the public. Codes and standards are usually copyrighted and may be purchased from the originating organization or, if they are American National Standards, from—

American National Standards Institute
11 West 42nd Street
New York, NY 10036–8002
www.ansi.org
212–642–4900

Legally binding regulatory requirements are stated only in laws; NRC regulations; licenses, including technical specifications; or orders, not in NUREG-series publications. The views expressed in contractor-prepared publications in this series are not necessarily those of the NRC.

The NUREG series comprises (1) technical and administrative reports and books prepared by the staff (NUREG–XXXX) or agency contractors (NUREG/CR–XXXX), (2) proceedings of conferences (NUREG/CP–XXXX), (3) reports resulting from international agreements (NUREG/IA–XXXX), (4) brochures (NUREG/BR–XXXX), and (5) compilations of legal decisions and orders of the Commission and Atomic and Safety Licensing Boards and of Directors' decisions under Section 2.206 of NRC's regulations (NUREG–0750).

NUREG-1804
Revision 2

Yucca Mountain Review Plan

Final Report

Manuscript Completed: July 2003
Date Published: July 2003

Prepared by
Center for Nuclear Waste Regulatory Analyses
6220 Culebra Road
San Antonio, TX 78228-0510

Division of Waste Management
Office of Nuclear Material Safety and Safeguards
U.S. Nuclear Regulatory Commission
Washington, DC 20555-0001

J. A. Ciocco, NRC Project Manager

Prepared for
Division of Waste Management
Office of Nuclear Material Safety and Safeguards
U.S. Nuclear Regulatory Commission
Washington, DC 20555-0001

ABSTRACT

The Yucca Mountain Review Plan provides guidance for the U.S. Nuclear Regulatory Commission staff to evaluate a U.S. Department of Energy license application for a geologic repository. It is not a regulation and does not impose regulatory requirements. The licensing criteria are contained in the U.S. Code of Federal Regulations (CFR) Title 10, Part 63 (10 CFR Part 63), "Disposal of High-Level Radioactive Wastes in a Proposed Geologic Repository at Yucca Mountain, Nevada." The Secretary of Energy has recommended the Yucca Mountain site to the President for the development of a Yucca Mountain repository. The President has notified Congress that he considers the Yucca Mountain site qualified for application for a construction authorization for a repository. Nevada filed a notice of disapproval of the President's recommendation; however, Congress later approved the site recommendation. The U.S. Department of Energy may now submit a license application to the U.S. Nuclear Regulatory Commission. The principal purpose of the Yucca Mountain Review Plan is to ensure the quality, uniformity, and consistency of U.S. Nuclear Regulatory Commission staff reviews of the license application and any requested amendments. The Yucca Mountain Review Plan has separate sections for reviews of general information, repository safety before permanent closure, repository safety after permanent closure, the research and development program to resolve safety questions, the performance confirmation program, and administrative and programmatic requirements. Each section addresses determining compliance with specific regulatory requirements from 10 CFR Part 63. The regulations and the Yucca Mountain Review Plan are risk-informed, performance-based to the extent practical.

Draft Revision 2 of the Yucca Mountain Review Plan was made available for public comment in March 2002. This Final Revision 2 reflects revisions to address those comments, as appropriate.

CONTENTS

CONTENTS (continued)

Section Page

CONTENTS (continued)

Section Page

CONTENTS (continued)

CONTENTS (continued)

Section Page

CONTENTS (continued)

Section Page

CONTENTS (continued)

CONTENTS (continued)

Section Page

CONTENTS (continued)

FIGURES

EXECUTIVE SUMMARY

Disposal of high-level radioactive waste requires a U.S. Nuclear Regulatory Commission license. Part 63 under Title 10 of the U.S. Code of Federal Regulations ("Disposal of High-Level Radioactive Wastes in a Proposed Geologic Repository at Yucca Mountain, Nevada") contains the governing regulations. U.S. Nuclear Regulatory Commission authority to regulate a high-level waste repository comes from the Atomic Energy Act of 1954, as amended; the Energy Reorganization Act of 1974, as amended; and the Nuclear Waste Policy Act of 1982, as amended. The Yucca Mountain Review Plan is guidance to the U.S. Nuclear Regulatory Commission staff for review of any license application from the U.S. Department of Energy for a geologic repository for disposal of high-level radioactive waste at Yucca Mountain, Nevada. The U.S. Nuclear Regulatory Commission has directed the staff to carry out risk-informed, performance-based regulatory programs. 10 CFR Part 63 is risk-informed and performance-based, because risk of health effects to the reasonably maximally exposed individual is the basis for its performance objectives. 10 CFR Part 63 also requires protection of ground water by limiting the radioactivity in a representative volume of ground water and an assessment of repository performance under conditions of human intrusion. The U.S. Nuclear Regulatory Commission will base its licensing decision on whether the U.S. Department of Energy has demonstrated compliance with the performance objectives. Therefore, the Yucca Mountain Review Plan is risk-informed and performance-based.

The principal purpose of the Yucca Mountain Review Plan is to ensure the quality and uniformity of U.S. Nuclear Regulatory Commission staff licensing reviews. Yucca Mountain Review Plan sections present the areas of review, review methods, acceptance criteria, evaluation findings, and references the staff will use for its review. There are sections for reviews of general information, repository safety before permanent closure, repository safety after permanent closure, the research and development program to resolve safety questions, the performance confirmation program, and administrative and programmatic requirements. A summary of the risk-informed, performance-based foundation for each section follows.

An acceptance review is the first screening of the U.S. Department of Energy license application. The application must be complete and provide enough information to demonstrate compliance with the regulations. The reviewer will evaluate whether the information is sufficient to support a detailed review, and will assess the schedule for any later U.S. Nuclear Regulatory Commission milestones. The acceptance review does not determine the technical adequacy of the submitted information; that will be accomplished during a subsequent detailed technical review. U.S. Nuclear Regulatory Commission staff will send the results of the acceptance review, with a projected schedule for the rest of the review, to the U.S. Department of Energy within 90 days of receiving the license application. If the license application fails the acceptance review,, the staff will inform the U.S. Department of Energy in writing, that the application is not acceptable for processing or docketing and will inform the U.S. Department of Energy as to how the document is deficient.

During the detailed technical review, the general information section of the license application will be evaluated to determine if it provides a broad overview of the U.S. Department of Energy engineering design concept for the repository and allows the U.S. Department of Energy to demonstrate its understanding of which aspects of the Yucca Mountain site and its environs influence repository design and performance. More detailed technical information about the repository is in the Safety Analysis Report sections of the license application. The assessment of safety before permanent closure evaluates compliance with performance objectives to limit doses to workers and the public to acceptable risk levels. 10 CFR Part 63 requires the U.S. Department of Energy to demonstrate compliance using a preclosure safety analysis. A preclosure safety analysis systematically examines the site, the design, the potential hazards, initiating events and their consequences, and the potential dose consequences to workers and the public. The preclosure safety analysis considers the probabilities and uncertainties associated with potential hazards. The preclosure review will focus on information supplied by the U.S. Department of Energy to demonstrate that repository design, construction, and operation will meet the performance objectives (exposure limits). The U.S. Nuclear Regulatory Commission staff review will include risk-significant systems, structures, and components important to safety.

10 CFR Part 63 requires the U.S. Department of Energy to conduct a performance assessment to demonstrate compliance with postclosure performance objectives. A performance assessment systematically analyzes what can happen, the likelihood, and the consequences. The U.S. Nuclear Regulatory Commission staff will use risk information to focus its review on those items most important to waste isolation. The staff will examine the U.S. Department of Energy identification of natural and engineered barriers important to waste isolation. The staff will use risk insights from previous performance assessments for the Yucca Mountain site, detailed process-level modeling efforts, laboratory and field experiments, and natural analog studies. The staff will then evaluate the U.S. Department of Energy scenario analysis. The scenario analysis must consider the risk information from identified barriers and include the identification and screening of features, events, and processes, and the construction of scenarios from the retained features, events, and processes of the Yucca Mountain site. Finally, the performance assessment review will examine information on 14 model abstractions. The abstractions arose from engineered, geosphere, and biosphere subsystems shown to be most important to waste isolation, based on previous performance assessments, knowledge of site characteristics, and repository design. The staff review will focus on those models and abstracts that are most risk significant to repository safety. For the postclosure period, "important to safety" means important to meeting the radiation exposure performance objective. The risk of radiation health effects is the basis for the radiation exposure limit. The postclosure performance objectives also protect ground water by limiting the radioactivity in a representative volume of ground water and require an assessment of performance under conditions of human intrusion.

The review of the research and development program for resolving safety questions applies to systems, structures, and components important to safety, and engineered and natural barriers important to waste isolation. The program identifies, describes, and discusses safety features or components that require further information to confirm the adequacy of design. This will be a risk-informed review, because it focuses on those items most important to safety or waste isolation.

The review of the performance confirmation program examines the program of tests, experiments, and analyses the U.S. Department of Energy will conduct to evaluate the adequacy of the information used to demonstrate compliance with the performance objectives in 10 CFR Part 63. A performance confirmation program addresses uncertainties in estimating repository performance over thousands of years. This section is risk-informed and performance-based because it focuses on parameters and engineered and natural barriers important to health and safety.

10 CFR Part 63 provides no performance objectives for the administrative and programmatic sections of the Yucca Mountain Review Plan. Existing regulatory programs are the basis for acceptance criteria and review methods in this section and are identified. The acceptance criteria and review methods were prepared based on expected operations and associated risks. The quality assurance section of the Yucca Mountain Review Plan contains review methods and acceptance criteria to support a review of either a graded or nongraded quality assurance program. The U.S. Nuclear Regulatory Commission staff will conduct a review of the quality assurance program proposed by the U.S. Department of Energy.

1 REVIEW PLAN FOR GENERAL INFORMATION

Chapter 1, "General Information," reviews the requirements specified in 10 CFR 63.21(b). The intent of providing general information in the license application is twofold. First, it allows the U.S. Department of Energy to provide an overview of its engineering design concept for the repository (Section 1.1). Second, it allows the U.S. Department of Energy to demonstrate its understanding of what aspects of the Yucca Mountain site and its environs (Section 1.5) influence repository design and performance. Understanding the performance of the design, in the context of the Yucca Mountain site and its environs, allows the U.S. Department of Energy to make risk-informed, performance-based judgments regarding compliance with the regulations, which are subsequently evaluated by the U.S. Nuclear Regulatory Commission staff elsewhere in the Safety Analysis Report (Chapter 2). Accordingly, the material to be reviewed by the staff is generally informational in nature, with the more detailed technical discussions and descriptions found elsewhere in the Safety Analysis Report section of the license application. Overall, there are five sections in Chapter 1.

1.1 General Description

Review Responsibilities—High-Level Waste Branch and Environmental and Performance Assessment Branch

1.1.1 Areas of Review

This section reviews the general information to be included in the license application for the proposed geologic repository at Yucca Mountain. Reviewers will evaluate the information required by 10 CFR 63.21(b)(1).

The "General Information" section of the license application is expected to contain a broad overview that describes the proposed geologic repository at Yucca Mountain, including its major structures, systems, and components, as well as a discussion of proposed geologic repository operations area operations and activities. The level of detail presented should be similar to that in an executive summary. The material to be reviewed is informational in nature, with the more detailed technical discussions and descriptions found elsewhere in the safety analysis report section of the license application. Therefore, no detailed technical analysis of the information addressed in this section of the Yucca Mountain Review Plan is required. The detailed review of the information covered by these other technical subjects will be conducted under other sections of this review plan.

This review will address the following:

(1) A description of the facilities and location of structures, systems, and components of the geologic repository operations area, both surface and subsurface;

(2) A discussion of the proposed geologic repository operations area operations and activities; and

(3) The delineation of the regulatory basis for proposed geologic repository operations.

Review Plan for General Information

The "General Information" to be reviewed will be evaluated using the review methods and acceptance criteria found in Sections 1.1.2 and 1.1.3, respectively, of the Yucca Mountain Review Plan. In general, these review methods and acceptance criteria are based on well-established and accepted U.S. Nuclear Regulatory Commission regulatory activities. Because some of the information contained in this portion of the license application is informational in nature and may not concern performance-related issues, the review methods used to evaluate this information are not risk-informed, performance-based. In instances such as these, there will be no performance measures against which the review methods can be compared.

1.1.2 Review Methods

Review Method 1 Location and Arrangement of Structures, Systems, and Components of the Geologic Repository Operations Area

Confirm that the U.S. Department of Energy has provided an accurate general description of the geologic repository operations area. This general description, at a minimum, should include:

(1) A general discussion of the physical characteristics of the proposed repository site and environs critical to repository health and safety;

(2) Scaled drawings or maps, showing the location of the geologic repository operations area and its associated structures, systems, and components, including but not limited to, barriers, roads and connecting transportation infrastructure, utility services, and natural and man-made boundaries;

(3) A summary of the major design features of the above- and below-ground structures, systems, and components, with a designation of whether they are temporary or permanent;

(4) Those geologic repository operations area structures, systems, and components to be dismantled for the purposes of decommissioning and permanent closure;

(5) The identification and description of each major structure, system, and component of the geologic repository operations area, including a definition of the purpose of each and a description of the interrelationships among these structures, systems, and components;

(6) A general discussion of the plans to restrict access to the geologic repository operations area and to regulate land uses around the geologic repository operations area including requirements for ownership and control of interests in land (the detailed technical review of this information is addressed in Section 2.5.8 of the Yucca Mountain Review Plan);

(7) The identification and description of radiological monitoring instrumentation and activities, including the U.S. Department of Energy plans for the mitigation of radiological impacts associated with the construction and operation of the proposed repository; and

(8) Information that is consistent with the U.S. Department of Energy Final Environmental Impact Statement for Yucca Mountain and relevant updated information, if any.

Review Method 2 General Nature of the Geologic Repository Operations Area Activities

The staff should confirm that the U.S. Department of Energy has provided a summary description of the proposed geologic repository operations area operations. An acceptable summary description would include:

(1) Information on the types, kinds, and amounts of spent nuclear fuel and other high-level radioactive waste to be disposed of at the proposed repository;

(2) Information on routine waste package receipt, handling, and emplacement operations (the detailed technical review of this information will take place in Section 2.5.6 of the Yucca Mountain Review Plan);

(3) Description of plans for the inspection and testing of waste forms and waste packages as they are received at the geologic repository operations area (the detailed technical review of this information will take place in Section 2.5.6 of the Yucca Mountain Review Plan);

(4) Description of plans for the retrieval, and the alternative storage of, waste packages from emplacement drifts (the detailed technical review of this information will take place in Section 2.1.2 of the Yucca Mountain Review Plan);

(5) Description of plans for decommissioning and permanent closure of the geologic repository operations area (the detailed technical review of this information will take place in Section 2.1.3 of the Yucca Mountain Review Plan);

(6) A general discussion of proposed uses of the geologic repository operations area for purposes other than the disposal of spent nuclear fuel and other types of high-level radioactive waste (the detailed technical review of this information will take place in Section 2.5.9 of the Yucca Mountain Review Plan); and

(7) Description of plans for responses to emergencies. (The detailed technical review of this information is addressed in Section 2.5.7 of the Yucca Mountain Review Plan.)

In general, the reviewer should verify that the aforementioned summaries include adequate plans and procedures for the movement of personnel, materiel, and equipment during construction and normal operations.

Review Method 3 Basis for the Commission's Licensing Authority

The staff should verify that the license application describes the basis for the Commission's licensing authority that applies to the proposed activities at the geologic repository.

1.1.3 Acceptance Criteria

The following acceptance criteria are based on meeting the requirements of 10 CFR 63.21(b)(1), relating to the description of the general information.

Acceptance Criterion 1 The Location and Arrangement of the Geologic Repository Operations Area are Adequately Defined.

(1) A general but accurate description of the geologic repository operations area is provided. This description includes:

(a) A discussion of the physical characteristics of the site and the natural setting;

(b) Scaled drawings or maps showing the location of the geologic repository operations area and its associated structures, systems, and components;

(c) A summary of the design features of the above- and below-ground structures, systems, and components, with a designation of whether they are permanent or temporary;

(d) A definition of the purpose of each geologic repository operations area structure, system, and component, and any interrelationships among them;

(e) Plans to restrict access to, and to regulate land uses around, the geologic repository operations area; and

(f) A description of radiological monitoring instrumentation and activities, including the U.S. Department of Energy plans for the mitigation of radiological impacts associated with the construction and operation of the proposed repository.

Acceptance Criterion 2 The General Nature of the Activities to be Conducted at the Geologic Repository is Adequately Described.

(1) A summary description of the types, kinds, and amounts of spent nuclear fuel and other high-level radioactive waste to be disposed of is provided;

(2) A summary description of the proposed operations is provided that includes receipt, handling, emplacement, retrieval, of waste and waste packages. This description includes basic plans for the movement of personnel, material, and equipment during construction and normal operations;

(3) A description of plans for the inspection and testing of waste forms and waste package is provided;

(4) A description of plans for the retrieval and the alternative storage of radioactive wastes, should retrieval be necessary is provided;

(5) A description of plans for decommissioning and permanent closure of the geologic repository operations area is provided;

(6) A general discussion of proposed uses of the geologic repository operations area for purposes other than the disposal of spent nuclear fuel and other types of high-level radioactive waste is incorporated; and

(7) A description of plans for responses to emergencies is provided.

Acceptance Criterion 3 An Adequate Basis for the Exercise of the U.S. Nuclear Regulatory Commission Licensing Authority is Provided.

(1) The license application describes the basis for the Commission's licensing authority that applies to the proposed activities at the geologic repository.

1.1.4 Evaluation Findings

If the license application provides sufficient information and the regulatory acceptance criteria in Section 1.1.3 are appropriately satisfied, the staff concludes that this portion of the staff evaluation is acceptable. The reviewer writes material suitable for inclusion in the safety evaluation report prepared for the entire application. The report includes a summary statement of what was reviewed and why the reviewer finds the submittal acceptable. The staff can document the review as follows.

U.S. Nuclear Regulatory Commission staff has reviewed the "General Information" and other information submitted in support of the license application and has found, with reasonable assurance, that the requirements of 10 CFR 63.21(b)(1) are satisfied. An adequate general description of the geologic repository has been provided that identifies the location of the geologic repository operations area, discusses the general character of the proposed activities at the geologic repository operations area, and provides the basis for the exercise of the Commission's licensing authority.

1.1.5 References

None.

1.2 Proposed Schedules For Construction, Receipt, and Emplacement of Waste

Review Responsibilities—High-Level Waste Branch and Environmental and Performance Assessment Branch

1.2.1 Areas of Review

This section reviews proposed schedules for construction, receipt, and emplacement of waste. Reviewers will evaluate the information required by 10 CFR 63.21(b)(2).

Review Plan for General Information

The staff will evaluate the following parts of proposed schedules for construction, receipt, and emplacement of waste, using the review methods and acceptance criteria in Sections 1.2.2 and 1.2.3.

The material to be reviewed is informational in nature, and no detailed technical analysis is required. Because some of the information contained in this portion of the license application is informational in nature and may not concern health and safety-related issues, some of the review methods used to evaluate this information are not risk-informed, performance-based. In instances such as these, there will be no performance measures against which the review methods can be compared. Also, the staff recognizes that schedules will evolve and become more detailed over time.

(1) Schedules for construction of structures, systems, and components of the geologic repository operations area (including development of requisite infrastructure both on- and off-site); and

(2) Proposed schedules for the receipt, handling, and emplacement of waste package canisters.

1.2.2 Review Methods

Review Method 1 Major Steps for the Completion of Each Significant Work Element

Verify that the schedules for each significant work element necessary for both on- and off-site construction (including infrastructure development) and the receipt and emplacement of waste provide an adequate description of planned project activities. Traditional project management techniques (i.e., critical-path method diagrams, Gantt charts) should be used to convey the necessary information. In evaluating the adequacy of project planning, recognize that scheduling will be a function of evolving circumstances and expect distant scheduling to be less detailed than near-term scheduling. This review of project planning schedules should include:

(1) Verifying that the schedules, time-scaled charts, or work progress flow charts are provided;

(2) Verifying that the scheduled time allocated for each work step and the identified interdependence of work steps are sufficient to provide an overall understanding of the geologic repository operations area and infrastructure construction and waste-emplacement operations; and

(3) Verifying that construction of geologic repository operations area facilities will be substantially complete before the proposed scheduled receipt and emplacement of wastes.

1.2.3 Acceptance Criteria

The following acceptance criterion is based on meeting the requirements of 10 CFR 63.21(b)(2) relating to proposed schedules for construction, receipt, and emplacement of waste.

Acceptance Criterion 1 Major Steps for the Completion of Each Significant Work Element are Adequately Described.

(1) Major steps for the completion of each significant work element during construction of geologic repository operations area facilities and the associated infrastructure are identified in the proposed schedule of activities;

(2) Major steps and activities associated with the receipt of and emplacement of wastes are identified in the proposed schedule of activities; and

(3) For each of the activities described in the various phases of geologic repository operations area operations and activities, an adequate description of planned overall project progress is provided. Specifically:

   (a) Schedules, work-flow diagrams, and other project-management planning tools are provided; and

   (b) The scheduled time allocated for each major activity and the identified interdependence of major activities are sufficient to provide an overall understanding of the geologic repository operations area and infrastructure construction and routine waste-emplacement operations.

1.2.4 Evaluation Findings

If the license application provides sufficient information and the regulatory acceptance criteria in Section 1.2.3 are appropriately satisfied, the staff concludes that this portion of the staff evaluation is acceptable. The reviewer writes material suitable for inclusion in the safety evaluation report prepared for the entire application. The report includes a summary statement of what was reviewed and why the reviewer finds the submittal acceptable. The staff can document the review as follows.

U.S. Nuclear Regulatory Commission staff has reviewed the “General Information” and other information submitted in support of the license application and has found, with reasonable assurance, that the requirements of 10 CFR 63.21(b)(2) are satisfied. The U.S. Department of Energy provides schedules for construction, receipt of waste, and waste emplacement at the geologic repository operations area that are sufficiently detailed to allow staff to evaluate the overall construction program for the geologic repository operations area and its infrastructure.

1.2.5 References

None.

1.3 Physical Protection Plan

This review determines with reasonable assurance whether the U.S. Department of Energy has provided a description of the detailed security measures for physical protection that is complete in the light of information that is reasonably available at the time of docketing and provides assurance that activities involving high-level radioactive waste do not present an unreasonable risk to the public health and safety. The description must include the design for physical protection, the licensee's safeguard contingency plan, and security organization and personnel training and qualification plan. The plan must list tests, inspections, audits and other means used to demonstrate compliance with requirements. The physical protection system should be designed to protect against a loss of control of the geologic repository operations area that could be sufficient to cause radiation exposure exceeding the dose defined in 10 CFR 72.106. Physical protection requirements for high-level radioactive waste at a geologic repository operations area are at 10 CFR 73.51. These regulations specify the requirements for a U.S. Nuclear Regulatory Commission-approved physical protection plan. In light of the terrorist attacks of September 11, 2001, the Commission has directed the staff to conduct a comprehensive reevaluation of U.S. Nuclear Regulatory Commission physical requirements. If this effort indicates that U.S. Nuclear Regulatory Commission regulations or requirements warrant revision, such changes would occur through a public rulemaking or other appropriate methods and, if necessary, the Yucca Mountain Review Plan would be revised accordingly.

Review Responsibilities—High-Level Waste Branch, Environmental and Performance Assessment Branch, and Office of Nuclear Security and Incident Responses

1.3.1 Areas of Review

This section reviews a description of the detailed security measures for the physical protection of high-level radioactive waste in accordance with 10 CFR 73.51. This must include the design for physical protection, the licensee's safeguards contingency plan, and security organization personnel training and qualification plan. It must list tests, inspections, audits, and other means to be used to demonstrate compliance with such requirements.

The reviewer should evaluate the U.S. Department of Energy submittal for an acceptable physical protection system that protects against a loss of control of the geologic repository operations area that could be sufficient to cause radiation exposure exceeding the dose as defined in 10 CFR 72.106. The reviewer should verify that the U.S. Department of Energy has described how the general performance requirements, the performance capabilities, and the specific measures included in 10 CFR 73.51 will be met through developing, implementing, and maintaining a physical protection system.

The staff will evaluate a description of the detailed security measures for the physical protection of high-level radioactive waste in accordance with 10 CFR 73.51. This must include the design for physical protection, the licensee's safeguards contingency plan, and security organization personnel training and qualification plan. It must list tests, inspections, audits, and other means to be used to demonstrate compliance with such requirements, using the review methods and acceptance criteria in Sections 1.3.2 and 1.3.3:

(1) Introduction and schedule for implementation;
(2) General performance objectives;
(3) Protection goal;
(4) Security organization;
(5) Physical barrier subsystems;

(6) Access control subsystems and procedures;
(7) Detection, surveillance, and alarm subsystems and procedures;
(8) Communication subsystems;
(9) Equipment operability and compensatory measures;
(10) Contingency and response plans and procedures; and
(11) Reporting of safeguards events.

1.3.2 Review Methods

Review Method 1 Geologic Repository Operations Area Description and Schedule for Implementation

Verify that the U.S. Department of Energy specifies the geologic repository operations area location. The U.S. Department of Energy should describe the geologic repository operations area facilities, the nature of the wastes to be disposed of, the geologic repository operations area layout, the surrounding area, and the surrounding terrain. Verify that the U.S. Department of Energy has included a map of the entire facility, and other maps and illustrations, to assess the physical protection plan. The U.S. Department of Energy should indicate on these maps the controlled area; the location of all buildings; the locations of physical protection systems, subsystems, and major components; the protected area; and all entry/exit points, entry/exit control points, alarm stations, and security posts.

Confirm that the U.S. Department of Energy has presented an adequate schedule for implementing the physical protection plan. High-level radioactive waste may not be stored or used at the geologic repository operations area until the physical protection system is implemented and operational.

Review Method 2 General Performance Objectives

Verify that the U.S. Department of Energy provides a description of the detailed security measures for the physical protection of high-level radioactive waste in accordance with 10 CFR 73.51. This must include the design for physical protection, the licensee's safeguards contingency plan, and security organization personnel training and qualification plan. It must list tests, inspections, audits, and other means to be used to demonstrate compliance with such requirements. Items to be verified include that:

(1) The U.S. Department of Energy has described, in general terms, how the physical protection system will provide assurance that activities involving high-level radioactive waste do not present an unreasonable risk to the public health and safety;

(2) The U.S. Department of Energy has adequately described how, through establishing, maintaining, and arranging a physical protection system, the general performance objective and requirements in 10 CFR 73.51 will be met;

(3) The U.S. Department of Energy has identified and adequately described those portions of the physical protection system for which redundant and diverse components and redundant and diverse subsystems and components are necessary to ensure

adequate performance, as required by 10 CFR 73.51(b)(2). In general terms, the U.S. Department of Energy should describe the subsystems and components to be used to provide this redundancy and diversity and the ways in which these subsystems and components are redundant and diverse; and

(4) The U.S. Department of Energy has adequately described how the physical protection system is designed, tested, and maintained to ensure its continual effectiveness, reliability, and availability.

Review Method 3 Protection Goal

Verify that the U.S. Department of Energy will protect against a loss of control of the geologic repository operations area that could cause radiation exposure exceeding the dose defined in 10 CFR 72.106. The U.S. Department of Energy should have established a physical protection strategy that would deny unauthorized access to areas of the geologic repository operations area which could result in a loss of control sufficient to cause radiation exposure exceeding the dose as described in 10 CFR 72.106. Confirm that the U.S. Department of Energy will maintain and update the physical protection plan to reflect any changes that are necessary to ensure the continual ability to protect against situations leading to loss of control of the geologic repository operations area.

Review Method 4 Security Organization

Verify that the U.S. Department of Energy has described an adequate security organization to manage, control, and implement the physical protection system, consistent with the physical protection plan and consistent with maintaining its effectiveness. The security organization will be acceptable if it is consistent with the requirements in 10 CFR 73.51(d); associated Appendixes B, C, and G of 10 CFR Part 73; and the following criteria:

(1) The U.S. Department of Energy has stated whether the security organization is employed directly by the U.S. Department of Energy or is a contractor to the U.S. Department of Energy. Verify, if the security organization is managed by a contractor, that the U.S. Department of Energy has described adequate written agreements, between the U.S. Department of Energy and contract guard force management, that will govern how the security force will meet requirements at 10 CFR 73.51(d) and Appendix B, "General Criteria for Security Personnel," to 10 CFR Part 73;

(2) The U.S. Department of Energy will provide adequate structure and management for the security organization. This should include both uniformed security personnel and other persons responsible for security-related functions, consistent with 10 CFR 73.51(d). The structure description should include each supervisory and management position with responsibilities and lines of authority to facility and corporate management. The security organization must provide for sufficient personnel each shift to monitor detection systems and to conduct surveillance, assessment, access control, and communications, to assure adequate response time against a security threat;

(3) The U.S. Department of Energy will review the physical protection program at least once every 24 months per 10 CFR 73.51(d)(12), by individuals who are independent of physical protection management and who have no direct responsibility for implementation of the physical protection program. The physical protection program review shall evaluate the effectiveness of the physical protection system, and of the liaison established with the designated response force or local law enforcement agency;

(4) The U.S. Department of Energy will have an approved Guard Force Training Plan, that meets 10 CFR Part 73, Appendix B, "General Criteria for Security Personnel," being in effect. The physical protection plan will train, equip, and qualify all members of the security organization to perform their security duties in accordance with 10 CFR Part 73, Appendix B, "General Criteria for Security Personnel," consistent with 10 CFR 73.51(d)(5); and

(5) The U.S. Department of Energy will maintain records, required by 10 CFR 73.51(d)(13), being maintained/retained and adequately describing how they will be maintained/retained.

Review Method 5 Physical Barrier Subsystems

A performance objective of physical barriers is to define areas within which authorized activities and conditions are permitted. Other barrier performance objectives are to channel persons, vehicles, and material to or from entry/exit control points; to delay or deny unauthorized penetration attempts by persons, vehicles, or material; to delay attempts to cause loss of control of the geologic repository operations area; to assist detection and assessment; and to permit a timely response by the security force or local law enforcement to prevent the intended act.

Verify that the U.S. Department of Energy has adequately described the physical barrier subsystems for the geologic repository operations area. This description will be acceptable if the U.S. Department of Energy physical protection plan is consistent with the following criteria:

(1) The U.S. Department of Energy will store high-level radioactive waste within a protected area. Access to material in the protected area shall require passage or penetration through two physical barriers—one barrier at the perimeter of the protected area, and one barrier offering substantial penetration resistance. The physical barrier at the perimeter of the protected area must be as defined in 10 CFR 73.2. The barrier offering substantial resistance to penetration must be adequately defined and described. The U.S. Department of Energy will install the protected area barrier fence, so that it cannot be lifted to allow an individual to crawl under it. The U.S. Department of Energy should describe any access points through the protected area barrier, the manner in which they are to be used, and the means to control and protect them to ensure the integrity of the barrier. Barriers designed to protect against the malevolent use of a vehicle are not required at the geologic repository operations area;

(2) The U.S. Department of Energy has adequately described the location and size of any geologic repository operations area isolation zones. The U.S. Department of Energy will establish isolation zones alongside physical barriers at the perimeter of the protected

area, being at least 6.1 meters [20 feet] wide and being maintained clear of obstacles or structures on either side of the barriers, to permit assessment consistent with 10 CFR 73.51(d)(1); and

(3) The U.S. Department of Energy has described the lighting system sufficiently to demonstrate that it will be adequate to ensure illumination for monitoring, observation, and assessment activities for exterior areas within the protected area. The illumination must be sufficient to assess unauthorized penetrations of, or activities within, the protected area, consistent with 10 CFR 73.51(d)(2). The U.S. Department of Energy should demonstrate acceptable emergency backup power for protected area lighting and security assessment if normal power is lost. Illumination should be maintained during all periods of darkness (not just during periods of assessment). The level of illumination should be sufficient for the security assessment means proposed; however, 10 CFR 73.51 defines no specific required illumination level. The reviewer should consider that the physical layout of the geologic repository operations area may complicate maintaining a consistent level of illumination throughout the protected area because of obstruction from such structures as storage casks.

Review Method 6 Access Control Subsystems and Procedures

The performance objectives of access authorization controls and procedures are to verify the identity of persons, vehicles, and materials, and to initiate timely response measures to deny unauthorized entries.

Verify that the U.S. Department of Energy will provide adequate access control subsystems for the geologic repository operations area. These subsystems will be acceptable if they are consistent with the requirements in 10 CFR 73.51(d)(9), and the following criteria:

(1) The U.S. Department of Energy will establish and maintain a personnel identification system to limit access only to authorized individuals. The personnel identification system should provide unique identification of individuals granted access to the protected area. A picture identification system using a driver's license photograph, a name badge system using a badge medium that is difficult to counterfeit, or facial recognition could be used. Use of facial recognition should be justified (e.g., long-term employment and small site population);

(2) The U.S. Department of Energy has described adequate procedures for control of points of personnel access into the protected area, consistent with 10 CFR 73.51(d)(9). These procedures should include a discussion of methods used to identify individuals and to verify individual authorization. Procedures should also describe techniques for conducting visual searches of individuals, vehicles, and hand-carried packages for explosives before entry into the protected area. If an individual can be positively identified, is authorized access, and has been searched for explosives without positive findings, then no escort is required. If the individual cannot meet any one of these three criteria, access to the protected area should be denied;

(3) The U.S. Department of Energy will establish and maintain a controlled lock system, to limit access to authorized individuals, consistent with 10 CFR 73.51(d)(7). Regulatory Guide 5.12, "General Use of Locks in the Protection and Control of Facilities and Special Nuclear Materials" (U.S. Nuclear Regulatory Commission, 1973) should be used as guidance for developing a controlled lock system; and

(4) The U.S. Department of Energy will retain the following documentation for 3 years after the record is made or until termination of the license: (I) a log of individuals granted access to the protected area; (ii) screening records of members of the security organization; (iii) a log of all patrols; (iv) a record of each alarm received, identifying the type of alarm, location, date, and time when received—and disposition of the alarm; and (v) the physical protection program review reports.

Review Method 7 Detection, Surveillance, and Alarm Subsystems and Procedures

The performance objectives of detection, surveillance, and alarm subsystems and procedures are to detect, assess, and communicate any unauthorized access or penetrations, or such attempts by persons, vehicles, or materials at the time of occurrence, so the response will prevent the unauthorized access or penetration.

Verify that the U.S. Department of Energy has adequate detection, surveillance, and alarm subsystems for the geologic repository operations area. These subsystems will be acceptable if they are consistent with the requirements in 10 CFR 73.51(d), and the following criteria:

(1) An adequate intrusion detection system will be installed in the isolation zone between the two barriers at the protected area perimeter, consistent with 10 CFR 73.51(d)(3). The U.S. Department of Energy will provide a volumetric intrusion-detection system capable of detecting an individual in the isolation zone of the protected area. The capabilities, installation, and testing of the intrusion-detection equipment should be consistent with Regulatory Guide 5.44, "Perimeter Intrusion Alarm Systems," Revision 3 (U.S. Nuclear Regulatory Commission, 1997);

(2) The location, construction, and characteristics of the central and secondary alarm stations are consistent with 10 CFR 73.51(d)(3). The U.S. Department of Energy will have all required alarms annunciate in a continuously manned central alarm station located within the protected area and in at least one other continuously manned independent on-site station. Continuous manning of alarm stations and methods used for annunciation of required alarms should be described, along with protection afforded the stations (both procedural and physical), so that a single act cannot remove the capability of calling for assistance or responding to an alarm. The reviewer should confirm that access to the alarm stations will be controlled on a need-to-know basis, and that the central alarm station will not contain any activities that would interfere with the alarm response. The annunciation systems at the alarm stations should indicate the status of all alarms and alarm zones in both alarm stations. The secondary location need only provide a summary indication that an alarm has been generated. The U.S. Department of Energy should follow the guidelines of Regulatory Guide 5.44,

"Perimeter Intrusion Alarm Systems," Revision 3 (U.S. Nuclear Regulatory Commission, 1997) for alarm annunciation;

(3) Detection systems and supporting subsystems must be tamper-indicating with line supervision. These systems and the surveillance/assessment and illumination systems must be maintained in operable condition; and

(4) The U.S. Department of Energy will monitor the protected area with daily random patrols, consistent with 10 CFR 73.51(d)(4). To evaluate the proposed frequency of random patrols, the reviewer should consider the remoteness of the geologic repository operations area, the nature of activities adjacent to the site, and the size of the geologic repository operations area.

Review Method 8 Communication Subsystems

The performance objective of communication subsystems is to notify of an attempted unauthorized intrusion, so response can prevent loss of control of the geologic repository operations area.

Verify that the U.S. Department of Energy will have adequate communications subsystems for the geologic repository operations area. The communications subsystems will be acceptable if they are consistent with the requirements in 10 CFR 73.51(d), and the following criteria:

(1) The individual in each continuously manned alarm station should be able to call for assistance from other guards and watchmen and from local law enforcement or designated response forces;

(2) Redundant systems should be used to ensure communications with the local law enforcement authority or designated response force, consistent with 10 CFR 73.51(d)(8); and

(3) The methods used to maintain communications systems in operable condition should be consistent with 10 CFR 73.51(d)(11).

Review Method 9 Equipment Operability and Compensatory Measures

The performance objective of test and maintenance procedures is to provide confidence that security equipment will be available and reliable to perform when needed.

Confirm that the U.S. Department of Energy will have adequate test and maintenance programs for the geologic repository operations area. The test and maintenance programs will be acceptable if they are consistent with the requirements in 10 CFR 73.51(d), and the U.S. Department of Energy will provide a testing program for the perimeter intrusion detection system consistent with Regulatory Guide 5.44, Revision 3 (U.S. Nuclear Regulatory Commission, 1997).

Review Method 10 Contingency and Response Plans and Procedures

The performance objective for contingency response plans and procedures is to provide predetermined response to safeguards contingency events, so the adversary will be engaged and impeded until off-site assistance arrives.

Verify that the U.S. Department of Energy has adequate contingency and response plans for the geologic repository operations area. The contingency and response plans will be acceptable if the U.S. Department of Energy plans are consistent with the requirements in 10 CFR 73.51(d)(10), Appendix C to 10 CFR Part 73, and the following criteria:

(1) The U.S. Department of Energy will develop a safeguards contingency plan for unauthorized penetrations of, or activities within, the protected area, that includes the Category 5, "Procedures," of Appendix C to 10 CFR Part 73, consistent with 10 CFR 73.51(d)(10); and

(2) The U.S. Department of Energy will have adequate documented response arrangements with designated response force or local law enforcement agencies, consistent with the requirements of 10 CFR 73.51(d)(6). The designated response force could be a privately contracted security force that meets the requirements of Appendix B to 10 CFR Part 73. If the designated response force cannot respond quickly enough, additional protective measures may be required, including the use of armed guards.

Review Method 11 Reporting of Safeguards Events

Verify that the U.S. Department of Energy will report safeguards events to the U.S. Nuclear Regulatory Commission, consistent with the criteria in 10 CFR Part 73, Appendix G, "Reportable Safeguards."

1.3.3 Acceptance Criteria

The following acceptance criteria are based on meeting the requirements of 10 CFR 63.21(b)(3), relating to the physical protection plan.

Acceptance Criterion 1 The Physical Protection Plan Contains an Adequate Geologic Repository Operations Area Description and Provides an Acceptable Schedule for Implementation.

(1) The physical protection plan adequately specifies the location of the geologic repository operations area, the geologic repository operations area facilities, the nature of the wastes to be disposed of, the geologic repository operations area layout, the surrounding area, and the surrounding terrain. Adequate maps are provided to support the physical protection plan; and

(2) An acceptable schedule is provided for implementing the physical protection plan. High-level radioactive waste will not be stored or used at the geologic repository

operations area facility, until the physical protection system is implemented and operational.

Acceptance Criterion 2 General Performance Objectives Will be Met.

(1) The physical protection system will provide assurance that activities involving high-level radioactive waste do not present an unreasonable risk to the public health and safety;

(2) Through establishing, maintaining, and arranging a physical protection system, the general performance objective and requirements prescribed in 10 CFR 73.51 will be met;

(3) Those portions of the physical protection system for which redundant and diverse components, and redundant and diverse subsystems and components, are necessary to ensure adequate performance, will meet the requirements of 10 CFR 73.51(b)(2); and

(4) The physical protection system will be designed, tested, and maintained to ensure its continual effectiveness, reliability, and availability.

Acceptance Criterion 3 The Protection Goal Will be Met.

The physical protection system will be designed to protect against a loss of control of the geologic repository operations area that could cause radiation exposure exceeding the dose defined in 10 CFR 72.106. The U.S. Department of Energy will have a physical protection strategy that will deny unauthorized access to areas of the geologic repository operations area which could result in a loss of control sufficient to cause radiation exposure exceeding the dose as described in 10 CFR 72.106. The U.S. Department of Energy will maintain and update the physical protection plan to reflect any changes that are necessary to ensure the continual ability to protect against situations leading to loss of control of the geologic repository operations area.

Acceptance Criterion 4 The Security Organization Will be Adequate.

The U.S. Department of Energy has an adequate security organization to manage, control, and implement the physical protection system consistent with the physical protection plan and will continually maintain its effectiveness.

(1) The U.S. Department of Energy has stated whether the security organization is employed directly by the U.S. Department of Energy or is a contractor to the U.S. Department of Energy. The U.S. Department of Energy has adequate written agreements between the U.S. Department of Energy and the contract guard force;

(2) The U.S. Department of Energy has an adequate structure and management for the security organization, including both uniformed security personnel and other persons responsible for security-related functions. The security organization provides for sufficient personnel each shift to monitor detection systems and to conduct surveillance, assessment, access control, and communications to assure adequate response time against security threats;

(3) The U.S. Department of Energy will review the physical protection program at least once every 24 months using individuals who are independent of physical protection management, and who have no direct responsibility for implementation of the physical protection program. The physical protection program review will evaluate the effectiveness of the physical protection system, and of the liaison established with the designated response force or local law enforcement agency;

(4) The U.S. Department of Energy will establish an adequate guard force training plan. The physical protection plan will properly train, equip, and qualify members of the security organization to perform their security duties; and

(5) The U.S. Department of Energy will adequately maintain the records required by 10 CFR 73.51(d)(13).

Acceptance Criterion 5 Physical Barrier Subsystems Will be Adequate.

The physical barriers will control areas within which authorized activities and conditions are permitted. The barriers will channel persons, vehicles, and material to or from entry/exit control points; will delay or deny unauthorized penetration attempts by persons, vehicles, or material; will delay any attempts to cause loss of control of the geologic repository operations area; will assist detection and assessment; and will permit a timely response by the security force or local law enforcement to prevent the intended act.

The U.S. Department of Energy has adequate physical barrier subsystems at the geologic repository operations area.

(1) High-level radioactive waste will be stored only within a protected area. Access to material in the protected area will require passage or penetration through two physical barriers; one barrier at the perimeter of the protected area, and one barrier offering substantial penetration resistance. The physical barrier at the perimeter of the protected area will be as defined in 10 CFR 73.2. The barrier offering substantial resistance to penetration is adequately defined and described. The U.S. Department of Energy will install the protected area barrier fence, so that it cannot be lifted to allow an individual to crawl under it. Access points through the protected area barrier, the manner in which they are to be used, and the means to control and protect them to ensure the integrity of the barrier are adequately described;

(2) The location and size of any geologic repository operations area isolation zones are adequately defined. The isolation zones adjacent to the physical barriers at the perimeter of the protected area will be at least 6.1 meters [20 feet] wide, and will be maintained clear of obstacles or structures on either side of the barriers, to permit assessment consistent with 10 CFR 73.51(d)(1); and

(3) The U.S. Department of Energy has described the lighting system sufficiently to demonstrate that it will be adequate to ensure illumination for monitoring, observation, and assessment activities for exterior areas within the protected area. The illumination will be sufficient to permit assessment of unauthorized penetrations of, or activities

within, the protected area, consistent with 10 CFR 73.51(d)(2). The U.S. Department of Energy demonstrates that there will be acceptable emergency backup power for protected area lighting and security assessment capability if normal power is lost. Illumination will be maintained during all periods of darkness. The level of illumination will be sufficient for the security assessment means proposed.

Acceptance Criterion 6 Access Control Subsystems and Procedures Will be Adequate.

Controls and procedures are adequate to verify the identity of persons, vehicles, and materials, and to initiate timely response measures to deny unauthorized entries.

The U.S. Department of Energy will provide adequate access control subsystems for the geologic repository operations area.

(1) The U.S. Department of Energy will establish and maintain an adequate personnel identification system to limit access only to authorized individuals. The personnel identification system will provide unique identification of individuals granted access to the protected area;

(2) The U.S. Department of Energy will provide adequate procedures for control of points of personnel access into the protected area. These procedures will include appropriate methods to identify individuals and to verify individual authorization, and techniques for conducting visual searches of individuals, vehicles, and hand-carried packages for explosives before entry into the protected area;

(3) The U.S. Department of Energy will employ an adequate controlled lock system to limit access to authorized individuals, consistent with 10 CFR 73.51(d)(7); and

(4) The U.S. Department of Energy will maintain adequate records of access control.

Acceptance Criterion 7 Detection, Surveillance, and Alarm Subsystems and Procedures Will be Adequate.

Detection, surveillance, and alarm subsystems and procedures will be adequate to detect, assess, and communicate any unauthorized access or penetrations, or such attempts by persons, vehicles, or materials at the time of the act or the attempt, so the response can prevent the unauthorized access or penetration.

The U.S. Department of Energy has adequate detection, surveillance, and alarm subsystems for the geologic repository operations area.

(1) An adequate intrusion-detection system will be installed in the isolation zone between the two barriers at the protected area perimeter;

(2) The location, construction, and characteristics of the central and secondary alarm stations are consistent with 10 CFR 73.51(d)(3). The U.S. Department of Energy will have all required alarms annunciate in a continuously manned central alarm station

located within the protected area, and in at least one other continuously manned independent on-site station. The U.S. Department of Energy will provide continuous manning of alarm stations, and methods used for annunciation of required alarms are adequate, so that a single act cannot remove the capability of calling for assistance or responding to an alarm. Access to the alarm stations will be controlled on a need-to-know basis, and the central alarm station will not contain any operational activities that would interfere with the execution of alarm response functions. The annunciation systems at the alarm stations will indicate the status of all alarms and alarm zones in both alarm stations;

(3) Detection systems and supporting subsystems will be tamper-indicating with line supervision. These systems and the surveillance/assessment and illumination systems will be maintained in operable condition; and

(4) The protected area will be monitored with adequate daily random patrols.

Acceptance Criterion 8 Communication Subsystems Will be Adequate.

The communication subsystems will provide adequate notification of an attempted unauthorized intrusion, so that response can prevent loss of control of the geologic repository operations area.

The U.S. Department of Energy will have adequate communications subsystems for the geologic repository operations area.

(1) The individual in each continuously manned alarm station will be capable of calling for assistance from other guards and watchmen and from local law enforcement authorities or a designated response force;

(2) Redundant systems will be used to ensure the capability of communications with the local law enforcement authority or a designated response force; and

(3) The methods used to maintain communications systems in operable condition are adequate.

Acceptance Criterion 9 Equipment Operability and Compensatory Measures are Adequate.

Test and maintenance procedures provide adequate confidence that security equipment will be available and reliable to perform when needed.

The U.S. Department of Energy will have adequate test and maintenance programs for the geologic repository operations area physical protection systems.

Acceptance Criterion 10 Contingency and Response Plans and Procedures Will be Adequate.

Contingency response plans and procedures will provide adequate predetermined response to safeguards contingency events, so that the adversary will be engaged and impeded until off-site assistance arrives.

The U.S. Department of Energy has adequate contingency and response plans for the geologic repository operations area.

(1) The U.S. Department of Energy will provide an adequate safeguards contingency plan for dealing with unauthorized penetrations of, or activities within, the protected area; and

(2) The U.S. Department of Energy will have adequate documented response arrangements with designated response force or local law enforcement agencies.

Acceptance Criterion 11 Reporting of Safeguards Events Will be Adequate.

The U.S. Department of Energy will provide adequate reporting of safeguards events to the U.S. Nuclear Regulatory Commission.

1.3.4 Evaluation Findings

If the license application provides sufficient information and the regulatory acceptance criteria in Section 1.3.3 are appropriately satisfied, the staff concludes that this portion of the staff evaluation is acceptable. The reviewer writes material suitable for inclusion in the safety evaluation report prepared for the entire application. The report includes a summary statement of what was reviewed and why the reviewer finds the submittal acceptable. The staff can document the review as follows.

U.S. Nuclear Regulatory Commission staff has reviewed the Safety Analysis Report and other information submitted in support of the license application and has found, with reasonable assurance, that the requirements of 10 CFR 63.21(b)(3) are satisfied. The U.S. Department of Energy will implement an adequate physical protection program for high-level radioactive waste that includes physical protection, a safeguards contingency plan, and a security organization personnel training and qualification plan that complies with 10 CFR 73.51 of this chapter.

1.3.5 References

U.S. Nuclear Regulatory Commission. Regulatory Guide 5.44, “Perimeter Intrusion Alarm Systems.” Revision 3. Washington, DC: U.S. Nuclear Regulatory Commission, Office of Standards Development. October 1997.

———. Regulatory Guide 5.12, “General Use of Locks in the Protection and Control of Facilities and Special Nuclear Materials.” Washington, DC: U.S. Nuclear Regulatory Commission, Office of Standards Development. November 1973.

1.4 Material Control and Accounting Program

This review is to verify the U.S. Department of Energy material control and accounting plan describes, establishes, implements, and maintains a program adequate to protect against, detect, and respond to loss of high-level radioactive waste. Material control and accounting requirements for high-level radioactive waste are required by 10 CFR 63.21(b)(4) and stipulated in 10 CFR 63.78. At the construction authorization stage, the U.S. Department of Energy is required to submit a description of the material control and accounting program to meet the requirements of 10 CFR 63.78 and, that includes design basis information, assesses any potential impact of the material control and accounting program on design features, and describes physical aspects of the material control and accounting program.

Review Responsibilities—High-Level Waste Branch, Division of Fuel Cycle Safety and Safeguards, and Environmental and Performance Assessment Branch

1.4.1 Areas of Review

This section reviews the material control and accounting program. Reviewers will evaluate the information required by 10 CFR 63.21(b)(4).

The staff will evaluate the following parts of the material control and accounting program, using the review methods and acceptance criteria in Sections 1.4.2 and 1.4.3.

(1) Material balance, inventory, and records and procedures for stored high-level radioactive waste;

(2) Procedures for preparing accidental criticality or loss of special nuclear material reports;

(3) Procedures for preparing material status reports; and

(4) Procedures for preparing nuclear material transfer reports.

1.4.2 Review Methods

Review Method 1 Material Balance, Inventory, and Record-Keeping Procedures

Verify that the material control and accounting plan establishes the bases for identifying, controlling, and accounting for the nuclear materials that the U.S. Department of Energy will be authorized to possess at the geologic repository operations area.

Verify records will adequately document the receipt, inventory (including location), disposal, acquisition, and transfer of spent nuclear fuel and high-level radioactive waste, including provision to maintain inventory during any retrieval operations. Information on the waste form, proposed waste package, characteristics of any encapsulation material, radionuclide characteristics, heat generation rate, and material control history should be provided in these records. Confirm procedures require that records be maintained for as long as the material is stored, and for 5 years after the repository is closed. Verify that the following minimum information will be included in the retained records per 10 CFR 72.72(a), unless the Commission directs otherwise:

(1) Name of shipper;

(2) Estimated quantity of radioactive material per item, including high-level radioactive waste;

(3) Item identification and seal number;

(4) Storage or emplacement location;

(5) On-site movement of each fuel assembly or storage canister; and

(6) Ultimate disposal.

Confirm that a physical inventory of spent nuclear fuel and high-level radioactive waste in storage will be made at intervals not to exceed 12 months (unless directed otherwise by the U.S. Nuclear Regulatory Commission). The U.S. Department of Energy will retain a copy of the current inventory until the U.S. Nuclear Regulatory Commission terminates the license.

Verify that policies, practices, and procedures are designed and implemented to ensure the quality of physical inventories, and the control and maintenance of records and documentation associated with the physical inventories. A copy of the current inventory should be maintained until the U.S. Nuclear Regulatory Commission terminates the license.
Confirm that written material control and accounting procedures, sufficient for the U.S. Department of Energy to account for the material in storage, will be established, maintained, and followed. The U.S. Department of Energy will retain a copy of the current material control and accounting procedures until the U.S. Nuclear Regulatory Commission terminates the license.

Verify that checks and balances in the material control and accounting system ensure that falsification of data and reports that could conceal a diversion of high-level radioactive waste by employees acting individually, or in collusion, will be readily detected.

Confirm that records of spent nuclear fuel or high-level radioactive waste in storage will be in duplicate. Duplicate sets of records should be at separate locations, so a single event will not destroy both sets. The U.S. Department of Energy will preserve records of spent nuclear fuel or high-level radioactive waste transferred out of the geologic repository operations area for a minimum of 5 years after transfer.

Review Method 2 Reports of Accidental Criticality or Loss of Special Nuclear Material

Verify that any loss is considered and incorporated in a collusion protection program designed to thwart attempts from an insider to divert special nuclear material.

Verify that procedures ensure that anomalies (off-normal or abnormal situations), suggesting a likelihood that special nuclear material may be missing (whether or not the cause is assumed deliberate), are promptly and accurately reported to the U.S. Nuclear Regulatory Commission.

Confirm that the anomaly reporting system is able to respond promptly to alarms indicating potential loss of special nuclear material and discrimination of actual loss or system error is readily determined. Verify appropriate remedial action is planned, verified, and reported after alarms are tripped.

Confirm adequate procedures for reporting accidental criticality or loss of special nuclear material to the U.S. Nuclear Regulatory Commission Operations Center, using the Emergency Notification System. If this system is inoperable, commercial telephone, other dedicated telephonic service, or any means that assures the U.S. Nuclear Regulatory Commission receipt of the report may be used. Reports should be made within one hour of the discovery of accidental criticality or any loss of special nuclear material.

Review Method 3 Procedures for Preparation of Material Status Reports

Verify whether procedures that require a material status report will be completed, in computer-readable format, and submitted to the U.S. Nuclear Regulatory Commission in accordance with instructions in NUREG/BR–0007 (U.S. Nuclear Regulatory Commission, 2000a) and Nuclear Materials Management and Safeguards Report D–24, "Personal Computer Data Input for U.S. Nuclear Regulatory Commission Licensees" (U.S. Nuclear Regulatory Commission, 1994). Information on special nuclear material contained in the spent nuclear fuel possessed, received, transferred, disposed of, or lost by the licensee should be reported. Confirm procedures require material status reports as of March 31 and September 30 of each year, to be filed within 30 days after the end of the period covered by the report, unless otherwise specified by the U.S. Nuclear Regulatory Commission or by 10 CFR 75.35, pertaining to implementation of the United States/International Atomic Energy Agency Safeguards Agreement.

Review Method 4 Procedures for Preparation of Nuclear Material Transfer Reports

Confirm that the U.S. Department of Energy establishes auditable records sufficient to demonstrate reporting requirements have been met. Verify procedures specify forms of records and adequate safeguards to ensure the integrity of records. Verify procedures require that whenever spent nuclear fuel is transferred or received, a Nuclear Material Transaction Report will be completed, in computer-readable format, in accordance with instructions in NUREG/BR–0006 (U.S. Nuclear Regulatory Commission, 2000b) and Nuclear Materials Management and Safeguards System Report D–24, "Personal Computer Data Input for U.S. Nuclear Regulatory Commission Licensees" (U.S. Nuclear Regulatory Commission, 1994), as required by 10 CFR 72.78.

1.4.3 Acceptance Criteria

The following acceptance criteria are based on meeting the requirements of 10 CFR 63.78, relating to the material control and accounting program achieving the system capabilities stipulated by 10 CFR 72.72, 72.74, 72.76, and 72.78.

Acceptance Criterion 1 Material Balance, Inventory, and Record-Keeping Procedures for Spent Nuclear Fuel and High-Level Radioactive Waste Are Adequate.

(1) The material control and accounting plan establishes the basis for identifying, controlling, and accounting for the nuclear materials that the U.S. Department of Energy will be authorized to possess;

(2) Records adequately document the receipt, inventory (including location), disposal, acquisition, and transfer of spent nuclear fuel and high-level radioactive waste, including provision to maintain inventory during any retrieval operations. Adequate information on the waste form, proposed waste package, characteristics of any encapsulation material, radionuclide characteristics, heat generation rate, and history is provided. The procedures require that records be maintained for as long as the material is stored and for 5 years after the repository is closed, per 10 CFR 72.72(a), unless the Commission directs otherwise. The information in the retained records will include:

  (a) Name of shipper;

  (b) Estimated quantity of radioactive material per item, including high-level radioactive waste;

  (c) Item identification and seal number;

  (d) Storage or emplacement location;

  (e) On-site movement of each fuel assembly or storage canister; and

  (f) Ultimate disposal.

(3) A physical inventory of spent nuclear fuel and high-level radioactive waste in storage will be made at intervals not to exceed 12 months (unless directed otherwise by the U.S. Nuclear Regulatory Commission);

(4) Adequate policies, practices, and procedures are designed and implemented to ensure the quality of physical inventories, and the control and maintenance of records and documentation associated with the physical inventories. A copy of the current inventory will be retained until the U.S. Nuclear Regulatory Commission terminates the license;

(5) Written material control and accounting procedures sufficient for the U.S. Department of Energy to account for the material in storage are established, maintained, and followed.

A copy of the current material control and accounting procedures will be retained until the U.S. Nuclear Regulatory Commission terminates the license;

(6) The material control and accounting system incorporates checks and balances sufficient to detect falsification of data and reports that could conceal a possible diversion of high-level radioactive waste by employees acting individually or in collusion; and

(7) Records of spent nuclear fuel or high-level radioactive waste in storage are in duplicate. Duplicate sets of records are kept at separate locations, so a single event will not destroy both sets. Records of spent nuclear fuel or high-level radioactive waste transferred out of the facility will be preserved for a minimum of 5 years after transfer.

Acceptance Criterion 2 Procedures Are Adequate to Ensure Timely Reports of Accidental Criticality or Loss of Special Nuclear Material.

(1) The U.S. Department of Energy will have an adequate collusion protection program to thwart attempts from an insider to divert special nuclear material;

(2) The U.S. Department of Energy will report to the U.S. Nuclear Regulatory Commission any anomalies (off-normal or abnormal conditions or situations) suggesting a likelihood that special nuclear material may be missing (whether or not the cause is deliberate);

(3) The U.S. Department of Energy anomaly reporting system is able to respond promptly to alarms indicating a potential loss of special nuclear material, and allows determination of whether the unusual observable condition is caused by an actual loss or by a system error. The reporting procedure and resolution program will identify the type of system error or innocent cause, so remedial action can be taken. The response will be timely to ensure that indicators that might result from diversion, loss or other misuse, are investigated and resolved promptly; and

(4) Procedures for reporting accidental criticality or loss of special nuclear material to the U.S. Nuclear Regulatory Commission Operations Center, using the Emergency Notification System, are adequate. If this system is inoperable, commercial telephone, other dedicated telephonic service, or any means that assures the U.S. Nuclear Regulatory Commission receipt of the report may be used. Reports should be made within 1 hour of the discovery of accidental criticality or any loss of special nuclear material.

Acceptance Criterion 3 Procedures for Preparation of Material Status Reports Are Adequate.

(1) Procedures require that a material status report be completed, in computer-readable format, and submitted to the U.S. Nuclear Regulatory Commission, in accordance with instructions in NUREG/BR–0007 (U.S. Nuclear Regulatory Commission, 2000a) and Nuclear Materials Management and Safeguards System Report D–24, "Personal Computer Data Input for U.S. Nuclear Regulatory Commission Licensees" (U.S. Nuclear Regulatory Commission, 1994). Information on the amount of spent nuclear fuel

possessed, received, transferred, disposed of, or lost by the licensee will be reported. Procedures require material status reports as of March 31 and September 30 of each year, to be filed within 30 days after the end of the period covered by the report, unless otherwise specified by the U.S. Nuclear Regulatory Commission or by 10 CFR 75.35, pertaining to implementation of the United States/International Atomic Energy Agency Safeguards Agreement.

Acceptance Criterion 4 Procedures for Preparation of Nuclear Material Transfer Reports Are Adequate.

(1) The U.S. Department of Energy will establish auditable records sufficient to demonstrate that reporting requirements have been met. In addition, each record pertaining to receipt and disposal of spent nuclear fuel will be retained until the Commission terminates the license;

(2) The procedures specify in what form those records will be kept;

(3) The procedures provide adequate safeguards against tampering with and loss of records; and

(4) Procedures require that whenever spent nuclear fuel is transferred or received, a Nuclear Material Transaction Report is completed, in computer-readable format, in accordance with instructions in NUREG/BR–0006 (U.S. Nuclear Regulatory Commission, 2000b) and Nuclear Materials Management and Safeguards System Report D–24, "Personal Computer Data Input for U.S. Nuclear Regulatory Commission Licensees" (U.S. Nuclear Regulatory Commission, 1994), as required by 10 CFR 72.78.

1.4.4 Evaluation Findings

If the license application provides sufficient information and the regulatory acceptance criteria in Section 1.4.3 are appropriately satisfied, the staff concludes that this portion of the staff evaluation is acceptable. The reviewer writes material suitable for inclusion in the safety evaluation report prepared for the entire application. The report includes a summary statement of what was reviewed and why the reviewer finds the submittal acceptable. The staff can document the review as follows.

U.S. Nuclear Regulatory Commission staff has reviewed the Safety Analysis Report and other information submitted in support of the license application and has found, with reasonable assurance, that the requirements of 10 CFR 63.78 are satisfied. The U.S. Department of Energy has established a material control and accounting program that meets the requirements of 10 CFR 72.72, 72.74, 72.76, and 72.78.

1.4.5 References

U.S. Nuclear Regulatory Commission. NUREG/BR–0006, "Instructions for Completing Nuclear Material Transfer Reports." Revision 4. Washington, DC: U.S. Nuclear Regulatory Commission. February 2000a.

———. NUREG/BR–0007, “Instructions for the Preparation and Distribution of Material Status Reports.” Revision 3. Washington, DC: U.S. Nuclear Regulatory Commission. February 2000b.

———. “Personal Computer Data Input for U.S. Nuclear Regulatory Commission Licensees.” Nuclear Materials Management and Safeguards System Report D–24. Washington, DC: U.S. Nuclear Regulatory Commission. May 1994.

1.5 Description of Site Characterization Work

Review Responsibilities—High-Level Waste Branch and Environmental and Performance Assessment Branch

1.5.1 Areas of Review

This section reviews the description and results of site characterization work performed at Yucca Mountain that support the technical discussions and descriptions found elsewhere in the Safety Analysis Report. The reviewers will evaluate the information required by 10 CFR 63.21(b)(5).

The level of detail presented in this section of the license application should be similar to that in an executive summary. The material to be reviewed is informational in nature, with the more detailed technical discussions and descriptions found elsewhere in the Safety Analysis Report section of the license application. Therefore, no detailed technical analysis of the information contained in this section of the license application is required. The detailed review of the information covered by these technical subjects will be conducted using other sections of the Yucca Mountain Review Plan.

The staff will review the following parts of the description and results of site characterization work, using the review methods and acceptance criteria in Sections 1.5.2 and 1.5.3:

(1) Geology;
(2) Hydrology;
(3) Geochemistry;
(4) Geotechnical properties and conditions of the host rock;
(5) Climatology, meteorology, and other environmental sciences; and
(6) Reference biosphere definition.

Because the information reviewed using this section of the Yucca Mountain Review Plan is generally informational in nature, the review methods are not risk-informed, performance-based. In instances such as these, there are no performance measures against which the information can be compared.

1.5.2 Review Methods

Review Method 1 Description of Site Characterization Activities

Confirm that site characterization activities have been described in the "General Information" section of the license application. This general description, at a minimum, should include site-specific information in the following areas:

(1) Geology;
(2) Hydrology;
(3) Geochemistry;
(4) Geotechnical properties and conditions of the host rock;
(5) Climatology, meteorology, and other environmental sciences; and
(6) Reference biosphere definition.

Review Method 2 Summary of Site Characterization Results

Confirm that the results of site characterization activities have been described in the "General Information" section of the license application. An acceptable summary description should include areas such as:

(1) An overview of geology, consistent with other site characterization summaries, that includes:

 (a) A description of the physical setting of the site, including the major physiographic and geological features;

 (b) A description of the principal rock units at the surface and in the subsurface, and their stratigraphic relationships;

 (c) A description and location of potentially important stratigraphic and structural features (such as faults, fractures, and joint sets and systems);

 (d) A description of geotechnical properties of stratigraphic units involved in the operation and safety of the proposed repository;

 (e) The delineation of the proposed geologic system to be used in estimating the performance of the proposed repository;

 (f) A summary of regional geomorphic, tectonic, seismic, and volcanic models (i.e., conceptual, technical basis, interpretation of data), with particular emphasis on those features, events, and processes that may have an effect on repository operations and safety;

 (g) The identification of potential geologic hazards requiring complex engineering measures;

(h) A summary evaluation of seismic seismicity; and

(i) A summary evaluation of volcanic activity.

(2) An overview of hydrology consistent with other site characterization summaries that includes:

(a) A description of hydrogeologic (aquifers and confining units) features, including those occurring at the location of the reasonably maximally exposed individual, with emphasis on known or inferred hydrologic significance: this description should include information on hydraulic conductivity, transmissivity, porosities, permeability, and other important hydrogeologic parameters of the major hydrostratigraphic units, as appropriate;

(b) An interpretation of the regional ground-water flow system, including a discussion of the major features and controls that affect local and regional ground-water supply;

(c) The delineation of the proposed hydrogeologic system (saturated and unsaturated) to be used in estimating the performance of the proposed repository;

(d) A description and discussion of local climate, including precipitation, temperature, and surface runoff;

(e) A discussion of ground-water quality;

(f) A discussion of current water-use patterns, including ground-water withdrawals by aquifer source;

(g) An estimated water budget for the respective aquifer systems; and

(h) The identification of surface hydrologic features (including impoundments and stream channels (either continuous or intermittent), or other geomorphic features, that could potentially affect the geologic repository operations area operations or safety.

(3) An overview of geochemistry consistent with other site characterization summaries that includes:

(a) A delineation of the proposed geochemical environment to be used in estimating the safety of the repository;

(b) Evaluation of ground water to determine characteristics such as water chemistry, radionuclide solubility, and radionuclide sorption capability; and

(c) A description of the anticipated geochemical environment in the vicinity of emplaced waste packages.

(4) An overview of geotechnical properties and conditions consistent with other site characterization summaries that includes:

(a) A discussion of the results of site investigations necessary to characterize the engineering properties of the soils present at the site;

(b) A discussion of the results of site investigations necessary to characterize the engineering properties of the rock types present at the site, with particular emphasis on the host rock and its immediate environs necessary for the underground excavation of the geologic repository; and

(c) A discussion and description of other site characterization work conducted to define the relevant geotechnical properties and anticipated response/ performance of both surface and subsurface facilities.

(5) An overview of climatological, meteorological, and other environmental information for the site. This overview should also include a description of paleoclimate features, events, and processes; and

(6) An overview of the reference biosphere. The biosphere pathways selected for dose assessments should be consistent with arid or semi-arid conditions found in a mid-latitude desert. The location and representative local diet and living style of the reasonably maximally exposed individual must be proposed by the U.S. Department of Energy in the license application. The detailed review of information on the characteristics of the reasonably maximally exposed individual is conducted using Section 2 ("Review Plan for Safety Analysis Report") of the Yucca Mountain Review Plan.

1.5.3 Acceptance Criteria

The following acceptance criteria are based on meeting the requirements of 10 CFR 63.21(b)(5), relating to the description of site characterization work provided in the "General Information" section of the license application. (In general, the detailed technical review of this information is addressed in Section 2, "Review Plan for Safety Analysis Report," of the Yucca Mountain Review Plan.)

Acceptance Criterion 1 The "General Information" Section of the License Application Contains an Adequate Description of Site Characterization Activities.

(1) An adequate overview is provided of the site characterization activities related to geology; hydrology; geochemistry; geotechnical properties and conditions of the host rock; climatology, meteorology, and other environmental sciences; and the reference biosphere.

Acceptance Criterion 2 The "General Information" Section of the License Application Contains an Adequate Description of Site Characterization Results.

(1) A sufficient understanding is provided of current features and processes present in the Yucca Mountain region;

(2) Adequate information is provided for evolution of future events and processes likely to be present in the Yucca Mountain region that could effect repository safety; and

(3) The description of the reference biosphere is consistent with present knowledge of natural processes in and around the Yucca Mountain site, including the location of the reasonably maximally exposed individual.

1.5.4 Evaluation Findings

If the license application provides sufficient information and the regulatory acceptance criteria in Section 1.5.3 are appropriately satisfied, the staff concludes that this portion of the staff evaluation is acceptable. The reviewer writes material suitable for inclusion in the safety evaluation report prepared for the entire application. The report includes a summary statement of what was reviewed and why the reviewer finds the submittal acceptable. The staff can document the review as follows.

U.S. Nuclear Regulatory Commission staff has reviewed the general information and other information submitted in support of the license application on site characterization and has found, with reasonable assurance, that the requirements of 10 CFR 63.219(b)(5) are satisfied. There is an adequate summary description of the work done to characterize the Yucca Mountain site and a summary of the results from that work, to allow staff to evaluate the overall sufficiency of the license application.

1.5.5 References

None.

2 REVIEW PLAN FOR SAFETY ANALYSIS REPORT

Dose Projections: Determination of compliance with the preclosure and postclosure dose limits involves the use of computer programs for estimating potential exposures. The regulations specify a total effective dose equivalent as the measure to be used in estimating the dose. The staff should use the sum of the committed effective dose equivalent from internal doses resulting from one year's exposure to radioactive materials, and the effective dose equivalent from external radiation exposure during the year to calculate potential exposures. Additionally, the staff should use organ weighting factors, from Federal Guidance Report 12 and International Commission on Radiological Protection in its Publication 26, for external dose calculations. (Note: The Statement of Considerations to 10 CFR Part 63, 66 FR 55734 and 55735, describe the method to be used for calculating the total effective dose equivalent.)

Occupational Dose Monitoring: Actual exposures to radiation workers at the site will be measured to assure compliance with 10 CFR Part 20 requirements.

2.1 Repository Safety Before Permanent Closure

2.1.1 Preclosure Safety Analysis

Risk-Informed Review Process for Preclosure Safety Analysis—This section provides for review of compliance with the performance objectives in 10 CFR Part 63, which are based on permissible levels of doses to workers and the public, established on the basis of acceptable levels of risk. 10 CFR 63.21(c)(5) requires a preclosure safety analysis of the geologic repository operations area for the period before permanent closure, to ensure compliance with the performance objectives. Preclosure safety analysis is a systematic examination of the site; the design; the potential hazards, initiating events, and their consequences; and the potential dose consequences to workers and the public. Preclosure safety analysis considers the probability of potential hazards, taking into account the range of uncertainty associated with the data that support the probability calculations. Event sequences are defined, and these sequences of human-induced and natural events are used as inputs to calculate consequences of potential failures of structures, systems, and components, in terms of doses to workers and the public. These calculated doses are compared to allowable doses in establishing compliance with performance objectives. The structures, systems, and components that must be functional to comply with the performance objective dose limits are identified as structures, systems, and components important to safety. Preclosure safety analysis also identifies and describes the controls that are relied on to prevent potential event sequences from occurring or to mitigate their consequences, and identifies measures taken to ensure the availability of the safety systems. The end products of the preclosure safety analysis are a list of structures, systems, and components important to safety (also known as the Q-List) and the associated design criteria and technical specifications necessary to keep them functional and to meet the performance objectives. The structures, systems, and components important to safety may also be further categorized, based on relative safety significance, using risk information from the preclosure safety analysis. This distinction may be used to focus on the level of design details to be provided in the license application and the application of quality assurance controls through a graded quality assurance program. The U.S. Department of Energy plans on categorizing structures, systems, and components based on safety/risk-significance and implementing a graded quality assurance program commensurate with safety significance.

Accordingly, the Yucca Mountain Review Plan has included appropriate criteria to evaluate the U.S. Department of Energy technical basis for categorizing structures, systems, and components and grading quality assurance requirements.

The staff review is focused on items that the preclosure safety analysis has determined to be important to safety. The rigor of review for the design items on the Q-List, and the level of attention to detail, depend on relative safety significance. No prescriptive design criteria are imposed in the Yucca Mountain Review Plan, because 10 CFR Part 63 allows the U.S. Department of Energy to develop the design criteria and demonstrate their appropriateness. Thus, the U.S. Department of Energy has flexibility to use any codes, standards, and methodologies it demonstrates to be applicable and appropriate. This flexibility is necessary when implementing a risk-informed, performance-based regulation. The risk-informed, performance-based review process in the Yucca Mountain Review Plan focuses on determining compliance with performance objectives as demonstrated by the U.S. Department of Energy preclosure safety analysis. In summary, the review philosophy is based on the following premises: (i) the U.S. Department of Energy must demonstrate, through its preclosure safety analysis, that the repository will be designed, constructed, and operated to meet the specified exposure limits (performance objectives) throughout the preclosure period; (ii) the staff must focus the review on the design of the structures, systems, and components important to safety in the context of the design's ability to meet the performance objectives; and finally, (iii) the staff resources will be focused proportionately on the inspection and review of high-risk significant structures, systems, and components important to safety.

2.1.1.1 Site Description as it Pertains to Preclosure Safety Analysis

Review Responsibilities—High-Level Waste Branch and Environmental and Performance Assessment Branch

2.1.1.1.1 Areas of Review

This section provides guidance on the review of site description, as it pertains to preclosure safety analysis and geologic repository operations area design. The reviewers will also evaluate the information required by 10 CFR 63.21(c)(1)(i)–(iii).

The adequacy of the site description should be assessed in the context of the information required to conduct the preclosure safety analysis and design the geologic repository operations area. The reviewers of this section should coordinate their reviews with the reviewers of Sections 2.1.1.3 ("Identification of Hazards and Initiating Events") and 2.1.1.7 ("Design of Structures, Systems, and Components Important to Safety and Safety Controls") of the Yucca Mountain Review Plan.

The staff will evaluate the following parts of the site description, as they pertain to preclosure safety analysis and geologic repository operations area design, using the review methods and acceptance criteria in Sections 2.1.1.1.2 and 2.1.1.1.3.

(1) Site geography;

(2) Regional demography;

(3) Local meteorology and regional climatology;

(4) Regional and local surface and ground-water hydrology;

(5) Site geology and seismology, including geoengineering properties that are relevant to design of surface and subsurface facilities;

(6) Igneous activity;

(7) Site geomorphology;

(8) Site geochemistry; and

(9) Land use, structures and facilities, and residual radioactivity within the entire land withdrawal area.

2.1.1.1.2 Review Methods

Review Method 1 Description of Site Geography

Verify that the site location is adequately defined and is specified relative to prominent natural and man-made features, such as mountains, streams, military bases, civilian and military airports, population centers, roads, railroads, transmission lines, wetlands, surface water bodies, and potentially hazardous commercial operations and manufacturing centers, that may be significant for the review of the preclosure safety analysis and geologic repository operations area design.

Confirm that the characteristics of natural and man-made features, within the restricted area of the site, that may be significant for evaluation of the preclosure safety analysis and geologic repository operations area design, have been acceptably defined.

Ascertain that maps of the site and nearby facilities are included, and are of sufficient detail and of appropriate scale to provide information needed to review the preclosure safety analysis and geologic repository operations area design. A site map should clearly indicate the site boundary and the controlled area, controlled area access points, and distances from the boundary to significant features of the installation. Maps should describe the site topography and surface drainage patterns, as well as roads, railroads, transmission lines, wetlands, and surface-water bodies.

Review Method 2 Description of Regional Demography

Verify that regional demographic information is based on current census data, and presents the population distribution as a function of distance from the geologic repository operations area. The demographic information should be in sufficient detail to determine the location of real members of the public. The demographic information should be projected for the preclosure period.

Review Method 3 Description of Local Meteorology and Regional Climatology

Evaluate the adequacy of the license application data on local meteorology and regional climatology that may be significant for the review of the preclosure safety analysis and geologic repository operations area design, including items such as:

(1) Temperature extremes;
(2) Atmospheric stability;
(3) Average wind speeds and prevailing wind direction;
(4) Extreme winds; and
(5) Tornadoes.

Confirm that data collection techniques are based on accepted methods [e.g., those described in NUREG–0800 (U.S. Nuclear Regulatory Commission, 1987)], and that technical bases for data summaries are provided.

Assess the information provided on the annual amount and forms of precipitation, and the probable maximum precipitation at the site. Confirm that acceptable methods were used to develop this information.

Confirm that the license application adequately defines the type, frequency, magnitude, and duration of severe weather, such as tornados, lightning, and storms; and assess the validity of the design bases/criteria provided for the severe weather assessment.

Verify whether the U.S. Department of Energy conducted appropriate trending analyses supported by sufficient historical data.

Review Method 4 Description of Regional and Local Surface and Ground-Water Hydrology

Evaluate the description of the Yucca Mountain surface and ground-water hydrology, to verify that hydrologic features relevant to the preclosure safety analysis and geologic repository operations area design are adequately identified.

Verify that the analyses of the effects of any proposed changes to natural drainage features on geologic repository operations area design are acceptable. To make this determination, coordinate with the reviewer of Section 2.1.1.7 ("Design of Structures, Systems, and Components Important to Safety and Safety Controls") of the Yucca Mountain Review Plan.

Confirm that the calculation of probable maximum flood is supported by sufficient data, including actual storm data for the drainage basin. Section 2.4.3 of NUREG–0800 (U.S. Nuclear Regulatory Commission, 1987) may be used to conduct this review.

Review Method 5 Descriptions of Site Geology and Seismology

Verify that the U.S. Department of Energy has provided sufficient data on the geology of the site to support the preclosure safety analysis and geologic repository operations area design, including the stratigraphy and lithology for the entire surface and subsurface construction area. To make this determination, coordinate with the reviewers of Section 2.2.1.3 ("Model Abstraction") of the Yucca Mountain Review Plan.

Confirm that site characterization data include geomechanical properties and conditions of host rock, based on *in situ* and laboratory test results for the rock formations, where major construction activities will take place. Collection and processing of these data should be based on accepted industry techniques and standards. Verify that rock mechanics testing data support the license application analyses of the stability of subsurface materials. Note that evaluation of the sufficiency of data and appropriateness of design parameters will be conducted using the appropriate subsection of Section 2.1.1.7 ("Design of Structures, Systems, and Components Important to Safety and Safety Controls") of the Yucca Mountain Review Plan.

Confirm that the engineering properties provided for soils in the areas where surface facilities will be constructed are based on laboratory and *in situ* test results. Verify that the U.S. Department of Energy collected and processed these data using accepted industry techniques.

Confirm that detailed soil testing data support the license application analyses of the stability of surface materials, considering surface subsidence, previous loading histories, and liquefaction potential.

Consult with the reviewers of Section 2.2.1.3.2.3 ("Acceptance Criteria—Mechanical Disruption of Engineered Barriers") of the Yucca Mountain Review Plan, to confirm that the vibratory ground motion and surface and subsurface fault displacements of the site have been adequately characterized. This assessment should include a list of Type I faults, areal seismic source zones, earthquake parameters such as maximum magnitude and recurrence for each source, historical earthquake data, paleoseismic data, and ground motion attenuation models. Topical Report YMP/TR–002–NP: "Methodology to Assess Fault Displacement and Vibratory Ground Motion Hazards at Yucca Mountain," Revision 1 (U.S. Department of Energy, 1997) presents an acceptable method.

Verify that conversion of the characterized vibratory ground motion and surface and subsurface fault displacements of the site to engineering design parameters uses acceptable methods.

Evaluate the analyses of the static and dynamic stability of facility foundations, subsurface emplacement drifts, and natural and man-made slopes (both cut and fill), the failure of which could lead to radiological release. Verify that appropriate methods are used for the analyses, data used are appropriate for the methods, and results are properly interpreted.

Review Method 6 Site Igneous Activity Information

Consult with the reviewer of Section 2.2.1.2 ("Scenario Analysis and Event Probability") of the Yucca Mountain Review Plan to verify the license application adequately considers igneous activity at the site, including volcanic eruption, subsurface magmatic activity/flow, and volcanic ash flow/ash fall.

Review Method 7 Site Geomorphology Information

Evaluate the analysis of site geomorphology [using guidance such as NUREG/CR–3276 (Schumm and Chorley, 1983) and "Standard Format and Content for Documentation of Remedial Action Selection at Title I Uranium Mill Tailings Sites" (U.S. Nuclear Regulatory Commission, 1989), as appropriate]. Assess the extent of erosion of the land surface and the likelihood that mass wasting, such as landslides or rock avalanches, or rapid fluvial degradation in channels or interfluves, might affect site structures or operations.

Review Method 8 Site Geochemical Information

Evaluate the description of the geochemical information at Yucca Mountain that is relevant to the preclosure safety analysis and geologic repository operations area design, to confirm that it is adequate, including items such as:

(1) Geochemical composition of any subsurface water held within the rock matrix or perched water zones, or episodically flowing through fractures to determine corrosivity;

(2) Geochemical composition of rock strata within and above the repository horizon to identify minerals that might leach and increase the corrosivity of water flowing through the strata; and

(3) Any geochemical alterations to the rock fractures and rock matrix through heating or other processes that might significantly alter geomechanical rock mass properties.

Review Method 9 Land Use, Structures and Facilities, and Residual Radioactivity

Evaluate the description of previous uses of land within the land withdrawal area; the description, locations, and uses of man-made structures and facilities; and the identification of any residual sources of radiation within the land withdrawal area as they relate to the preclosure safety analysis. The evaluation should include such items as:

(1) Conflicts with uses of the land for a repository;

(2) Impacts on existing structures and facilities or potential contamination from these facilities; and

(3) Potential for public or worker exposures from residual radiation at the site.

2.1.1.1.3 Acceptance Criteria

The following acceptance criteria are based on meeting the requirements of 10 CFR 63.112(c) relating to the site description as it pertains to the preclosure safety analysis.

Acceptance Criterion 1 The License Application Contains a Description of the Site Geography Adequate to Permit Evaluation of the Preclosure Safety Analysis and the Geologic Repository Operations Area Design.

(1) The site location is adequately defined. The site location is specified relative to prominent natural and man-made features, such as mountains, streams, military bases, civilian and military airports, population centers, and potentially hazardous commercial operations and manufacturing centers, that may be significant for the review of the preclosure safety analysis and geologic repository operations area design;

(2) The characteristics of natural and man-made features, within the controlled area of the site, that may be significant for evaluation of the preclosure safety analysis and geologic repository operations area design, are adequately defined; and

(3) Maps of the site and nearby facilities are included, and are of sufficient detail and of appropriate scale to provide information needed to review the preclosure safety analysis and geologic repository operations area design. A site map clearly indicates the site boundary and the controlled area, controlled area access points, and distances from the boundary to significant features of the installation. Maps describe the site topography and surface drainage patterns, as well as roads, railroads, transmission lines, wetlands, and surface water bodies.

Acceptance Criterion 2 The License Application Contains a Description of the Regional Demography Adequate to Permit Evaluation of the Preclosure Safety Analysis and the Geologic Repository Operations Area Design.

(1) Regional demographic information is based on current census data and presents the population distribution as a function of distance from the geologic repository operations area.

Acceptance Criterion 3 The License Application Contains a Description of the Local Meteorology and Regional Climatology Adequate to Permit Evaluation of the Preclosure Safety Analysis and the Geologic Repository Operations Area Design.

(1) The license application data on local meteorology and regional climatology, that may be significant for the review of the preclosure safety analysis and geologic repository operations area design, are adequate;

(2) The data collection techniques are based on accepted methods, and the technical bases for data summaries are provided;

(3) Adequate information is provided on the annual amount and forms of precipitation, and the probable maximum precipitation at the site. Acceptable methods are used to develop this information;

(4) The license application adequately defines the type, frequency, magnitude, and duration of severe weather. Valid design bases/criteria are provided for the severe weather assessment; and

(5) Trending analyses are appropriately conducted and supported by sufficient historical data presented in the license application.

Acceptance Criterion 4 The License Application Contains Sufficient Local and Regional Hydrological Information to Support Evaluation of the Preclosure Safety Analysis and the Geologic Repository Operations Area Design.

(1) The description of the Yucca Mountain surface and ground-water hydrology, adequately identifies hydrologic features relevant to the preclosure safety analysis and geologic repository operations area design;

(2) The analyses of the effects of any proposed changes to natural drainage features on geologic repository operations area design are acceptable; and

(3) The calculation of probable maximum flood is supported by sufficient data, including actual storm data for the drainage basin.

Acceptance Criterion 5 The License Application Contains Descriptions of the Site Geology and Seismology Adequate to Permit Evaluation of the Preclosure Safety Analysis and the Geologic Repository Operations Area Design.

(1) The license application provides sufficient data on the geology of the site to support the preclosure safety analysis and geologic repository operations area design, including the stratigraphy and lithology for the entire surface and subsurface construction area;

(2) Site characterization data adequately include rock mechanics properties based on *in situ* and laboratory test results for the rock formations where major construction activities will take place. Collection and processing of these data are based on accepted industry techniques;

(3) Rock mechanics testing data adequately support the license application analyses of the stability of subsurface materials;

(4) The engineering properties provided for soils in the areas where surface facilities will be constructed are based on laboratory and *in situ* test results. These data are collected and processed using accepted industry techniques;

(5) Detailed soil testing data support the license application analyses of the stability of surface materials, considering surface subsidence, previous loading histories, and liquefaction potential;

(6) The vibratory ground motion and surface and subsurface fault displacements of the site are adequately characterized, taking into account the assessment in Section 2.2.1.3.2.3 ("Mechanical Disruption of Engineered Barriers") of the Yucca Mountain Review Plan and considering a list of Type 1 faults, areal seismic source zones, earthquake parameters such as maximum magnitude and recurrence for each source, historical earthquake data, paleoseismic data, and ground motion attenuation models. Topical report YMP/TR–002–NP (U.S. Department of Energy, 1997) presents an acceptable methodology for evaluating vibratory ground motion and fault displacement hazards;

(7) Acceptable methods are used to develop seismic design data using the characterized vibratory ground motion and surface and subsurface fault displacement; and

(8) The license application provides adequate analyses of the stability of the facility foundations, subsurface emplacement drifts, and natural and man-made slopes (both cut and fill), the failure of which could result in radiological release. Appropriate methods are used for the analyses, data used are appropriate for the methods, and results are properly interpreted.

Acceptance Criterion 6 The License Application Contains Descriptions of the Historical Regional Igneous Activity Adequate to Permit Evaluation of the Preclosure Safety Analysis and the Geologic Repository Operations Area Design.

(1) The license application adequately considers igneous activity at the site, including volcanic eruption, subsurface magmatic activity/flow, and volcanic ash flow/ash fall.

Acceptance Criterion 7 The License Application Provides Analysis of Site Geomorphology Adequate to Permit Evaluation of the Preclosure Safety Analysis and Geologic Repository Operations Area Design.

(1) The license application adequately considers the extent of erosion of the land surface and the likelihood that mass wasting, such as landslides or rapid fluvial degradation in channels or interfluves, might affect site structures or operations.

Acceptance Criterion 8 The License Application Contains Site-Sufficient Geochemical Information to Support Evaluation of the Preclosure Safety Analysis and the Geologic Repository Operations Area Design.

(1) Information on the geochemical composition of subsurface water held within the rock matrix or perched water zone, or from episodic flows through fractures, is sufficient to determine corrosivity;

(2) The geochemical composition of the rock strata, within and above the repository horizon, is adequately defined to identify minerals that might leach and add to the corrosivity of water flowing through the strata; and

(3) Potential geochemical alterations to the rock fractures and the rock matrix, through heating or other processes that might significantly alter geomechanical rock mass properties, are adequately characterized.

Acceptance Criterion 9 The License Application Contains Adequate Evaluations of Previous Land Use, Impacts on Existing Structures and Facilities; and the Potential for Exposures from Residual Radiation.

(1) Information on previous land uses within the land withdrawal area is sufficient to identify any potential conflicts;

(2) Locations and descriptions of existing man-made structures or facilities are adequate to determine impacts on or from these structures and facilities; and

(3) Identification of any residual radiation is sufficient to determine the potential for exposures to workers or the public.

2.1.1.1.4 Evaluation Findings

If the license application provides sufficient information and the regulatory acceptance criteria in Section 2.1.1.1.3 are appropriately satisfied, the staff concludes that this portion of the staff evaluation is acceptable. The reviewer writes material suitable for inclusion in the safety evaluation report prepared for the entire application. The report includes a summary statement of what was reviewed and why the reviewer finds the submittal acceptable. The staff can document the review as follows.

U.S. Nuclear Regulatory Commission staff has reviewed the Safety Analysis Report and other information submitted in support of the license application and has found, with reasonable assurance, that the requirements of 10 CFR 63.112(c) are satisfied. Requirements for conducting an adequate preclosure safety analysis and evaluation of geologic repository operations area design have been met in that adequate data from the Yucca Mountain site and the surrounding region have been provided to identify naturally occurring and human-induced hazards and geomechanical properties and conditions of the host rock.

2.1.1.1.5 References

U.S. Department of Energy. "Topical Report YMP/TR–002–NP: Methodology to Assess Fault Displacement and Vibratory Ground Motion Hazards at Yucca Mountain." Revision 1. Washington, DC: U.S. Department of Energy. 1997.

U.S. Nuclear Regulatory Commission. "Standard Format and Content for Documentation of Remedial Action Selection at Title I Uranium Mill Tailings Sites." Washington, DC: U.S. Nuclear Regulatory Commission. 1989.

———. NUREG–0800, "Standard Review Plan for the Review of Safety Analysis Reports for Nuclear Power Plants." LWR Edition. Washington, DC: U.S. Nuclear Regulatory Commission. 1987.

Schumm, S.A. and R.J. Chorley. NUREG/CR–3276, "Geomorphic Controls on the Management of Nuclear Waste." Washington, DC: U.S. Nuclear Regulatory Commission. 1983.

2.1.1.2 Description of Structures, Systems, Components, Equipment, and Operational Process Activities

Review Responsibilities—High-Level Waste Branch and Environmental and Performance Assessment Branch

2.1.1.2.1 Areas of Review

This section provides guidance on the review of the description of structures, systems, components, equipment, and operational process activities. The reviewers will also evaluate the information required by 10 CFR 63.21(c)(2), (c)(3)(i), and (c)(4).

The description of structures, systems, and components, equipment, and operational process activities should be sufficient for the reviewer to understand the design of geologic repository operations area facilities, and to identify hazards and event sequences. The reviewers of this section should coordinate their reviews with the reviews under Sections 2.1.1.3 ("Identification of Hazards and Initiating Events") and 2.1.1.4 ("Identification of Event Sequences") of the Yucca Mountain Review Plan.

The staff will evaluate the following parts of the description of structures, systems, and components, equipment, and operational process activities, using the review methods and acceptance criteria in Sections 2.1.1.2.2 and 2.1.1.2.3.

(1) Descriptions of locations of surface facilities and their functions, including structures, systems and components, and equipment;

(2) Descriptions of, and design details for, structures, systems, and components, equipment, and utility systems of surface facilities;

(3) Descriptions of, and design details for, structures, systems, and components, equipment, and utility systems of the subsurface facility;

(4) Description of high-level radioactive waste characteristics;

(5) Descriptions of engineered barrier system components (e.g., waste package, drip shield, and backfill); and

(6) Descriptions of geologic repository operations area processes activities and procedures, including interfaces and interactions between structures, systems, and components.

2.1.1.2.2 Review Methods

Review Method 1 Description of Location of Surface Facilities and their Functions

Confirm that the license application describes all surface facilities, including their locations and arrangements at the site, and their distances from the site boundary. This description should include drawings of sufficient detail and appropriate scale.

Verify that the description of the design of the surface facilities is adequate to permit an evaluation of the preclosure safety analysis.

Verify that descriptions of the functional requirements for all the facilities provide an understanding of geologic repository operations area operational activities, sequences, and locations, sufficient for evaluation of the preclosure safety analysis and geologic repository operations area design.

Verify that the license application has descriptions of the capabilities of the equipment, training of the operators, and testing/maintenance plans, sufficient for evaluation of the preclosure safety analysis. Make this verification in collaboration with the reviewers for Sections 2.5.3 ("Training and Certification of Personnel") and 2.5.6 ("Plans for Conduct of Normal Activities Including Maintenance, Surveillance, and Periodic Testing") of the Yucca Mountain Review Plan.

Review Method 2 Descriptions of, and Design Details for, Structures, Systems, and Components, and Equipment of Surface Facilities

Confirm the license application has provided adequate descriptions and design information for the structures, systems, and components, equipment, and the utility systems that support the structures, systems, and components of the surface facilities, such as:

(1) Design codes and standards employed;
(2) General arrangement drawings of buildings;
(3) Materials of construction;
(4) Equipment layout;
(5) Process flow diagrams;
(6) Piping and instrumentation diagrams;

(7) Electrical systems;
(8) Pressure relief systems;
(9) Crane systems;
(10) Welding systems;
(11) Heating, ventilation, air conditioning, and filtration systems;
(12) On-site transportation systems;
(13) Confinement system;
(14) Decontamination system;
(15) Safety systems (e.g., interlocks, radiation detection, and fire suppression systems);
(16) Waste package and cask receipt, transfer, and handling systems;
(17) Loading and unloading systems (including remote operations);
(18) Emergency and radiological safety systems;
(19) Criticality and radiological monitoring systems;
(20) Criticality safety program;
(21) Communication and control systems;
(22) Power distribution systems, including any backup power supplies;
(23) Shielding and criticality control systems; and
(24) Water supply systems.

Focus on systems used for radiological waste handling, packaging, transfer, containment, or storage, and on any other structures, systems, and components important to safety. Identification of structures, systems, and components important to safety is reviewed using Section 2.1.1.6 ("Identification of Structures, Systems, and Components Important to Safety; Safety Controls; and Measures to Ensure Availability of the Safety Systems") of the Yucca Mountain Review Plan.

Confirm the license application has provided adequate descriptions of potential interactions among support systems and structures, systems, and components.

Verify the license application has provided adequate descriptions of the location and functional arrangement of the structures, systems, and components within each facility.

Confirm the license application has provided adequate discussion of design information about the capability of the surface facilities to withstand natural phenomena (e.g., seismic ground motions). The appropriateness and adequacy of the design will be reviewed using Section 2.1.1.7 ("Design of Structures, Systems, and Components Important to Safety and Safety Controls") of the Yucca Mountain Review Plan.

Review Method 3 Descriptions of, and Design Details for, Structures, Systems, and Components, and Equipment of the Subsurface Facility

Confirm the license application has provided adequate descriptions and design information for the structures, systems, components, equipment, and utility systems that support the structures, systems, and components of the subsurface facility, such as:

(1) Design codes and standards employed;

(2) The layout of the subsurface facility in relation to any constraints imposed by natural conditions (geologic and hydrologic) and generic design goals (e.g., maximum rock temperature allowable);

(3) Materials of construction;

(4) Ground control/support systems;

(5) Process flow diagrams;

(6) Power distribution systems;

(7) Subsurface ventilation systems;

(8) Subsurface filtration systems;

(9) Communication and inspection/monitoring systems;

(10) Transportation systems;

(11) Safety, detection, and suppression systems for fire and radiological emergencies;

(12) Waste package emplacement system;

(13) Emergency and radiological safety systems;

(14) Air seal systems to separate the waste emplacement area from the emplacement drift construction area;

(15) Waste package support/invert systems;

(16) Drip shield and drip shield placement systems;

(17) Backfill emplacement systems;

(18) Drainage system;

(19) Instrumentation and control systems; and

(20) Limits and interlocks.

Review Method 4 Description of Spent Nuclear Fuel and High-Level Radioactive Waste Characteristics

Verify that the license application has adequately characterized the ranges of parameters that describe the spent nuclear fuel, such as:

(1) Reactor type (e.g., boiling water, pressurized water);

(2) Fuel assembly manufacturer and model designation;

(3) Fuel assembly physical characteristics and dimensions;

(4) Fuel cladding material (including crud deposits, oxide layer, hydride content, and extent of failure and damage);

(5) Thermal characteristics;

(6) Heat generation rate and dose rate;

(7) Radionuclide inventory;

(8) Radiochemical characteristics; and

(9) History (enrichment, burnup, and postirradiation storage).

Confirm that the license application has adequately characterized properties of the high-level radioactive waste besides spent nuclear fuel, such as:

(1) Waste form composition and amount;
(2) Waste form characteristics (phase stability and product consistency);
(3) Canister and characteristics of any waste encapsulation;
(4) Radionuclide inventory;
(5) Radiochemistry;
(6) Heat generation rate and dose rate;
(7) Proposed storage unit of material; and
(8) History.

Review Method 5 Description of Engineered Barrier System and Its Components

Confirm that the principal characteristics of the waste package, including dimensions, weights, materials, fabrications, and weldings, have been provided.

Confirm that adequate discussion on analyses and characterization of functional features of the waste package and canister, such as containment, criticality control, shielding, drop fracture resistance, and confinement, has been provided.

Verify that the discussion of analyses and characterization of engineered barrier system components, such as drip shields, backfill (if used in the license application design), supports/inverts, and sorption barrier, is sufficient for evaluation in the preclosure safety analysis and geologic repository operations area design review.

Review Method 6 Description of Geologic Repository Operations Area Operational Processes and Procedures

Evaluate the descriptions of operational processes and procedures to confirm that they provide an adequate understanding of the component and facility functions and sequences of activities.

Verify that information provided on operational process design, equipment design and specifications, and instrumentation and control systems is sufficient to assess the preclosure safety analysis.

The descriptions and information should include:

(1) Modes of operations, for example, normal process operations, maintenance;

(2) The purpose of each operational process and its relationship to overall geologic repository operations area operations;

(3) Basic operational process function and theory, including a discussion of the basic theory of operational processes and an adequate discussion of ranges and limits for measured variables used to ensure safe operation of processes;

(4) Diagrams or flow charts that demonstrate the interfaces and interactions of parts of the operational processes, such as schematics or descriptions showing the inventory, location, and geometry of nuclear materials, moderators, and other materials associated with processes;

(5) Procedures for startup, shutdown, normal operations, and emergency operations;

(6) Hazardous material locations and quantities;

(7) Locations and types of interlocks and controls;

(8) Process block diagrams, including decontamination and monitoring;

(9) Safety equipment; control systems; and instrumentation locations, characteristics, and functions;

(10) Maximum intended inventories of radioactive materials; and

(11) Criticality safety program.

2.1.1.2.3 Acceptance Criteria

The following acceptance criteria are based on meeting the requirements of 63.112(a), relating to the description of structures, systems, and components, equipment, and operational process activities.

Acceptance Criterion 1 The License Application Contains a Description of the Location of the Surface Facilities and Their Designated Functions Sufficient to Permit Evaluation of the Preclosure Safety Analysis and the Geologic Repository Operations Area Design.

(1) The license application has a description of surface facilities that includes their location and arrangement at the site and their distance from the site boundary. This description includes drawings of sufficient detail and appropriate scale;

(2) The description of the design of the surface facilities is sufficient to permit an evaluation of the preclosure safety analysis;

(3) The descriptions of the functional requirements for the facilities provide an understanding of geologic repository operations area operational activities, sequences, and locations, sufficient for evaluation of the preclosure safety analysis and geologic repository operations area design; and

(4) The descriptions of the capabilities of the equipment, training of the operators, and testing/maintenance plan are sufficient for evaluation of the preclosure safety analysis.

Acceptance Criterion 2 The License Application Contains Descriptions and Design Details for Structures, Systems, and Components, and Equipment, of the Surface Facilities, Sufficient to Permit Evaluation of the Preclosure Safety Analysis and the Geologic Repository Operations Area Design.

(1) The license application provides adequate descriptions and design information for the structures, systems, and components, and equipment, of the surface facilities;

(2) The license application provides adequate discussion on potential interactions among support systems and structures, systems, and components;

(3) The license application provides adequate descriptions of the location and functional arrangement of the structures, systems, and components within each facility; and

(4) The license application provides adequate discussion of design information, regarding the capability of the surface facilities to withstand the effects of natural phenomena.

Acceptance Criterion 3 The License Application Contains Descriptions and Design Details for Structures, Systems, and Components, and Equipment of the Subsurface Facility, Sufficient to Permit Evaluation of the Preclosure Safety Analysis and the Geologic Repository Operations Area Design.

(1) The license application provides adequate descriptions and design information for the structures, systems, and components, and equipment, of the subsurface facility.

Acceptance Criterion 4 The License Application Describes the Characteristics of the Spent Nuclear Fuel and High-Level Radioactive Waste, Sufficiently to Permit Evaluation of the Preclosure Safety Analysis and the Waste Package Design.

(1) The license application adequately characterizes the ranges of parameters that describe the spent nuclear fuel; and

(2) The license application adequately characterizes the properties of the high-level radioactive waste besides spent nuclear fuel.

Acceptance Criterion 5 The License Application Provides a General Description of the Engineered Barrier System and Its Components, Sufficient to Support Evaluation of the Preclosure Safety Analysis and the Engineered Barrier System Design.

(1) The principal characteristics of the waste package, including dimensions, weights, materials, fabrications, and weldings, are defined;

(2) Adequate characterization of functional features of the waste package and canister, such as criticality control, shielding, drop fracture resistance, and confinement, is provided; and

(3) The discussion of analyses and characterization of engineered barrier system components, such as drip shields, backfill, supports/inverts, and sorption barrier, is sufficient to support evaluations in the preclosure safety analysis and geologic repository operations area design reviews.

Acceptance Criterion 6 The Description of the Operational Processes to be Used at the Geologic Repository Operations Area is Sufficient for Review of the Preclosure Safety Analysis.

(1) Descriptions of geologic repository operations area operational processes provide an adequate understanding of the component and facility functions and sequences of activities; and

(2) Information provided on interfaces and interactions of parts of the operational processes design, equipment design and specifications, and instrumentation and control systems, is sufficient to assess the preclosure safety analysis.

2.1.1.2.4 Evaluation Findings

If the license application provides sufficient information and the regulatory acceptance criteria in Section 2.1.1.2.3 are appropriately satisfied, the staff concludes that this portion of the staff evaluation is acceptable. The reviewer writes material suitable for inclusion in the safety evaluation report prepared for the entire application. The report includes a summary statement

of what was reviewed and why the reviewer finds the submittal acceptable. The staff can document the review as follows.

U.S. Nuclear Regulatory Commission staff has reviewed the Safety Analysis Report and other information submitted in support of the license application and has found, with reasonable assurance, that the requirements of 10 CFR 63.112(a) are satisfied in that an adequate general description of the structures, systems, and components, equipment, and process activities of the geologic repository operations area, has been provided.

2.1.1.2.5 References

None.

2.1.1.3 Identification of Hazards and Initiating Events

Review Responsibilities—High-Level Waste Branch and Environmental and Performance Assessment Branch

2.1.1.3.1 Areas of Review

This section provides guidance on the review of the identification of hazards and initiating events. The reviewers of this section should coordinate their reviews with the reviewers of Sections 2.1.1.1 ("Site Description as it Pertains to Preclosure Safety Analysis") and 2.1.1.2 ("Description of Structures, Systems, Components, Equipment, and Operational Process Activities") of the Yucca Mountain Review Plan. Reviewers will also evaluate the information required by 10 CFR 63.21(c)(5).

The staff will evaluate the following parts of the identification of hazards and initiating events, using the review methods and acceptance criteria in Sections 2.1.1.3.2 and 2.1.1.3.3.

(1) Technical basis and assumptions for methods used for identification of hazards and initiating events;

(2) Use of relevant data for identification of hazards and initiating events;

(3) Determination of frequency or probability of occurrence of hazards and initiating events;

(4) Technical basis for inclusion or exclusion of specific hazards and initiating events; and

(5) List of hazards and initiating events to be considered in the preclosure safety analysis.

2.1.1.3.2 Review Methods

Review Method 1 Technical Basis and Assumptions for Methods Used for Identification of Hazards and Initiating Events

Confirm that methods used to identify hazards and initiating events are consistent with Agency guidance or standard industry practices. Examples include NUREG/CR–2300 (U.S. Nuclear Regulatory Commission, 1983a); NUREG–1513 (U.S. Nuclear Regulatory Commission, 2001); NUREG–1520 (U.S. Nuclear Regulatory Commission, 2000a); and the American Institute of Chemical Engineers (1992), Appendixes A and B. If expert elicitation was used, review the expert elicitation process, using Section 2.5.4 ("Expert Elicitation") of the Yucca Mountain Review Plan.

If Agency guidance or standard industry practices are not used by the U.S. Department of Energy, evaluate whether the U.S. Department of Energy basis and justification for choosing a particular hazard and initiating event identification method are defensible.

Confirm that methods selected for hazard and initiating event identification are appropriate for the available data on the site and geologic repository operations area. Review descriptions of the site and its structures, systems, and components using Sections 2.1.1.1 ("Site Description as it Pertains to Preclosure Safety") and 2.1.1.2 ("Description of Structures, Systems, Components, Equipment, and Operational Process Activities") of the Yucca Mountain Review Plan for this purpose.

Confirm that assumptions used to identify naturally occurring and human-induced hazards and initiating events are well-defined and have adequate technical basis, and are supported by information in Section 2.1.1.1 ("Site Description as it Pertains to Preclosure Safety Analysis") and Section 2.1.1.2 ("Description of Structures, Systems, Components, Equipment, and Operational Process Activities") of the Yucca Mountain Review Plan.

Review Method 2 Use of Relevant Data for Identification of Site-Specific Hazards and Initiating Events

Verify that appropriate site-specific data (including frequency of occurrence, where relevant) have been used to identify naturally occurring and human-induced hazards and initiating events, such as:

(1) Seismicity and faulting;
(2) Winds and tornadoes;
(3) Volcanic activity;
(4) Slope instability;
(5) Other extreme meteorological or geological conditions; and
(6) Human-induced events.

Coordinate with the reviewers for Section 2.2 ("Repository Safety after Permanent Closure") of the Yucca Mountain Review Plan, to verify that naturally occurring hazards (e.g., seismicity, faulting, and igneous activity) identified in this section are consistent with the list of features, events, and processes.

Verify that the appropriate properties and factors are considered in determining the adequacy of the hazard and initiating event identification, such as:

(1) Heat generation from the high-level radioactive waste;
(2) Flammable, corrosive, pressurized, and toxic materials;
(3) Conditions under which available fissionable material could pose a criticality hazard; and
(4) Potential interactions among hazardous materials and conditions.

Confirm that the identification of human-induced hazards encompasses relevant aspects of the geologic repository operations area radiological systems. In particular, consider the list of such systems evaluated, using Section 2.1.1.2 ("Description of Structures, Systems, Components, Equipment, and Operational Process Activities") of the Yucca Mountain Review Plan. Confirm that the identification of hazards encompasses all modes of operation. Modes of operation include normal process operations; maintenance (e.g., shutting down critical equipment); and backfilling operations (if included in the license application design) within waste emplacement drifts.

Consult with reviewers of Section 2.1.1.2.3 ("Acceptance Criteria—Description of Structures, Systems, Components, Equipment, and Operational Process Activities") of the Yucca Mountain Review Plan, to verify that system descriptions used to support hazard and initiating event identification are adequate.

Review Method 3 Determination of Frequency or Probability of Occurrence of Hazards and Initiating Events

Confirm that methods selected for determining probability or frequency of occurrence for hazards and initiating events are appropriate. Also, verify that uncertainties associated with the frequency or probability estimates are quantified.

If Agency guidance or standard industry practices are not used by the U.S. Department of Energy, evaluate whether the U.S. Department of Energy basis and justification for choosing the method(s) used to determine the frequency or probability of occurrence of hazards and initiating events are defensible.

If relevant frequency or probability data are insufficient or not available, verify that appropriate probability estimates are used, and defensible technical bases are provided. Also, evaluate the adequacy of the associated probability estimation method (e.g., bounding Bayesian, expert elicitation). If expert elicitation was used, review the expert elicitation process using Section 2.5.4 ("Expert Elicitation") of the Yucca Mountain Review Plan.

Consult with the reviewers of Section 2.2.1.2.1 ("Identification of Features, Events, and Processes Affecting Compliance with the Overall Performance Objectives") of the Yucca Mountain Review Plan, to confirm the validity of the frequencies and/or probabilities established for naturally occurring events. Also, assess the validity of the frequencies and/or probabilities established for human-induced hazards and initiating events.

Verify that human errors that may lead to radiological consequences are adequately identified, and that adequate human reliability analyses are performed. Confirm that the U.S. Department of Energy provides an adequate technical basis for any human reliability method used, its range of applicability, and its assumptions and uncertainties. Guidance from documents such as NUREG–1278 (U.S. Nuclear Regulatory Commission, 1983b); NUREG–1624 (U.S. Nuclear Regulatory Commission, 2000b); and NUREG–2300 (U.S. Nuclear Regulatory Commission, 1983a), can assist the review.

Review Method 4 Technical Basis for Inclusion or Exclusion of Specific Hazards and Initiating Events

Verify that adequate technical bases for the inclusion and exclusion of hazards and initiating events are provided.

Confirm that technical bases are defensible and consistent with site and system information reviewed in Sections 2.1.1.1 ("Site Description as it Pertains to Preclosure Safety Analysis") and 2.1.1.2 ("Description of Structures, Systems, Components, Equipment, and Operational Process Activities") of the Yucca Mountain Review Plan.

Confirm that technical bases include consideration of uncertainties.

Review Method 5 List of Hazards and Initiating Events To Be Considered in the Preclosure Safety Analysis

Verify that the U.S. Department of Energy list of hazards and initiating events contains the credible natural and human-induced events.

Perform limited independent assessment to confirm that the list of hazards and initiating events that may result in radiological releases is acceptable.

2.1.1.3.3 Acceptance Criteria

The following acceptance criteria are based on meeting the requirements of 10 CFR 63.112(b) and (d), relating to the identification of hazards and initiating events.

Acceptance Criterion 1 Technical Bases and Assumptions for Methods Used for Identification of Hazards and Initiating Events are Adequate.

(1) Methods used for hazard and initiating event identification are consistent with standard industry practices;

(2) If Agency guidance or standard industry practices are not used, the U.S. Department of Energy basis and justification for choosing particular hazard and initiating event identification method(s) are defensible;

(3) Methods selected for hazard and initiating event identification are appropriate for the available data on the site and geologic repository operations area; and

(4) Assumptions used to identify naturally occurring and human-induced hazards and initiating events are well-defined, have adequate technical basis, and are supported by information on the site and its structures, systems, components, equipment and operational processes.

Acceptance Criterion 2 Site Data and System Information Are Appropriately Used in Identification of Hazards and Initiating Events.

(1) Appropriate site-specific data are used to identify naturally occurring hazards and initiating events;

(2) In determining the adequacy of the hazard and initiating event identification, the appropriate properties and factors are considered; and

(3) The identification of human-induced hazards encompasses relevant aspects of the geologic repository operations area radiological systems. The identification of hazards encompasses all geologic repository operations area modes of operation.

Acceptance Criterion 3 Determination of Frequency or Probability of Occurrence of Hazards and Initiating Events is Acceptable.

(1) Methods selected for determining probability or frequency of occurrence for hazards and initiating events are appropriate, and uncertainties are adequately quantified;

(2) An appropriate basis and justification are provided for any use of nonstandard practices for determining frequency or probability estimates;

(3) The frequencies and/or probabilities established for naturally occurring events and human-induced hazards and initiating events are valid; and

(4) Human errors that may lead to radiological consequences are adequately identified, and adequate human reliability analyses are performed.

Acceptance Criterion 4 Adequate Technical Bases for the Inclusion and Exclusion of Hazards and Initiating Events are Provided.

(1) The technical bases are defensible and consistent with site and system information; and

(2) The technical bases include adequate consideration of uncertainties, associated with frequency or probability of the hazards and initiating events.

Acceptance Criterion 5 The List of Hazards and Initiating Events That May Result in Radiological Releases is Acceptable.

(1) The U.S. Department of Energy list of hazards and initiating events contains the credible natural and human-induced events; and

(2) Independent assessment confirms that the list of hazards and initiating events that may result in radiological releases is acceptable.

2.1.1.3.4 Evaluation Findings

If the license application provides sufficient information and the regulatory acceptance criteria in Section 2.1.1.3.3 are appropriately satisfied, the staff concludes that this portion of the staff evaluation is acceptable. The reviewer writes material suitable for inclusion in the safety evaluation report prepared for the entire application. The report includes a summary statement of what was reviewed and why the reviewer finds the submittal acceptable. The staff can document the review as follows.

U.S. Nuclear Regulatory Commission staff has reviewed the Safety Analysis Report and other information submitted in support of the license application and has found, with reasonable assurance, that the requirements of 10 CFR 63.112(b) are satisfied, related to identification of hazards and initiating events. The naturally occurring and human-induced hazards and potential initiating events have been adequately identified. The identification of the initiating events and the associated probabilities of occurrence is acceptable.

U.S. Nuclear Regulatory Commission staff has reviewed the Safety Analysis Report and other information submitted in support of the license application and has found, with reasonable assurance, that the requirements of 10 CFR 63.112(d) are satisfied. An adequate technical basis for either inclusion or exclusion of specific naturally occurring or human-induced hazards and initiating events has been provided.

2.1.1.3.5 References

American Institute of Chemical Engineers. “Guidelines for Hazard Evaluation Procedures, Second Edition with Worked Examples.” New York City, New York: American Institute of Chemical Engineers, Center for Chemical Process Safety. 1992.

U.S. Nuclear Regulatory Commission. NUREG–1513, “Integrated Safety Analysis Guidance Document.” Washington, DC: U.S. Nuclear Regulatory Commission. 2001.

———. NUREG–1520, “Standard Review Plan for the Review of a License Application for a Fuel Cycle Facility, Draft Report.” Washington, DC: U.S. Nuclear Regulatory Commission. 2000a.

———. NUREG–1624, "Technical Basis and Implementation Guidelines for A Technique for Human Event Analysis (ATHEANA)." Revision 1. Washington, DC: U.S. Nuclear Regulatory Commission. 2000b.

———. NUREG/CR–2300, "PRA Procedures Guide–A Guide to the Performance of Probabilistic Risk Assessments for Nuclear Power Plants." Washington, DC: U.S. Nuclear Regulatory Commission. 1983a.

———. NUREG–1278, "Handbook of Human Reliability Analysis with Emphasis on Nuclear Power Plant Application." Washington, DC: U.S. Nuclear Regulatory Commission. 1983b.

2.1.1.4 Identification of Event Sequences

Review Responsibilities—High-Level Waste Branch and Environmental and Performance Assessment Branch

2.1.1.4.1 Areas of Review

This section provides guidance on the review of the identification of event sequences considered in the preclosure safety analysis. Reviewers will also evaluate the information required by 10 CFR 63.21(c)(5).

The staff will evaluate the following parts of the identification of event sequences, using the review methods and acceptance criteria in Sections 2.1.1.4.2 and 2.1.1.4.3.

(1) Technical bases for methods used and assumptions made for identification of event sequences, and

(2) Categories 1 and 2 event sequences.

2.1.1.4.2 Review Methods

Review Method 1 Technical Basis and Assumptions for Methods Used for Identification of Event Sequences

Verify that methods selected for event sequence identification are appropriate, and are consistent with standard practices [e.g., NUREG/CR–2300 (U.S. Nuclear Regulatory Commission, 1983); and American Institute of Chemical Engineers (1992), Appendixes A and B].

Verify that an appropriate basis and justification are provided for use of any nonstandard method.

Confirm that methods selected are consistent with, and supported by, site-specific data.

Verify that assumptions made in identifying the event sequences are justified and valid.

Review Method 2 Categories 1 and 2 Event Sequences

Verify that the U.S. Department of Energy has properly considered the hazards and initiating events reviewed in Section 2.1.1.3 ("Identification of Hazards and Initiating Events") of the Yucca Mountain Review Plan. Confirm that the U.S. Department of Energy has appropriately applied the methods for identification of event sequences.

Verify that the potentially relevant human factors reviewed using Section 2.1.1.3 ("Identification of Hazards and Initiating Events") of the Yucca Mountain Review Plan have been appropriately considered in event sequence identification. Verify that the analysis has applied methods for human reliability analysis similar to those shown to be acceptable in licensing other facilities [e.g., NUREG–1624 ("Technical Basis and Implementation Guidelines for a Technique for Human Event Analysis (ATHEANA)"] (U.S. Nuclear Regulatory Commission, 2000a).

Verify that the U.S. Department of Energy has considered reasonable combinations of initiating events and the associated event sequences that could lead to exposure of individuals to radiation.

Verify that Category 1 event sequences include those sequences that are expected to occur one or more times before permanent closure of the geologic repository operations area.

Verify that Category 2 event sequences include those event sequences that have at least one chance in 10,000 of occurring before permanent closure. Confirm that the methods and technical bases for determining those event sequence probabilities are consistent with the applicable governing regulation, policy, and guidance.

Perform limited independent assessments as appropriate to confirm that possible event sequences that may lead to radiological releases have been adequately identified, and to verify that the U.S. Department of Energy analyses and calculations were performed properly.

2.1.1.4.3 Acceptance Criteria

The following acceptance criteria are based on meeting the requirements of 10 CFR 63.112(b), relating to the identification of event sequences.

Acceptance Criterion 1 Adequate Technical Basis and Justification are Provided for the Methodology Used and Assumptions Made to Identify Preclosure Safety Analysis Event Sequences.

(1) Methods selected for event sequence identification are appropriate, and are consistent with Agency guidance or standard industry practices or are adequately justified;

(2) The methods selected are consistent with, and supported by, site-specific data; and

(3) Assumptions made in identifying event sequences are valid and reasonable.

Acceptance Criterion 2 Categories 1 and 2 Event Sequences are Adequately Identified.

(1) The U.S. Department of Energy has adequately considered the relevant hazards and initiating events. Methods selected for identification of event sequences have been applied properly;

(2) The potentially relevant human factors are appropriately considered in the event sequence identification;

(3) The U.S. Department of Energy considers reasonable combinations of initiating events and the associated event sequences that could lead to exposure of individuals to radiation;

(4) Category 1 event sequences are identified on the basis that they could occur one or more times before permanent closure of the geologic repository operations area, and the technical methods used to determine the event sequences are acceptable;

(5) Category 2 event sequences include all those event sequences that have at least 1 chance in 10,000 of occurring during the preclosure period, and the technical methods used to determine the probabilities of occurrence are acceptable; and

(6) Limited independent assessments confirm that possible event sequences that may cause radiological releases are adequately identified, and related U.S. Department of Energy analyses and calculations are performed properly.

2.1.1.4.4 Evaluation Findings

If the license application provides sufficient information and the regulatory acceptance criteria in Section 2.1.1.4.3 are appropriately satisfied, the staff concludes that this portion of the staff evaluation is acceptable. The reviewer writes material suitable for inclusion in the safety evaluation report prepared for the entire application. The report includes a summary statement of what was reviewed and why the reviewer finds the submittal acceptable. The staff can document the review as follows.

U.S. Nuclear Regulatory Commission staff has reviewed the Safety Analysis Report and other information submitted in support of the license application and has found, with reasonable assurance, that, in part, the requirements of 10 CFR 63.112(b) are satisfied. An identification and analysis of potential event sequences has been provided.

2.1.1.4.5 References

American Institute of Chemical Engineers. "Guidelines for Hazard Evaluation Procedures, Second Edition with Worked Examples." New York City, New York: American Institute of Chemical Engineers, Center for Chemical Process Safety. 1992.

Electric Power Research Institute. "PSA Applications Guide." EPRI TR–105396. Walnut Creek, California: Electric Power Research Institute. 1995.

U.S. Nuclear Regulatory Commission. NUREG–1624, "Technical Basis and Implementation Guidelines for a Technique for Human Event Analysis (ATHEANA)." Revision 1. Washington, DC: U.S. Nuclear Regulatory Commission. 2000a.

———. NUREG/CR–2300, "PRA Procedures Guide–A Guide to the Performance of Probabilistic Risk Assessments for Nuclear Power Plants." Washington, DC: U.S. Nuclear Regulatory Commission. 1983.

2.1.1.5 Consequence Analyses

2.1.1.5.1 Consequence Analysis Methodology and Demonstration that the Design Meets 10 CFR Parts 20 and 63 Numerical Radiation Protection Requirements for Normal Operations and Category 1 Event Sequences

Review Responsibilities—High-Level Waste Branch and Environmental and Performance Assessment Branch

2.1.1.5.1.1 Areas of Review

This section provides guidance on the review of the consequence analysis methodology and demonstration that the design meets 10 CFR Parts 20 and 63 numerical radiation protection requirements for normal operations and Category 1 event sequences. The reviewers will also evaluate the information required by 10 CFR 63.21(c)(5).

The staff will evaluate the following parts of consequence analysis methodology and demonstration that the design meets 10 CFR Parts 20 and 63 numerical radiation protection requirements for normal operations and Category 1 event sequences, using the review methods and acceptance criteria in Sections 2.1.1.5.1.2 and 2.1.1.5.1.3.

(1) Consequence evaluations for normal operations and Category 1 event sequences;

(2) On-site and off-site doses during normal operations and Category 1 event sequences; and

(3) Compliance with performance objectives.

2.1.1.5.1.2 Review Methods

Review Method 1 Consequence Analyses of Normal Operations, Category 1 Event Sequences, and Factors that Allow an Event Sequence to Propagate within the Geologic Repository Operations Area

Confirm that the U.S. Department of Energy has conducted consequence analyses for normal operations and Category 1 event sequences, that were reviewed using Section 2.1.1.4

("Identification of Event Sequences") of the Yucca Mountain Review Plan. Verify that consequence analyses consider the following:

(1) Hazard event sequences that could lead to radiological consequences (including the controls used to prevent or mitigate the event sequences);

(2) Interactions of identified hazards and proposed controls;

(3) Modes of geologic repository operations area operation, including normal process operations; maintenance (e.g, shutting down critical equipment); removal of damaged waste disposal containers from subsurface to surface facilities; and backfilling operations (if included in the license application design) within waste emplacement drifts. Analyses should assume that operations will be carried out at the maximum capacity and rate of receipt of radioactive waste stated in the license application; and

(4) Descriptions of event sequences for which consequences (radiation dose) will be determined, including information on the hazard, structures, systems, and components that take part in the event sequences, and controls relied on to prevent or mitigate the event sequences.

Review Method 2 Assessment of Calculations of Consequences to Workers and Members of the Public from Normal Operations and Category 1 Event Sequences

Evaluate methods used to perform the consequence (radiation dose) calculations. Verify that adequate technical bases for selecting these methods have been provided. Confirm that adequate technical bases have been provided for assumptions used for the calculations and methods. Confirm methods are consistent with site-specific data and system design and process information that were evaluated using Sections 2.1.1.1 ("Site Description as it Pertains to Preclosure Safety Analysis") and 2.1.1.2 ("Description of Structures, Systems, Components, Equipment, and Operational Process Activities") of the Yucca Mountain Review Plan.

Evaluate the identification of the real members of the public likely to receive doses from geologic repository operations area normal operations or Category 1 event sequences, and the rationale for this identification. Confirm that the annual dose to any real member of the public located beyond the site boundary is below the limits at 10 CFR 63.111(a)(2).

Verify that input data and information used for the consequence analyses are identified, and are consistent with site-specific data and system design and process information. Verify that adequate technical bases are provided for selection of this input data and information and that uncertainty in the input data is appropriately considered in the consequences analyses.

Evaluate the calculation of the source term, and confirm the following:

(1) Characteristics of the high-level radioactive waste used in the source term calculation (e.g., enrichment, burnup, and decay time) reasonably represent or bound the range of characteristics of waste that will be handled at the geologic repository operations area, as reviewed using Section 2.1.1.2 ("Description of Structures, Systems, Components,

Equipment, and Operational Process Activities") of the Yucca Mountain Review Plan; and

(2) The type, quantity, and concentration of airborne radionuclides released during normal operations and Category 1 event sequences are supported by appropriate data, or are in accordance with appropriate U.S. Nuclear Regulatory Commission guidance documents, such as NUREG–1567 (U.S. Nuclear Regulatory Commission, 2000).

Evaluate the calculations of on-site and off-site direct exposures, during normal operations and Category 1 event sequences, and confirm the following:

(1) The analyses are consistent with commonly acceptable shielding calculations, such as the guidance in NUREG–1567 (U.S. Nuclear Regulatory Commission, 2000), and are provided in sufficient detail to allow independent confirmatory calculations;

(2) Credit taken for shielding materials that reduce direct exposure dose rates is appropriate, and accounts for any degradation that may occur as a result of the event sequences;

(3) Methodologies used in shielding analyses are appropriate for the radiation types and geometries and materials modeled, and have been validated using dose rate measurements from similar facilities; and

(4) Flux-to-dose conversion factors, atmospheric dispersion data, and cross-section data used in the analyses are consistent with accepted practice, such as is documented in American National Standards Institute/American Nuclear Society 6.1.1 and American National Standards Institute/American Nuclear Society 6.1.2 (American Nuclear Society Standards Committee Working Group, 1977, 1989).

Evaluate the calculations of dose to workers and members of the public from airborne radionuclides, during normal operations and after Category 1 event sequences, and confirm the following:

(1) Credit taken for the use of ventilation and filtration systems in mitigating the release of airborne radioactive materials is appropriate;

(2) For the calculation of dose to the public from airborne radionuclides:

    (a) Airborne transport modeling uses acceptable methods, such as those outlined in Regulatory Guide 1.109 (U.S. Nuclear Regulatory Commission, 1977) for routine releases; and

    (b) Appropriate exposure pathways are considered, such as direct exposure to airborne radionuclides, inhalation of airborne radionuclides, and pathways associated with radionuclides deposited on the ground in the receptor location, for potential long-term exposure of the receptor.

(3) For the calculation of dose to workers from airborne radionuclides:

  (a) The calculation of airborne radioactivity concentrations within the geologic repository operations area uses times and levels of elevated airborne radioactivity concentrations that are reasonable or conservative, based on technically defensible data; and

  (b) The times that workers are assumed to be exposed to elevated radiation fields and airborne concentrations of radioactivity are reasonable or conservative, based on technically defensible data.

(4) The inhalation dose conversion factors used in the analyses are standard for dose assessments, such as those in Federal Guidance Report #11 (Eckerman, et al., 1992).

Review Method 3 Limitation of Dose to Workers and Members of the Public from Normal Operations and Category 1 Event Sequences to Within Limits Specified in 10 CFR 63.111(a)

Confirm that normal operations and Category 1 event sequences that could adversely affect radiological exposures have been considered.

Verify that an appropriate method has been used to aggregate the doses from normal operations and annualized doses from Category 1 event sequences.

Verify that the dose to workers and members of the public from normal operations and Category 1 event sequences will not exceed the limits specified in 10 CFR 63.111(a).

Confirm that the doses to workers and members of the public will be as low as is reasonably achievable, as evaluated using Section 2.1.1.8 ("Meeting the 10 CFR Part 20 As Low As Is Reasonably Achievable Requirements for Normal Operations and Category 1 Event Sequences") of the Yucca Mountain Review Plan.

2.1.1.5.1.3 Acceptance Criteria

The following acceptance criteria are based on meeting the requirements of 10 CFR 63.111(a)(1), (a)(2), (b)(1), (c)(1), and (c)(2), relating to consequence analysis methodology and demonstration that the design meets 10 CFR Parts 20 and 63 numerical radiation protection requirements for normal operations and Category 1 event sequences.

Acceptance Criterion 1 Consequence Analyses Adequately Assess Normal Operations and Category 1 Event Sequences, as Well as Factors That Allow an Event Sequence to Propagate Within the Geologic Repository Operations Area.

(1) The U.S. Department of Energy conducts consequence analyses for normal operations and Category 1 event sequences that adequately consider hazard event sequences that could lead to radiological consequences, interactions of identified hazards and proposed

controls, and all modes of geologic repository operations area operation. Analyses assume that operations are carried out at the maximum capacity and rate of receipt of radioactive waste stated in the license application. The consequence analyses provide details on the related hazard and the structures, systems, and components, and controls that are relied on to prevent or mitigate event sequences.

Acceptance Criterion 2 Consequence Calculations Adequately Assess the Consequences to Workers and Members of the Public From Normal Operations and Category 1 Event Sequences.

(1) Adequate methods are used to perform the consequence calculations, and adequate technical bases are provided for selecting these methods. Adequate technical bases are also provided for assumptions used for the calculations and methods. The selected methods are consistent with site-specific data and system design and process information;

(2) The identification of the member of the public likely to receive the highest dose from geologic repository operations area normal operations or Category 1 event sequences is adequate, and the rationale for this identification is adequate. The dose to this individual bounds the annual dose to any real member of the public located beyond the site boundary;

(3) Input data and information used for the consequence analysis are identified, and are consistent with site-specific data and system design and process information. Adequate technical bases are provided for selection of this data and information and uncertainty in this input data is appropriately considered in the consequence analyses;

(4) The calculation of the source term is acceptable, and is based on the following:

  (a) Characteristics of the high-level radioactive waste used in the source term calculation reasonably represent or bound the range of characteristics of waste that will be handled at the geologic repository operations area; and

  (b) The type, quantity, and concentration of airborne radionuclides released during normal operations and Category 1 event sequences are supported by appropriate data, or are in accordance with U.S. Nuclear Regulatory Commission guidance documents.

(5) The calculations of on-site and off-site direct exposures, during normal operations and Category 1 event sequences, are based on the following:

  (a) The analyses are consistent with commonly acceptable shielding calculations, and are provided in sufficient detail to allow independent confirmatory calculations;

(b) Credit taken for shielding materials that reduce direct exposure dose rates is appropriate and accounts for any degradation that may occur as a result of the event sequences;

(c) Methodologies used in any shielding analyses are appropriate for the radiation types and geometries and materials modeled, and are validated using dose rate measurements from similar facilities; and

(d) Flux-to-dose conversion factors, atmospheric dispersion data, and cross-section data used in the analyses are consistent with accepted practice.

(6) The calculations of dose to workers and members of the public from airborne radionuclides, during normal operations and after Category 1 event sequences, are adequate and are based on the following:

(a) Credit taken for the use of ventilation and filtration systems in mitigating the release of airborne radioactive materials is appropriate;

(b) For the calculation of dose to the public from airborne radionuclides, airborne transport modeling is conducted using acceptable methods, and the U.S. Department of Energy considers appropriate exposure pathways;

(c) For the calculation of dose to workers from airborne radionuclides, the calculation of airborne radioactivity concentrations within the geologic repository operations area uses times and levels of elevated airborne radioactivity concentrations that are reasonable, based on technically defensible data. The times that workers are assumed to be exposed to elevated radiation fields and airborne concentrations of radioactivity are reasonable, based on technically defensible data; and

(d) The inhalation dose conversion factors used in the analyses are appropriate for dose assessments.

Acceptance Criterion 3 The Dose to Workers and Members of the Public From Normal Operations and Category 1 Event Sequences is Within the Limits Specified in 10 CFR 63.111(a).

(1) Normal operations and Category 1 event sequences that could adversely affect radiological exposures are adequately considered;

(2) An appropriate method is used to aggregate annual doses from normal operations and annualized doses from Category 1 event sequences;

(3) Doses to workers and members of the public from normal operations and Category 1 event sequences will not exceed the limits in 10 CFR 63.111(a); and

(4) Doses to workers and members of the public will be as low as is reasonably achievable.

2.1.1.5.1.4 Evaluation Findings

If the license application provides sufficient information and the regulatory acceptance criteria in Section 2.1.1.5.1.3 are appropriately satisfied, the staff concludes that this portion of the staff evaluation is acceptable. The reviewer writes material suitable for inclusion in the safety evaluation report prepared for the entire application. The report includes a summary statement of what was reviewed and why the reviewer finds the submittal acceptable. The staff can document the review as follows.

U.S. Nuclear Regulatory Commission staff has reviewed the Safety Analysis Report and other information submitted in support of the license application and has found, with reasonable assurance, that the requirements of 10 CFR 63.111(a)(1) are satisfied. Performance objectives for the geologic repository operations area, up to the time of permanent closure, have been met in that the radiation exposure limits in 10 CFR Part 20 will not be exceeded.

U.S. Nuclear Regulatory Commission staff has reviewed the Safety Analysis Report and other information submitted in support of the license application and has found, with reasonable assurance, that the requirements of 10 CFR 63.111(a)(2) are satisfied. Performance objectives for the geologic repository operations area up to the time of permanent closure have been met in that, during normal operations and for Category 1 event sequences, the annual dose to any real member of the public, located beyond the boundary of the site, will not exceed 0.15 mSv [15 mrem].

U.S. Nuclear Regulatory Commission staff has reviewed the Safety Analysis Report and other information submitted in support of the license application and has found, with reasonable assurance, that the requirements of 10 CFR 63.111(b)(1) are satisfied. The geologic repository operations area has been designed such that, taking into consideration normal operation and Category 1 event sequences, radiation exposures, radiation levels, and releases of radioactive materials will be maintained, within the limits of 10 CFR 63.111(a).

U.S. Nuclear Regulatory Commission staff has reviewed the Safety Analysis Report and other information submitted in support of the license application and has found, with reasonable assurance, that the requirements of 10 CFR 63.111(c)(1) are satisfied. The preclosure safety analysis meets the requirements specified at 10 CFR 63.112, and demonstrates that the radiation protection limits of 10 CFR Part 20 will be met. During normal operations and Category 1 event sequences, the annual dose to any real member of the public, located beyond the boundary of the site, will not exceed 0.15 mSv [15 mrem].

U.S. Nuclear Regulatory Commission staff has reviewed the Safety Analysis Report and other information submitted in support of the license application and has found, with reasonable assurance, that the requirements of 10 CFR 63.111(c)(2) are satisfied. The preclosure safety analysis meets the requirements specified at 10 CFR 63.112, and demonstrates that the preclosure numerical radiation protection requirements will be met for geologic repository operations area normal operations and Category 1 event sequences.

2.1.1.5.1.5 References

American Nuclear Society Standards Committee Working Group. "Neutron and Gamma-Ray Cross Sections for Nuclear Radiation Protection Calculations for Nuclear Power Plants." American Nuclear Society–6.1.2–1989. Washington, DC: American National Standards Institute. 1989.

———. "Neutron and Gamma Ray Flux-to-Dose-Rate Factors." American National Standards Institute/American Nuclear Society–6.1.1–1977. Washington, DC: American National Standards Institute. 1977.

Eckerman, K.F., A.B. Wolbarst, and A. Richardson. Federal Guidance Report #11, "Limiting Values of Radionuclide Intake and Air Concentration and Dose Conversion Factors for Inhalation, Submersion, and Ingestion." Springfield, Virginia: U.S. Department of Commerce. 1992.

U.S. Nuclear Regulatory Commission. NUREG–1567, "Standard Review Plan for Spent Fuel Storage Facilities, Final Report." Washington, DC: U.S. Nuclear Regulatory Commission. 2000.

———. Regulatory Guide 1.109, "Calculation of Annual Doses to Man From Routine Releases of Reactor Effluents for the Purpose of Evaluating Compliance with 10 CFR Part 50." Appendix I. Washington, DC: U.S. Nuclear Regulatory Commission, Office of Standards Development. 1977.

2.1.1.5.2 Demonstration That the Design Meets 10 CFR Part 63 Numerical Radiation Protection Requirements for Category 2 Event Sequences

Review Responsibilities—High-Level Waste Branch and Environmental and Performance Assessment Branch

2.1.1.5.2.1 Areas of Review

This section provides guidance on the review of the design meeting 10 CFR Part 63 numerical radiation protection requirements for Category 2 event sequences. Reviewers will also evaluate the information required by 10 CFR 63.21(c)(5).

The staff will evaluate the following parts of the design meeting 10 CFR Part 63 numerical radiation protection requirements for Category 2 event sequences, using the review methods and acceptance criteria in Sections 2.1.1.5.2.2 and 2.1.1.5.2.3.

(1) Consequence evaluations for Category 2 event sequences;

(2) Off-site doses for Category 2 event sequences; and

(3) Compliance with performance objectives.

2.1.1.5.2.2 Review Methods

Review Method 1 Consequence Analyses of Category 2 Event Sequences and Factors That Allow An Event Sequence to Propagate Within the Geologic Repository Operations Area

Verify that the U.S. Department of Energy has conducted consequence analyses for Category 2 event sequences that were reviewed using Section 2.1.1.4 ("Identification of Event Sequences") of the Yucca Mountain Review Plan. Verify that consequence analyses consider the following:

(1) Hazard event sequences that could result in radiological consequences (including the controls used to prevent or mitigate the event sequences);

(2) Interactions of identified hazards and proposed controls;

(3) Whether the U.S. Department of Energy analyses assume that operations will be carried out at the maximum capacity and rate of receipt of radioactive waste stated in the license application; and

(4) Descriptions of event sequences for which consequences (radiation dose) will be determined, including information on the hazard; structures, systems, and components that take part in the event sequences; and controls relied on to prevent or mitigate the event sequences.

Review Method 2 Assessment of Calculations of Consequences to Members of the Public from Category 2 Event Sequences

Evaluate the methods used to perform consequence calculations, and verify that adequate technical bases for selecting these methods have been provided. Verify that adequate technical bases have also been provided for assumptions used for the calculations and methods. Confirm that methods are consistent with site-specific data and system design and process information that was evaluated using Sections 2.1.1.1 ("Site Description as it Pertains to Preclosure Safety Analysis") and 2.1.1.2 ("Description of Structures, Systems, Components, Equipment, and Operational Process Activities") of the Yucca Mountain Review Plan.

Evaluate the identification of the hypothetical member of the public, located on or beyond the site boundary, likely to receive the highest dose from the geologic repository operations area during a Category 2 event sequence, and the rationale for this identification.

Confirm that input data and information used for the consequence analyses are identified, and are consistent with site-specific data and system design and process information. Verify that adequate technical bases are provided for selection of this input data and information and that uncertainty in the input data is appropriately considered in the consequence analyses.

Evaluate the calculation of the source term, and confirm the following:

(1) Characteristics of the high-level radioactive waste used in the source term calculation (e.g., enrichment, burnup, and decay time) reasonably represent or bound the range of characteristics of waste that will be handled at the geologic repository operations area, as reviewed using Section 2.1.1.2 ("Description of Structures, Systems, Components, Equipment, and Operational Process Activities") of the Yucca Mountain Review Plan; and

(2) The type, quantity, and concentration of airborne radionuclides that could be released during Category 2 event sequences are supported by appropriate data and analyses, or are estimated in accordance with guidance documents, such as NUREG–1567 (U.S. Nuclear Regulatory Commission, 2000).

Evaluate the calculations of off-site dose from direct exposure after Category 2 event sequences, and confirm the following:

(1) The analyses are consistent with commonly acceptable shielding calculations, such as the guidance in NUREG–1567 (U.S. Nuclear Regulatory Commission, 2000). The analyses are provided in sufficient detail to allow independent confirmatory calculations;

(2) Credit taken for shielding materials that reduce direct exposure dose rates is appropriate, and accounts for any degradation that may occur as a result of the event sequences;

(3) Methodologies used in shielding analyses are appropriate for the radiation types and geometries and materials modeled, and have been validated using dose rate measurements from similar facilities;

(4) The time a member of the public is assumed to be exposed to elevated levels of radiation from Category 2 event sequences is reasonable. This time is based on the amount of time required for the facility to recover from the event sequence; and

(5) Flux-to-dose conversion factors and cross-section data used in the analyses are consistent with accepted practice, such as is documented in "American National Standards Institute/American Nuclear Society 6.1.1" and "American National Standards Institute/American Nuclear Society 6.1.2," (American Nuclear Society Standards Committee Working Group, 1977, 1989).

Evaluate the calculation of dose to members of the public from airborne radionuclides after Category 2 event sequences, and confirm that:

(1) Credit taken for the use of ventilation and filtration systems in mitigating the release of airborne radioactive materials is appropriate. The analyses consider credible damage to the ventilation system that may result from event sequences, such as ventilation duct collapse, fan failure, or filter blowout;

(2) Airborne transport modeling uses an acceptable method, such as that outlined in Regulatory Guide 1.145, "Atmospheric Dispersion Models for Potential Accident Consequence Assessments at Nuclear Power Plants" (U.S. Nuclear Regulatory Commission, 1983);

(3) The U.S. Department of Energy has considered appropriate exposure pathways, such as:

 (a) Direct exposure to airborne radionuclides;

 (b) Inhalation of airborne radionuclides;

 (c) Pathways associated with radionuclides deposited on the ground in the receptor location for potential long-term exposure of the receptor. This pathway may be omitted if the site emergency plan [reviewed using Section 2.5.7 ("Emergency Plan") of the Yucca Mountain Review Plan] has provisions to mitigate doses to members of the public after any accident that releases significant quantities of radioactive material;

(4) The time that a member of the public is assumed to be exposed to airborne radioactive materials from Category 2 event sequences is reasonable, and is based on the time that radioactive effluents are released from the geologic repository operations area, and

(5) The inhalation dose conversion factors used in the analyses are standard for dose assessments, such as those in Federal Guidance Report #11 (Eckerman, et al., 1992).

Review Method 3 Limitation of Dose to Hypothetical Members of the Public from Category 2 Event Sequences to Limits Specified in 10 CFR 63.111(b)(2)

Confirm that Category 2 event sequences that could adversely affect radiological exposures have been considered. Also, verify that no identified Category 2 event sequence will lead to a dose to a member of the public that exceeds the dose limit in 10 CFR 63.111(b)(2).

2.1.1.5.2.3 Acceptance Criteria

The following acceptance criteria are based on meeting the requirements of 10 CFR 63.111(b)(2) and (c), relating to the design meeting 10 CFR Part 63 numerical radiation protection requirements for Category 2 event sequences.

Acceptance Criterion 1 Consequence Analyses Include Category 2 Event Sequences as Well as Factors That Allow an Event Sequence to Propagate Within the Geologic Repository Operations Area.

(1) The U.S. Department of Energy conducts consequence analyses for Category 2 event sequences that adequately consider hazard event sequences that could lead to radiological consequences, interactions of identified hazards and proposed controls, and the maximum capacity and rate of receipt of radioactive waste. The consequence

analyses incorporate performance of the structures, systems, and components and controls that are relied on to prevent or mitigate event sequences.

Acceptance Criterion 2 Consequence Calculations Adequately Assess the Consequences to Members of the Public from Category 2 Event Sequences.

(1) Adequate methods are used to perform the consequence calculations, and adequate technical bases are provided for selecting these methods. Adequate technical bases are also provided for assumptions used for the calculations and methods. The selected methods are consistent with site-specific data and system design and process information;

(2) The identification of the hypothetical member of the public, located on or beyond the site boundary, likely to receive the highest dose from the geologic repository operations area during a Category 2 event sequence, is adequate, and the rationale for this identification is adequate;

(3) Input data and information used for the consequence analysis are identified, and are consistent with site-specific data and system design and process information. Adequate technical bases are provided for their selection and uncertainty in the input data is appropriately considered in the consequence analyses;

(4) The calculation of the source term is based on the following:

  (a) Characteristics of the high-level radioactive waste used in the source term calculation reasonably represent or bound the range of characteristics of waste that will be handled at the geologic repository operations area; and

  (b) The type, quantity, and concentration of airborne radionuclides that could be released during Category 2 event sequences are supported by appropriate data and analyses, or are estimated in accordance with U.S. Nuclear Regulatory Commission guidance documents.

(5) The calculations of off-site dose from direct exposure after Category 2 event sequences are adequate, and are based on the following:

  (a) The analyses are consistent with commonly acceptable shielding calculations, and are provided in sufficient detail to allow independent confirmatory calculations;

  (b) Credit taken for shielding materials that reduce direct exposure dose rates is appropriate, and accounts for any degradation that may occur as a result of the event sequence;

(c) Methodologies used in any shielding analyses are appropriate for the radiation types and geometries and materials modeled, and are validated using dose rate measurements from similar facilities;

(d) The time that a member of the public is assumed to be exposed to elevated levels of radiation from Category 2 event sequences is reasonable. The time is based on the amount of time required for the facility to recover from the event sequence; and

(e) Flux-to-dose conversion factors and cross-section data used in the analyses are consistent with accepted practice.

(6) The calculation of dose to members of the public from airborne radionuclides after Category 2 event sequences is adequate, and is based on the following:

(a) Credit taken for the use of ventilation and filtration systems in mitigating the release of airborne radioactive materials is appropriate. The analyses consider credible damage to the ventilation system that may result from event sequences;

(b) Airborne transport modeling uses an acceptable method;

(c) The U.S. Department of Energy considers appropriate exposure pathways;

(d) The time that a member of the public is assumed to be exposed to airborne radioactive materials from Category 2 event sequences is reasonable, and is based on the time that radioactive effluents are released from the facility; and

(e) The inhalation dose conversion factors used in the analyses are standard for dose assessments.

Acceptance Criterion 3 The Dose to Hypothetical Members of the Public from Category 2 Event Sequences is Within the Limits Specified in 10 CFR 63.111(b)(2).

(1) Category 2 event sequences that could adversely affect radiological exposures are adequately considered; and

(2) No identified Category 2 event sequence will lead to a dose to a member of the public that exceeds the dose limit in 10 CFR 63.111(b)(2).

2.1.1.5.2.4 Evaluation Findings

If the license application provides sufficient information and the regulatory acceptance criteria in Section 2.1.1.5.2.3 are appropriately satisfied, the staff concludes that this portion of the staff evaluation is acceptable. The reviewer writes material suitable for inclusion in the safety evaluation report prepared for the entire application. The report includes a summary statement

of what was reviewed and why the reviewer finds the submittal acceptable. The staff can document the review as follows.

U.S. Nuclear Regulatory Commission staff has reviewed the Safety Analysis Report and other information submitted in support of the license application, and has found, with reasonable assurance, that the requirements of 10 CFR 63.111(b)(2) are satisfied. The design of the geologic repository operations area is such that, taking into consideration Category 2 event sequences, no individual located on, or beyond, any point on the boundary of the site will receive, as a result of the single Category 2 event sequence, the more limiting of a total effective dose equivalent of 0.05 Sv [5 rem], or the sum of the deep dose equivalent and the committed dose equivalent to any individual organ or tissue (other than the lens of the eye) of 0.5 Sv [50 rem]. The lens dose equivalent will not exceed 0.15 Sv [15 rem], and the shallow dose equivalent to skin will not exceed 0.5 Sv [50 rem].

U.S. Nuclear Regulatory Commission staff has reviewed the Safety Analysis Report and other information submitted in support of the license application, and has found, with reasonable assurance, that the requirements of 10 CFR 63.111(c) are satisfied. The preclosure safety analysis meets the requirements specified at 10 CFR 63.112, including a demonstration that the numerical guides for design objectives for Category 2 events in 10 CFR 63.111(b)(2), are met.

2.1.1.5.2.5 References

American Nuclear Society Standards Committee Working Group. "Neutron and Gamma-Ray Cross Sections for Nuclear Radiation Protection Calculations for Nuclear Power Plants." American Nuclear Society–6.1.2–1989. Washington, DC: American National Standards Institute. 1989.

———. "Neutron and Gamma Ray Flux-to-Dose-Rate Factors." American National Standards Institute/American Nuclear Society–6.1.1–1977. Washington, DC: American National Standards Institute. 1977.

Eckerman, K.F., A.B. Wolbarst, and A. Richardson. Federal Guidance Report #11, "Limiting Values of Radionuclide Intake and Air Concentration and Dose Conversion Factors for Inhalation, Submersion, and Ingestion." Springfield, Virginia: U.S. Department of Commerce. 1992.

U.S. Nuclear Regulatory Commission. NUREG–1567, "Standard Review Plan for Spent Fuel Storage Facilities, Final Report." Washington, DC: U.S. Nuclear Regulatory Commission. 2000.

———. Regulatory Guide 1.145, "Atmospheric Dispersion Models for Potential Accident Consequence Assessments at Nuclear Power Plants." Washington, DC: U.S. Nuclear Regulatory Commission, Office of Standards Development. 1983.

2.1.1.6 Identification of Structures, Systems, and Components Important to Safety, Safety Controls, and Measures to Ensure Availability of the Safety Systems

Review Responsibilities—High-Level Waste Branch and Environmental and Performance Assessment Branch

2.1.1.6.1 Areas of Review

This section provides guidance on the review of the identification of structures, systems, and components important to safety, safety controls, and measures to ensure availability and reliability of the safety systems. Items on this list (Q-List) are important to safety and are subject to quality assurance requirements in Subpart G of 10 CFR Part 63. Reviewers will also evaluate the information required by 10 CFR 63.21(c)(5). The quality assurance program must control activities affecting the quality of Q-List items to an extent consistent with their importance to safety. The Q-List items are categorized consistent with their importance to safety.

The staff will evaluate the following parts of the identification and categorization of structures, systems, and components important to safety, safety controls, and measure, to ensure availability of the safety systems, using the review methods and acceptance criteria in Sections 2.1.1.6.2 and 2.1.1.6.3.

(1) Structures, systems, and components important to safety and measures to ensure availability and reliability of safety systems;

(2) Administrative or engineered safety controls for structures, systems, and components important to safety; and

(3) Risk-significance categorization of structures, systems, and components important to safety.

2.1.1.6.2 Review Methods

Review Method 1 List of Structures, Systems, and Components Important to Safety; Technical Bases for Identification of Structures, Systems, and Components and Safety Controls; and List and Analysis of Measures to Ensure Availability and Reliability of Safety Systems

Verify that analysis and classification of structures, systems, and components for the geologic repository operations area used the results of the iterative preclosure safety analysis reviewed in Sections 2.1.1.3 ("Identification of Hazards and Initiating Events"), 2.1.1.4 ("Identification of Event Sequences"), and 2.1.1.5 ("Consequence Analyses") of the Yucca Mountain review Plan. The identification of hazards, initiating events, event sequences, and consequence analysis should form the basis to identify structures, systems, and components that are important to safety that should be functional to meet the performance objectives. All structures, systems, and components and controls assumed to be functional in the consequence analyses should be considered in the list. Confirm that structures, systems, and components are classified as important to safety according to the definition specified in 10 CFR 63.2.

Confirm that analyses used to identify structures, systems, and components important to safety, safety controls, and measures to ensure the availability and reliability of the safety systems include adequate consideration of:

(1) Means to limit concentration of radioactive material in air, such as:

   (a) Appropriately designed ventilation systems;

   (b) Use of seals and/or air locks as part of geologic repository operations area design; and

   (c) Installation of radiation-monitoring systems that provide information on the dose rate and concentration of airborne radioactive material in selected areas.

(2) Means to limit time required to perform work in the vicinity of radioactive materials, such as:

   (a) Features that minimize the time that maintenance, health physics, or inspection personnel must remain in restricted areas; and

   (b) Use of remotely operated or robotic equipment, such as welders, wrenches, cutting tools, and radiation monitors, and means to remotely place temporary shielding.

(3) Suitable shielding, such as:

   (a) Shielding provided by the radioactive material being stored;

   (b) Neutron capture provided by borated water in casks and waste transfer pools, and by borated materials incorporated into casks;

   (c) Gamma and neutron shielding provided by the structural and nonstructural materials in the walls and ends of storage/transfer casks;

   (d) Temporarily positioned shielding used during operations for preparing casks, and/or during transfer of casks, and shielding provided by any pool facility interior and exterior walls; and

   (e) Selection of appropriate shielding materials, and that the design analysis of the shielding performance for normal and Categories 1 and 2 event sequences. Coordinate with the reviewer of the repository design for Section 2.1.1.7 ("Design of Structures, Systems, and Components Important to Safety and Safety Controls") of the Yucca Mountain Review Plan.

(4) Means to monitor and control dispersal of radioactive contamination;

(5) Means to control access to high radiation areas, very high radiation areas, or airborne radioactivity areas, to ensure compliance with the requirements of Subparts G and H of 10 CFR Part 20, such as:

(a) Analyses that identify airborne radioactivity areas. These analyses should provide a technical basis for any inability to practically apply process or other engineering controls, to restrict the concentrations of radioactive material in air to values below those that define an airborne radioactivity area;

(b) A plan for monitoring and limiting intakes of radiation (e.g., controlling access, limiting individual exposure times, using individual respiratory protection equipment); and

(c) Consistency with guidance such as Regulatory Guide 8.38, "Control of Access to High and Very High Radiation Areas of Nuclear Power Plants" (U.S. Nuclear Regulatory Commission, 1993).

(6) Means to prevent or control criticality, such as complying with American National Standards Institute/American Nuclear Society–8 nuclear criticality safety standard documents listed in Regulatory Guide 3.71 (U.S. Nuclear Regulatory Commission, 1998a);

(7) A radiation alarm system designed to warn of significant increases in radiation levels, concentrations of radionuclides in air, and increased radioactivity in effluents. This system should be designed to provide prompt notification to personnel both in the work area where an increase in radiation is detected and in control centers. Features of control centers should include:

(a) Appropriate installation of radiation alarms in areas where waste is being stored, transferred, or processed/repackaged;

(b) Availability of backup power systems for radiation alarm systems; and

(c) Design and operation of interior evacuation signals and signs consistent with Regulatory Guide 8.5, "Criticality and Other Interior Evacuation Signals" (U.S. Nuclear Regulatory Commission, 1981).

(8) The ability of structures, systems, and components to perform their intended safety functions, assuming the occurrence of event sequences, considering results from the review of consequence analyses using Section 2.1.1.5 ("Consequence Analyses") of the Yucca Mountain Review Plan;

(9) Explosion and fire detection systems and appropriate suppression systems, features of which may include:

(a) Installation of detection and suppression systems near probable sources of fire or explosion; and

(b) Designs to accommodate the interactions of ventilation systems with potential fires or explosions.

(10) Means to control radioactive waste and radioactive effluents, and to permit prompt termination of operations and evacuation of personnel during an emergency, such as:

(a) Design and operation of the geologic repository operations area to reduce the quantity of radioactive waste generated;

(b) Off-gas treatment, filtration, and ventilation systems for control of airborne radioactive effluents;

(c) Liquid waste management systems to handle the expected volume of potentially radioactive liquid waste generated during normal operations and Categories 1 and 2 event sequences. Design features and procedures for these systems should minimize generation of liquid waste and the possibility of spills, and should provide for control of spills, overflows, or leakage during packaging and transfer of site-generated radioactive liquid waste; and

(d) Solid waste management systems to handle the expected volume of potentially radioactive solid waste (e.g., contaminated equipment and personnel clothing) generated during normal operations and Categories 1 and 2 event sequences.

(11) Means to provide reliable and timely emergency power to instruments, utility service systems, and operating systems important to safety, such as:

(a) Instrumentation and/or monitoring systems with back-up battery power, for which the duration of backup battery life should be consistent with reasonable time periods of primary power loss;

(b) Uninterruptible power supplies on process control computers; and

(c) Standby diesel generators that should start on demand if primary power is lost, and should continue to operate for the required period of time.

(12) Means to provide redundant systems necessary to maintain, with adequate capacity, the capability of utility services important to safety, such as electrical systems, ventilation systems, air supply systems, water supply systems for fire suppression, and communication systems. Examples of design features for consideration in these systems may include electrical systems, ventilation systems, water supply systems, and communication systems; and

(13) Means to inspect, test, and maintain structures, systems, and components important to safety, as necessary, to ensure their continued function and readiness. This assessment should take into account the review of plans for conduct of normal activities, including maintenance, surveillance, and periodic testing conducted using Section 2.5.6

("Plans for Conduct of Normal Activities, Including Maintenance, Surveillance, and Periodic Testing") of the Yucca Mountain Review Plan.

Review Method 2 Administrative or Procedural Safety Controls to Prevent Event Sequences or Mitigate Their Effects

Confirm that management systems and procedures are sufficient to ensure that administrative or procedural safety controls will function properly. Coordinate with the reviewer for Sections 2.5.5 ("Plans for Startup Activities and Testing") and 2.5.6 ("Plans for Conduct of Normal Activities Including Maintenance, Surveillance, and Periodic Testing") of the Yucca Mountain Review Plan. Examples of such management systems are:

(1) Procedures;
(2) Training;
(3) Maintenance, calibration, and surveillance plans and schedules;
(4) Configuration controls for structures, systems, and components;
(5) Human factor evaluations;
(6) Audits and self-assessments;
(7) Emergency planning; and
(8) Accident/incident investigation requirements.

Confirm that administrative or procedural safety controls required for the structures, systems, and components to be functional and to meet the dose requirements are included in the list of structures, systems, and components important to safety.

Review Method 3 Risk Significance Categorization of Structures, Systems, and Components Important to Safety

Evaluate the methodology used for risk significance categorization of structures, systems, and components important to safety in the geologic repository operations area to confirm that the methodology is technically sound and defensible. Verify categorization methodology for structures, systems, and components important to safety is supported by appropriate qualitative descriptions and quantitative or semi-quantitative methods. Verify that the risk significance categorization of structures, systems, and components important to safety is consistent with the governing regulation and applicable policy and guidance.

Confirm that the identification of structures, systems, and components important to safety (Q-List generation) is done using a preclosure safety analysis methodology consistent with the requirements in 10 CFR 63.112. Verify the categorization methodology incorporates both event sequence frequencies and consequences in its consideration of risk.

Confirm the categorization methodology provides due consideration of uncertainties and sensitivity analyses for event sequence frequencies in a manner that is consistent with the applicable portions of existing U.S. Nuclear Regulatory Commission policy and guidance.

Verify that the categorization of structures, systems, and components important to safety is consistent with their relative importance to safety, as required in 10 CFR 63.142(c)(1). Verify that the distinctions between quality levels have a well defined and well documented technical basis. Verify that the frequencies and consequences of failures of structures, systems, and components important to safety identified for the various quality levels are well defined and consistent with applicable portions of existing U.S. Nuclear Regulatory Commission policy and guidance. Confirm that all structures, systems, and components important to safety identified in Review Methods 1 and 2 of this section are properly categorized and technical bases are adequately documented.

Verify that the categorization methodology is flexible enough to accommodate multiple revisions of the preclosure safety analysis and the subsequent re-evaluation of risk significance. Confirm that the categorization methodology permits revision to the categorization of structures, systems, and components important to safety as a result of the introduction of new data or design changes.

Verify the documentation, analysis, and criteria used for risk significance categorization of structures, systems, and components important to safety is transparent and traceable with a well defined technical basis. Verify that the categorization methodology is presented in such a manner that the reviewer can gain a clear understanding of the results at each step, and the technical bases for these results. Verify that the categorization methodology includes an unambiguous and comprehensive record of the decisions, criteria, and assumptions made, and the process used in arriving at a given conclusion or result.

2.1.1.6.3 Acceptance Criteria

The following acceptance criteria are based on meeting the requirements of 10 CFR 63.112(e), relating to the identification of structures, systems, and components important to safety, safety controls, and measures to ensure availability of the safety systems.

Acceptance Criterion 1 An Adequate List of Structures, Systems, and Components Identified as Being Important to Preclosure Radiological Safety; the Technical Bases for the Approaches Used to Identify Structures, Systems, and Components Important to Safety, and Safety Controls Based on Analysis of Their Performance; and a List and Analysis of the Measures to be Taken to Ensure That the Safety Systems are Available.

(1) The analysis and classification of structures, systems, and components for the geologic repository operations area uses results of the hazard assessment, identification of event sequences, and consequence analyses as a basis to identify those structures, systems, and components that are important to safety; and

(2) The analyses used to identify structures, systems, and components important to safety, safety controls, and measures to ensure the availability and reliability of the safety systems, include adequate consideration of:

   (a) Means to limit concentration of radioactive material in air;

(b) Means to limit time required to perform work in the vicinity of radioactive materials;

(c) Suitable shielding;

(d) Means to monitor and control dispersal of radioactive contamination;

(e) Means to control access to high radiation areas, very high radiation areas, or airborne radioactivity areas;

(f) Means to prevent or control criticality;

(g) A radiation alarm system designed to warn of significant increases in radiation levels, concentrations of radionuclides in air, and increased radioactivity in effluents;

(h) Ability of structures, systems, and components to perform their intended safety functions, assuming the occurrence of event sequences;

(i) Explosion and fire detection systems and appropriate suppression systems;

(j) Means to control radioactive waste and radioactive effluents, and to permit prompt termination of operations and evacuation of personnel during an emergency;

(k) Means to provide reliable and timely emergency power to instruments, utility service systems, and operating systems important to safety;

(l) Means to provide redundant systems necessary to maintain, with adequate capacity, the capability of utility services important to safety; and

(m) Means to inspect, test, and maintain structures, systems, and components important to safety, as necessary, to ensure their continued function and readiness.

Acceptance Criterion 2 Administrative or Procedural Safety Controls Needed to Prevent Event Sequences, or Mitigate Their Effects, Are Adequate, and Are Included in the List of Structures, Systems, and Components Important to Safety.

(1) Management systems and procedures are sufficient to ensure that administrative or procedural safety controls will function properly; and

(2) Administrative or procedural safety controls required for structures, systems, and components to be functional, and to meet dose requirements, are included in the list of structures, systems, and components important to safety.

Acceptance Criterion 3 The Approach and Criteria for Risk Significance Categorization of Structures, Systems, and Components Important to Safety Are Defensible and the Structures, Systems, and Components Important to Safety Are Adequately Categorized.

(1) Methodology for categorization of structures, systems, and components important to safety in the geologic repository operations area is technically sound and defensible;

(a) The risk significance categorization of structures systems and components important to safety are technically sound with a well supported technical basis and is consistent with regulatory framework;

(b) The categorization methodology for structures, systems, and components important to safety is supported by appropriate qualitative descriptions and quantitative or semi-quantitative methods;

(c) The identification of structures, systems, and components important to safety are consistent with the governing regulation and applicable policy and guidance;

(d) The identification of structures, systems, and components important to safety (Q-List generation) is done using a preclosure safety analysis methodology that is consistent with and fulfills the requirements in 10 CFR 63.112;

(e) The categorization methodology considers the frequency of event sequences (Categories 1 and 2) defined in 10 CFR 63.2;

(f) The categorization methodology considers the dose limits in 10 CFR 63.111 (including 10 CFR Part 20); and

(g) The categorization methodology provides due consideration of uncertainties and sensitivity analyses for event sequence frequencies in a manner that is consistent with the applicable portions of existing U.S. Nuclear Regulatory Commission policy and guidance.

(2) The risk significance categorization of structures, systems, and components important to safety is consistent with their relative importance to safety:

(a) The categorization methodology ensures that structures, systems, and components important to safety are categorized consistent with their risk significance and relative importance to safety [10 CFR 63.142(c)(1)];

(b) The distinctions between quality levels has a well defined and well documented technical basis;

(c) The frequencies and consequences of failures of structures, systems, and components important to safety identified for the various quality levels are well defined and consistent with applicable portions of existing U.S. Nuclear Regulatory Commission policy and guidance; and

(d) All structures, systems, and components important to safety are properly categorized and technical bases are adequately documented.

(3) The risk significance categorization of structures, systems, and components important to safety demonstrates flexibility:

(a) The categorization methodology is flexible enough to accommodate multiple revisions of the integrated safety analysis and the subsequent reevaluation of risk significance; and

(b) The categorization methodology permits the revision of the categorization level of individual and groups of structures, systems, and components important to safety as a result of the introduction of new data or design changes.

(4) The documentation and analysis for the risk significance categorization of structures, systems, and components important to safety is transparent and traceable:

(a) The categorization methodology is developed and presented in such a manner that the reviewer can gain a clear understanding of every step of what has been done, what the results are, and the technical bases for the results; and

(b) The categorization methodology includes an unambiguous and complete record of the decisions and assumptions made, and the process used in arriving at a given conclusion or result.

2.1.1.6.4 Evaluation Findings

If the license application provides sufficient information and the regulatory acceptance criteria in Section 2.1.1.6.3 are appropriately satisfied, the staff concludes that this portion of the staff evaluation is acceptable. The reviewer writes material suitable for inclusion in the safety evaluation report prepared for the entire application. The report includes a summary statement of what was reviewed and why the reviewer finds the submittal acceptable. The staff can document the review as follows.

U.S. Nuclear Regulatory Commission staff has reviewed the Safety Analysis Report and other information submitted in support of the license application and has found, with reasonable assurance, that the requirements of 10 CFR 63.112(e) are satisfied. An adequate analysis of the performance of the structures, systems, and components important to safety has been provided. In particular, this analysis demonstrates that:

(1) Structures, systems, and components important to safety are identified;

(2) Criteria for categorization of structures, systems, and components important to safety are adequately developed and categorization of items is acceptable;

(3) Controls that will be relied on to limit or prevent potential event sequences, or mitigate their consequences, are acceptable; and

(4) Measures are adequate to ensure the availability of structures, systems, and components important to safety.

2.1.1.6.5 References

Electric Power Research Institute. "PSA Applications Guide." EPRI TR–105396. Walnut Creek, California: Electric Power Research Institute. 1995.

U.S. Nuclear Regulatory Commission. NUREG–1513, "Integrated Safety Analysis Guidance Document." Draft Report. Washington, DC: U.S. Nuclear Regulatory Commission. 2000a.

———. NUREG–1520, "Standard Review Plan for the Review of a License Application for a Fuel Cycle Facility." Draft Report. Washington, DC: U.S. Nuclear Regulatory Commission. 2000b.

———. Regulatory Guide 3.71, "Nuclear Criticality Safety Standards for Fuels and Material Facilities." Washington, DC: U.S. Nuclear Regulatory Commission, Office of Standards Development. 1998a.

———. Regulatory Guide 8.38, "Control of Access to High and Very High Radiation Areas in Nuclear Power Plants." Washington, DC: U.S. Nuclear Regulatory Commission, Office of Standards Development. 1993.

———. Regulatory Guide 8.5, "Criticality and Other Interior Evacuation Signals." Washington, DC: U.S. Nuclear Regulatory Commission, Office of Standards Development. 1981.

———. Regulatory Guide 3.32, "General Design Guide for Ventilation Systems for Fuel Reprocessing Plants." Washington, DC: U.S. Nuclear Regulatory Commission, Office of Standards Development. 1975.

2.1.1.7 Design of Structures, Systems, and Components Important to Safety and Safety Controls

Review Responsibilities—High-Level Waste Branch and Environmental and Performance Assessment Branch

2.1.1.7.1 Areas of Review

This section provides guidance on the review of the design of structures, systems, and components important to safety and safety controls. Reviewers will also evaluate the information required by 10 CFR 63.21(c)(2) and (c)(3), and coordinate review of information, such as the geologic media, general arrangement, and dimensions, as specified in 10 CFR 63.21(c)(2), with the review of Sections 2.1.1.1 and 2.1.1.2 of the Yucca Mountain Review Plan.

Review Plan for Safety Analysis Report

The staff will evaluate the following parts of the design of structures, systems, and components important to safety and safety controls using the review methods and acceptance criteria in Sections 2.1.1.7.2 and 2.1.1.7.3.

(1) Design criteria and design bases;
(2) Design methodologies; and
(3) Geologic repository operations area design and design analyses.

The determination of the geologic repository operations area structures, systems, and components important to safety will depend largely on the final design and preclosure safety analysis results. The review methods and acceptance criteria provided in the following sections are examples. These structures, systems, and components may, or may not, be important to safety. If some structures, systems, and components listed below are determined not to be important to safety, based on the review conducted using Section 2.1.1.6 of the Yucca Mountain Review Plan, the reviewer may not have to review these structures, systems, and components. Similarly, for structures, systems, and components that are identified to be important to safety, but are not included in the Yucca Mountain Review Plan, the general aspects of the review methods and acceptance criteria provided below may still be applicable. However, for the remaining aspects not addressed in the following sections, the reviewer should exercise professional judgment to conduct the review.

2.1.1.7.2 Review Methods

2.1.1.7.2.1 Design Criteria and Design Bases

Review Method 1 Definitions of Relationship between Design Criteria and 10 CFR 63.111(a) and (b) Requirements; Relationship between Design Bases and Design Criteria; and Design Criteria and Design Bases for Structures, Systems, and Components Important to Safety

Verify that design criteria and bases have been identified for structures, systems, and components important to safety. Confirm these design criteria and bases are derived from the site characteristics and consequence analyses reviewed using Sections 2.1.1.1 ("Site Description as it Pertains to Preclosure Safety Analysis"), and 2.1.1.5 ("Consequence Analyses") of the Yucca Mountain Review Plan. Verify that these design criteria and bases are consistent with analyses used to identify structures, systems, and components as reviewed using Section 2.1.1.6 ("Identification of Structures, Systems, and Components Important to Safety; Safety Controls; and Measures to Ensure Availability of the Safety Systems") of the Yucca Mountain Review Plan.

Confirm that the design criteria for normal operating conditions are adequately developed. Verify that design criteria do not permit degradation of the geologic repository operations area structures, systems, and components, important to safety, which will reduce:

(1) Design capability of radioactive material handling and waste processing;

(2) Design capability to withstand further occurrence of Categories 1 and 2 event sequences without remedial action; or

(3) Capability to perform design functions as intended or as available for the full system lifetime.

Verify that design criteria adequately consider preclosure safety analysis results. Verify that structures, systems, and components important to safety will continue to prevent consequences, such as unacceptable releases of radioactive material, unacceptable radiation doses for workers or the public, and loss of removal capability.

Confirm that structural design criteria and bases for structures, systems, and components important to safety are consistent with relevant U.S. Nuclear Regulatory guidance for tornado protection, seismic design, explosion protection, and flood protection.

Verify that the design criteria and bases for thermal considerations are consistent with U.S. Nuclear Regulatory Commission and American National Standards Institute/American Nuclear Society guidance or standards for fire protection.

Verify that the design criteria and bases for shielding and confinement systems use appropriate guidance or standards.

Confirm that design criteria for fixed-area radiation monitors and continuous airborne monitoring instrumentation for radiological protection are consistent with appropriate guidance or standards.

Verify that criticality design criteria are developed based on the consequence analysis results from the preclosure safety analysis. Confirm that criticality design criteria are factored into models and assumptions used for criticality analysis. These criteria should be consistent with those given in NUREG–1567 (U.S. Nuclear Regulatory Commission, 2000) and those American National Standards Institute/American Nuclear Society–8 nuclear criticality standards adopted by the U.S. Nuclear Regulatory Commission as listed in Regulatory Guide 3.71 (U.S. Nuclear Regulatory Commission, 1998).

Verify that design bases and design criteria are based on the above listed, or other guidance documents and standards, considering the normal geologic repository operations area operating conditions and Categories 1 and 2 event sequences. For example, these design bases should incorporate:

(1) Thermal design bases and criteria that include temperatures for those temperature-sensitive structures, systems, and components important to safety that consequence analyses (reviewed using Section 2.1.1.5 of the Yucca Mountain Review Plan) indicate a potential radiological hazard if the design temperatures are not met. In reviewing adequacy of the structural design bases and criteria, the staff should:

    (a) Verify that thermal design criteria for the surface and subsurface facilities have been adequately developed, based on the maximum design waste inventory;

(b) Verify that design criteria for fire protection (e.g., fire ratings, fire barriers) are adequate, based on the maximum credible geologic repository operations area fire (duration and temperature), if determined to be of design importance from the consequence analyses (reviewed using Section 2.1.1.5 of the Yucca Mountain Review Plan); and

(c) Verify that design criteria for the surface and subsurface ventilation systems have been adequately developed, based on thermal and fire protection design criteria, in addition to airborne radiological dose limits.

(2) Structural design bases and criteria, that include maximum loads, stress/pressure loadings (static and/or dynamic), and displacements for structures, systems, and components important to safety, that consequence analyses (reviewed using Section 2.1.1.5 of the Yucca Mountain Review Plan) indicate a potential radiological hazard if the design loads and displacements are violated. In reviewing adequacy of the structural design bases and criteria:

(a) Verify that event sequences are properly converted into structural loads, pressures, and/or displacements, based on accepted methods; and

(b) Verify that the use of factored loads and load combinations is based on accepted methods or codes and standards.

(3) Shielding design bases and criteria, that include maximum dose rates and annual dose rates to workers and the public from the exterior of shielding surfaces, for structures, systems, and components important to safety;

(4) Criticality design bases and criteria, that include fuel geometry configurations, moderators, and waste forms effective neutron multiplication factor limits, to ensure that nuclear fuel remains subcritical during handling, transfer, repackaging, storage, and retrieval; and

(5) Operating design bases and criteria, that include the maximum limits of travel, vertical lift, and/or velocity, for structures, systems, and components important to safety for handling and transfer of high-level radioactive waste or containers that consequence analyses (reviewed using Section 2.1.1.5 of the Yucca Mountain Review Plan) show present a potential radiological hazard if operating limits are violated.

2.1.1.7.2.2 Design Methodologies

Review Method 1 Geologic Repository Operations Area Design Methodologies

Confirm that proposed design methodologies are supported by adequate technical bases, and are consistent with established industry practice. Verify that uncertainties associated with the proposed methodologies have been adequately addressed.

If the design methodologies depend on site-specific test data, confirm that such data are available. Also, verify that any analytical or numerical models used to support the design methodologies have been verified, calibrated, and validated. Verify that any assumptions or limitations relating to the proposed methodologies are identified, and that their implications for the design have been adequately analyzed and documented.

If the design methodologies depend on data from expert elicitations, coordinate with the reviewer of Section 2.5.4 ("Expert Elicitation") of the Yucca Mountain Review Plan, to verify that these elicitations are conducted and documented consistent with NUREG–1563 (Kotra, et al., 1996).

Confirm that seismic design methodologies use ground motion information that is consistent with proposed U.S. Department of Energy methodologies for hazard assessment and preclosure design criteria, and that, taken together, they provide adequate input for seismic design and for performance assessments.

2.1.1.7.2.3 Geologic Repository Operations Area Design and Design Analyses

I. Designs and Design Analyses for Structures, Systems, and Components, Equipment, and Safety Controls That are Safety Related for Surface Facilities

Review Method 1 Design Codes and Standards

Confirm that applicable design codes and standards are specified for the structural, thermal, shielding and confinement, criticality, and decommissioning designs. This review should include:

(1) Confirmation that structural design, fabrication, and testing of waste packages for storage of spent nuclear fuel is in accordance with the Boiler and Pressure Vessel Code of the American Society of Mechanical Engineers;

(2) Verification that prestressed and reinforced concrete structures, within the geologic repository operations area, that are used for containment of radioactive material are designed in accordance with American Concrete Institute and American Society of Mechanical Engineers Standards or other appropriate standards;

(3) Determination that steel structures and components are designed and constructed in accordance with applicable steel design codes and standards;

(4) Determination that foundations supporting structures, systems, and components important to safety are constructed in accordance with the applicable American Construction Institute code standards, and that site-related geotechnical parameters are obtained based on guidelines such as those provided in American National Standards Institute/American Nuclear Society;

(5) Verification that applicable standards and codes have been used for design and construction of processing equipment and facility power systems, instrumentation, control, and other operations systems. For example:

 (a) Crane systems [Nuclear Standard NOG–1–1995 (American Society of Mechanical Engineers, 1995); Crane Manufacturers' Association of America standards];

 (b) National Electrical Manufacturers' Association codes and Institute of Electrical and Electronics Engineers, Inc. nuclear standards;

 (c) Nuclear safety criteria for control air systems;

 (d) International Society for Measurement and Control and Institute of Electrical and Electronics Engineers codes);

 (e) National Fire Protection Association codes; and

 (f) Applicable standards or guides for heating, refrigeration, and air conditioning systems.

If other methods, standards, or guides are used for design, verify that the license application has provided adequate technical bases for their usage.

Review Method 2 Consistency of Materials with Design Methodologies

Verify that materials used for structures, systems, and components important to safety in surface facility design are consistent either with the accepted design criteria, codes, standards, and specifications, or with those specifically developed by the U.S. Department of Energy. For example, if American Society of Mechanical Engineers Boiler and Pressure Vessel Code, is used for waste package design criteria, the materials should be consistent with those prescribed by the associated paragraphs of the American Society of Mechanical Engineers Boiler and Pressure Vessel Code, or their equivalent. Other examples include:

(1) For concrete and steel design, applicable American Society for Testing and Materials standard specifications; and

(2) For shielding materials, American National Standards Institute/American Nuclear Society specifications may be used, and the geometric arrangement and the potential for shielding material to experience changes in material properties and geometry at high temperatures should be assessed. Confirm, based on review of the license application shielding analyses/design, that any temperature-sensitive shielding materials will not be subject to temperatures at or above their design limitations during normal operations and Categories 1 and 2 event sequences.

Evaluate the material properties and allowable stresses and strains for the design to verify the adequacy of the materials.

Confirm that the materials and their properties are appropriate for the expected design loading conditions. Also, confirm that anticipated stress limits for each material are based on maximum temperatures established in the thermal analysis evaluation in the license application.

Verify that the U.S. Department of Energy has considered the potential for creep or brittle fracture and drop fracture resistance of materials, to ensure that structures, systems, and components important to safety are adequate to perform their safety functions, as appropriate. For components governed by some codes and standards, providing only safety margins may be sufficient.

Review Method 3 Load Combinations Used for Normal and Categories 1 and 2 Event Sequence Conditions

Verify that loads used in the design analyses are consistent with loads under normal operations and Categories 1 and 2 event sequences for structures, systems, and components important to safety.

Evaluate load combinations used in the design analyses for consistency with those accepted by the U.S. Nuclear Regulatory Commission for the design of similar types of nuclear facilities, and for steel and reinforced concrete structures designed in accordance with accepted standards and codes.

Verify that design analyses use appropriate techniques that were correctly applied to provide design temperatures, mechanical loads, and pressures for the structures, systems, and components important to safety.

Review Method 4 Performance and Documentation of Design Analyses

Verify that design analyses include the relevant structural, thermal, shielding, criticality, confinement, and decommissioning factors, such as:

(1) For all analyses:

- (a) Computational models, data, assumptions, and results are adequately documented;
- (b) Computational models are validated;
- (c) Data are derived from relevant site and system information;
- (d) Assumptions having adequate technical justifications or bases are provided;
- (e) Normal operations and Categories 1 and 2 event sequences are considered in developing system loadings and environments; and
- (f) Analyses are based on the maximum capacity and rate of receipt of radioactive waste.

Confirm these analyses using limited confirmatory calculations.

(2) For shielding design and design analyses:

- (a) Dose rate estimates are presented for representative areas; and
- (b) Bases for flux-to-dose conversions are adequately documented and are acceptable to the U.S. Nuclear Regulatory Commission.

(3) For criticality design and design analyses:

- (a) Calculations determine the highest expected waste forms effective neutron multiplication factor that is likely to occur under the examined loading conditions;
- (b) Calculations are appropriate for the material properties;
- (c) Analyses are consistent with those for similar facilities, as appropriate; and
- (d) Any application for burnup credit is properly substantiated.

(4) For thermal design and design analyses:

- (a) Analyses are consistent with limiting fuel burnup and cooling times; and
- (b) Analyses specify the maximum and minimum expected temperatures for all components.

(5) For structural design and design analyses, loadings are correctly translated into either static or time-varying nodal forces, or element face pressures.

Confirm that values of material properties used for the design analyses have adequate technical bases, and are consistent with site-specific data.

Confirm that loads and load combinations used in the design analyses are consistent with defined normal operations and Categories 1 and 2 event sequences.

Verify that analytical methods, models, and codes used for the design analyses are appropriate for the conditions analyzed and are properly validated.

Confirm that technical bases for the assumptions used in the design analyses are adequately defined and based on accepted engineering practice.

Verify that designs and design analyses for structures, systems, and components important to safety are performed correctly. Also, verify that these structures, systems, and components have sufficient capability to withstand normal and Categories 1 and 2 event sequence loadings.

Conduct limited confirmatory checks or analyses using appropriate analytical methods, models, or codes.

II. Designs and Design Analyses for Structures, Systems, and Components, Equipment, and Safety Controls That are Safety Related for Subsurface Facility

Review Method 1 Design Assumptions, Codes, and Standards

Confirm the applicable design codes, standards, or other detailed criteria used for the design of the subsurface facility are specified. Codes and standards should be equivalent to, and consistent with, those accepted by the U.S. Nuclear Regulatory Commission for design of nuclear facilities with similar hazards and functions. If nonstandard approaches are used, confirm that the license application has provided adequate technical bases to justify why they are used.

Verify that the assumptions made for the design of the subsurface facility are technically defensible.

Confirm that geologic repository operations area subsurface facility designs for steel and concrete structures and components, air control systems, electrical power systems, and ventilation systems for the subsurface facility use applicable standards.

Review Method 2 Design of Subsurface Operating Systems

Verify that the methods, assumptions, and input data used in the ventilation design are consistent with proposed thermal loading performance goals in the emplacement drifts. Conduct limited confirmatory analyses to verify the results presented in the license application. Also, confirm that the analyses adequately address the thermal load in the ventilation tunnels, shafts, and raises.

Evaluate design analyses of control system functions, equipment, instrumentation, control links, and communication systems to confirm that the subsurface monitoring and control systems are appropriate for the structures, systems, and components important to safety during waste emplacement, and monitoring.

Assess the design of the waste emplacement system for compatibility with proposed waste emplacement procedures. Also, verify that interfaces with other systems are identified and assessed, and that continuity of operations and safety can be achieved.

Evaluate the layout of the subsurface facilities. Verify that emplacement drifts are located away from major faults, consistent with the seismic design, and that the subsurface layout is appropriate for the quantity of waste to be emplaced and satisfies the design thermal criteria.

Confirm that the geologic repository operations area design permits implementation of the performance confirmation plan provided in Section 4.4, as specified in 10 CFR 63.111(d).

Verify that standards and codes used for design of subsurface operating systems were properly applied.

Review Method 3 Materials and Material Properties Used for Subsurface Facility Design

Confirm that the selection of materials and the properties of these materials are appropriate for the anticipated subsurface environment.

Verify that materials and material properties are consistent with applicable design criteria, codes, standards, and specifications. If no standards are used, evaluate the technical bases provided to confirm that they are acceptable. Confirm that applicable American Society for Testing and Materials standard specifications are used.

Evaluate whether the selection of ground support materials accounts for degradation of such materials under elevated temperature and thermal loading. Also, confirm that plausible mechanisms for material degradation are identified and properly incorporated in assessments of subsystem structure, system, and component performance.

Verify that subsurface ventilation systems are constructed of fire-resistant materials (e.g., fire-resistant filters) to protect against fires occurring inside or outside the systems. Verify that ventilation equipment/components and materials, particularly those within or near waste emplacement drifts, are designed to withstand prolonged high temperature conditions, effects of blast cooling, and wet and corrosive environments, to minimize maintenance/ replacement of potentially contaminated ventilation components.

Review Method 4 Load Combinations Used for Normal and Categories 1 and 2 Event Sequences

Confirm that the arrangement of waste packages within the subsurface facility satisfies the thermal load design criteria.

Verify that the magnitude and time history of the applied thermal loading are consistent with the anticipated characteristics of the proposed waste, repository design configurations, and design areal mass loading.

Verify that thermal analyses have an appropriate technical basis; use site-specific thermal property data; consider temperature dependency and uncertainties of thermal property data; and use thermal models and analyses that are properly documented. If credit is taken for use of ventilation, confirm that assessments of the effects of ventilation are adequate.

Verify that design analyses consider appropriate *in situ* stresses, potential running ground conditions, and hydrologic changes to the rock mass, during the heating period, that might affect mechanical properties.

Confirm that dynamic loads used in design analyses are consistent with the seismic design ground motion parameters; consider faulting effects; and are consistent with accepted methodologies for assessing faulting hazards.

Review Method 5 Models and Site-Specific Properties of Host Rock Used in Design Analyses and Consideration of Spatial and Temporal Variation and Uncertainties in Such Properties

Confirm that appropriate combinations of continuum and discontinuum modeling of appropriate dimensionality have been used for assessing the behavior of a fractured rock mass under prolonged heated conditions and Categories 1 and 2 event sequences. Confirm that the bases for the choice of specific models and model combinations are adequate, and that appropriate bases for the assumptions and limitations of the modeling approach are provided.

Confirm that principles for the design analyses, the underlying assumptions, and the anticipated limitations are documented; are consistent with modeling objectives; and are technically sound.

Verify that values for the rock-mass thermal expansion coefficient are consistent with properly interpreted site-specific data, and that such interpretation accounts for likely scale effects and temperature dependency. Verify that uncertainty in the thermal expansion coefficient has been adequately assessed and considered in the thermal stress calculation.

For continuum rock-mass modeling, confirm that values for rock-mass elastic parameters (Young's modulus and Poisson's ratio) and strength parameters (friction angle and cohesion) are consistent with properly interpreted site-specific data. If the parameter values are obtained through empirical correlations with a rock quality index, verify that the empirical equations used are appropriate for the site, and are applied correctly. Confirm that the values of the index are consistent with site-specific data. If intact-rock-scale values are used, verify that the bases for application of the values to the rock-mass scale are adequate.

For discontinuum rock-mass modeling, verify that the selection of fracture patterns for numerical modeling is appropriate for the objectives of the design and analyses. Confirm that the interpretation of modeling results adequately considers effects of the representation of the characteristics of the modeled fracture network, compared with those of the *in situ* fracture network.

Confirm that the selection of stiffness and strength parameters for rock blocks between any fractures that are explicitly represented in the model is appropriate and accounts for fractures that are not explicitly represented.

Verify that the values for fracture stiffness and strength parameters are consistent with properly interpreted site-specific data.

If applicable, confirm that modeling time-dependent mechanical degradation of the rock mass, fractures, and ground support that may occur after the emplacement of waste is adequately accounted for in thermal-mechanical analyses. This may be based on extrapolations from the U.S. Department of Energy long-term exploratory studies facility's heated-drift and single-heater test studies, the cross-drift thermal test study, or other methods. Verify that the bases for the magnitude and rate of mechanical degradation applied in the analyses are appropriately established, and are technically defensible.

Confirm that uncertainties in rock mass and fracture mechanical properties are adequately estimated, and considered in both continuum and discontinuum modeling.

Verify that the models adequately address the stability of openings around drift intersections, considering the rock mass and its degraded properties and thermal loading. This information should be used in the design of ground support.

Conduct limited confirmatory continuum and discontinuum analyses to verify the rock mass behavior results presented in the license application, under design (normal) operating conditions and Categories 1 and 2 event sequences.

Review Method 6 Design Methodologies and Interpretations of Modeling Results for Ground Support Systems

Confirm that design methodologies or combinations of design methodologies are properly applied to the design of ground support systems. Confirm that, when used, the empirical design approach is consistent with accepted technology in the underground tunneling and mining industry. Also, verify that the evaluation and selection of ground support systems are supported by analyses that satisfy Acceptance Criteria 4 and 5 under Subsection II in Section 2.1.1.7.3.3 ("Geologic Repository Operations Area Design and Design Analyses") of the Yucca Mountain Review Plan. These analyses should provide mechanical evaluation of ground support systems under thermal and dynamic loads.

Confirm that the ground support system responses are adequately evaluated, based on the results of model analyses. If the ground support system is explicitly modeled, verify that the ground support responses include an adequate assessment of deformation and potential failure of the ground support systems. Verify that the interaction between the ground support system and the host rock units (e.g., interactions of rock bolts with lithophysae) is considered in the analysis. Review Method 5, and Acceptance Criterion 5, of this subsection should be used in assessing ground support system responses, where applicable. If the ground support system is not explicitly modeled, confirm that the anticipated ground support system responses from the modeling results are reasonably estimated, and that technical bases for these estimations are adequate.

Verify that geometrical, thermal, and mechanical characteristics of the ground support system used in the thermal-mechanical analyses are consistent with the design and construction specifications. Also, confirm that the time-dependent mechanical degradation of the ground support system under heated conditions is adequately accounted for in the analyses.

Verify that stability of emplacement drifts, ventilation tunnels, and shafts is adequately assessed, both with and without ground support. The assessment should identify rock blocks with potential to fall in the drifts; the potential for cave-in, collapse, or closure of the excavations; and the extent and severity of rock-mass disturbance near excavations. Confirm that selection of a ground support system is consistent with the anticipated rock-mass responses and potential failure mechanisms of the rock mass near the excavations.

Review Method 7 Design of Ventilation Systems

Confirm that the design of any subsurface ventilation systems important to safety has appropriate quality assurance classifications and conforms to appropriate codes and standards.

Confirm that any subsurface ventilation systems (including their power sources) important to safety (reviewed using Section 2.1.1.6 of the Yucca Mountain Review Plan) are designed to function under normal subsurface operating conditions (e.g., high temperature, potentially wet environments) and under Categories 1 and 2 event sequences. Coordinate with the reviewer using Sections 2.1.1.3, 4.1.1.4, and 2.1.1.5 to verify subsurface ventilation design has adequately considered event sequences that have radiological safety consequences.

For subsurface ventilation systems important to safety, confirm that the U.S. Department of Energy has an adequate periodic inspection, testing, and maintenance program to assure that ventilation requirements can be maintained, and that concentrations of radioactive materials within the subsurface worker operations areas, escape routes, and exhaust air are as low as is reasonably achievable. Verify that this program includes among others:

(1) Periodic replacement of any high-efficiency particulate air filters in the geologic repository operations area exhaust shafts, ramps, or other high radiation areas;

(2) Periodic testing/calibration of radiological monitoring devices that activate or deactivate any high-efficiency particulate air filter systems;

(3) Routine testing of any standby/backup ventilation equipment and emergency power to assure readiness to maintain ventilation functions and radiation safety; and

(4) Routine testing and calibration of airborne radiological monitoring devices, smoke detectors, and temperature sensors.

Verify that for any U.S. Department of Energy subsurface ventilation systems important to safety the design is adequate to seal off or isolate potential airborne radiological release areas (e.g., waste haulage routes, emplacement drifts) to limit the extent of radiological contamination and worker exposure.

Confirm that for any U.S. Department of Energy subsurface ventilation systems important to safety the design analysis is based on accepted industry codes or methods, incorporates site-specific data (i.e., resistance factors, humidity levels, time-varying waste package heat fluxes), and is based on an accurate representation of the subsurface drift structure (i.e., varying drift shapes and dimensions, varying flow rates between emplacement drifts and main drifts). Verify that subsurface ventilation flows from the least likely contaminated areas to the most contaminated areas, and meets design criteria (e.g., worker radiation exposure limits or other contaminant limits, air temperature limits, pressure differentials between high radiation/nonradiation areas).

Conduct limited confirmatory independent analyses to verify the U.S. Department of Energy analyses results for any subsurface ventilation systems important to safety.

Review Method 8 Design of Subsurface Power and Power Distribution Systems

Verify that the design of subsurface electric power supplies (e.g., electric transformers, electric substations) and power distribution systems, for structures, systems, and components important to safety, is consistent with accepted design criteria, codes, standards, and specifications for underground usage. Confirm these systems are suitable for the normal geologic repository operations area operating environment and those Categories 1 and 2 event sequences of radiological consequence reviewed using Section 2.1.1.5 of the Yucca Mountain Review Plan.

Confirm that the design incorporates proper grounding of electrical power sources/equipment to protect workers.

Verify that the design has sufficient emergency backup power capability to support equipment that is important to safety.

Verify that the U.S. Department of Energy design of electric power systems important to safety permits appropriate periodic inspection and testing.

Review Method 9 Maintenance Plan for Subsurface Facility Structures, Systems, and Components, Equipment, and Controls Important to Safety

Evaluate the adequacy of the maintenance plan developed to maintain drift stability before permanent closure of the repository. Verify that the maintenance plan considers the likely effects of uncertainties caused by high temperature and high radiation levels, and is based on an appropriate interpretation of modeling results that assesses the possibility of degradation of both the rock mass and the ground support system under sustained thermal load.

Verify that adequate maintenance plans for other subsurface facility structures, systems, and components, equipment, and controls important to safety are in place, and that they account for drift stability and accessibility during the period before permanent closure. Also, verify that the consideration of drift stability effects in the maintenance plan is based on an appropriate interpretation of modeling results. Plans for conduct of normal activities including maintenance, surveillance, and periodic testing are reviewed using Section 2.5.6 ("Plans for Conduct of Normal Activities Including Maintenance, Surveillance, and Periodic Testing") of the Yucca Mountain Review Plan.

III. Designs for Structures, Systems, and Components and Safety Controls That are Safety-Related for Waste Package/Engineered Barrier System

Review Method 1 Design of Waste Package and Engineered Barrier System Structures, Systems, and Components and Their Controls

Confirm that the waste package/engineered barrier system design adequately incorporates containment (considering corrosion resistance), criticality control, shielding, structural strength and drop fracture resistance of waste packages, thermal control, waste form degradation, drip shield, waste package support/invert, backfill, and sorption barrier, as appropriate.

Verify that the description and assessment of components for the waste packages include containers and internal structures, such as structural guides, baskets, fuel baskets, fuel basket plates with neutron absorbers, neutron absorber rods, canisters, fillers, and fill gas. The description and assessment should also consider specific components of the engineered barrier system, such as drip shield, backfill, and sorption barrier. Confirm that the design analyses for these components are adequate.

Verify that the materials, methods, and processes used in the fabrication of containers, internal waste package components, and engineered barrier system components are consistent with accepted design criteria, codes, standards, and specifications, such as American Society of Mechanical Engineers Boiler and Pressure Vessel Code. Confirm that processes specified for fabrication, assembly, closure, and inspection are based on accepted industry technology. Confirm that the license application evaluates and assesses the consequences of any significant uncertainties related to the corrosion and mechanical resistance of container materials and relevant engineered barrier system components ,such as the drip shield. If the U.S. Department of Energy uses design criteria, codes, standards, specifications, and industry technology, other than those mentioned above, evaluate the adequacy of the technical bases provided.

Confirm that specifications for the container and internal waste package materials are in agreement with those established in the final design. Verify that the specifications for closure welding, preparation for welding, materials to be used in welds, and inspection of welding comply with appropriate American Society of Mechanical Engineers codes, such as American Society of Mechanical Engineers Boiler and Pressure Vessel Code.

Verify that appropriate methods for nondestructive examination of fabricated containers and other structural components of waste packages have been identified to detect and evaluate fabrication defects and any other defects.

Confirm that criticality design criteria are consistent with those used in model calculations that support the design, and that isotopic enrichment of waste is properly characterized for these models. Verify the model configurations are appropriate for the postulated repository environments, and that appropriate computer models are used in design calculations.

Verify that the assessment of shielding provided by the containers is adequate. This assessment should include estimates of dose rates, a description of the source of data for the evaluation, and the methods for estimating dose rate, including the use of computational codes.

Confirm that the components of the waste package and internals have been designed to sustain loads from normal operation and Categories 1 and 2 event sequences.

Confirm that thermal control is such that the fuel cladding temperature is sufficiently low to prevent cladding failure.

Verify that the materials used in construction of the internal components of the waste package are compatible with the waste form, and that interactions among these materials will not be detrimental to the stability of the waste form. This verification should confirm that no pyrophoric, explosive, or chemically reactive materials are introduced in the waste package.

Confirm that the design of any drip shield, including materials of construction, configuration, and method of emplacement, is adequate. Confirm that the safety aspects of the engineered barrier system design and waste package handling are not impaired by the drip shield.

Verify that the design of backfill (if used in the license application design), including materials and physical characteristics, configuration, and methods of emplacement and compaction is adequate.

Confirm that the design of any sorption barrier is adequate.

2.1.1.7.3 Acceptance Criteria

The following acceptance criteria are based on meeting the requirements of 10 CFR 63.112(f), relating to the design of structures, systems, and components important to safety and safety controls.

2.1.1.7.3.1 Design Criteria and Design Bases

Acceptance Criterion 1 The Relationship Between the Design Criteria and the Requirements Specified in 10 CFR 63.111(a) and (b), the Relationship Between the Design Bases and the Design Criteria, and the Design Criteria and Design Bases for Structures, Systems, and Components Important to Safety are Adequately Defined.

(1) Design criteria and bases for structures, systems, and components important to safety and for those structures, systems, and components that affect the proper functioning of structures, systems, and components important to safety, are identified, and these criteria and bases are derived from the specific site characteristics and consequence analyses. The design criteria and bases are consistent with the analyses used in the identification of the structures, systems, and components;

(2) Design criteria for normal operating conditions are adequately developed. Design criteria do not permit degradation of the performance of geologic repository operations area structures, systems, and components important to safety;

(3) Design criteria adequately consider preclosure safety analysis results, to ensure that structures, systems, and components important to safety will continue to prevent unacceptable consequences;

(4) Structural design criteria and bases for structures, systems, and components important to safety meet relevant guidance or standards;

(5) Thermal design criteria and bases are consistent with relevant regulatory guidance or standards;

(6) Design criteria and bases for shielding and confinement systems use appropriate guidance or standards;

(7) Design criteria for fixed-area radiation monitors and continuous airborne monitoring instrumentation are consistent with relevant regulatory guidance;

(8) Criticality design criteria are developed, based on consequence analysis results from the preclosure safety analysis ,and are consistent with relevant regulatory guidance or standards. Design criteria are adequately factored into the models and assumptions used for criticality analysis; and

(9) Design bases and criteria are adequately identified for thermal, structural, shielding, criticality, and other operating limits for the geologic repository operations area facilities.

2.1.1.7.3.2 Design Methodologies

Acceptance Criterion 1 Geologic Repository Operations Area Design Methodologies Are Adequate.

(1) Proposed design methodologies are supported by adequate technical bases, are consistent with established industry practice, and address uncertainties.

(2) If the design methodologies depend on site-specific test data, such data are available. Analytical or numerical models used to support the design methodologies are verified, calibrated, and validated; assumptions or limitations relating to the proposed methodologies are identified, and their implications for the design are adequately analyzed and documented;

(3) Expert elicitations are properly conducted; and

(4) Seismic design methodologies use ground motion information that is consistent with proposed U.S. Department of Energy methodologies for hazard assessment and preclosure design criteria, and, taken together, they provide adequate input for seismic design and for performance assessments.

2.1.1.7.3.3 Geologic Repository Operations Area Design and Design Analyses

I. Designs and Design Analyses for Structures, Systems, and Components, Equipment, and Safety Controls That are Safety Related for Surface Facilities

Acceptance Criterion 1 Design Codes and Standards Used for the Design of Surface Facility Structures, Systems, and Components Important to Safety Are Identified, and Are Appropriate for the Design Methodologies Selected.

(1) Applicable design codes and standards are specified for structural, thermal, shielding and confinement, criticality, and decommissioning designs; and

(2) If other methods are used for design, the license application provides adequate technical bases for those methods.

Acceptance Criterion 2 The Materials to Be Used for Structures, Systems, and Components Important to Safety Related to Surface Facility Design Are Consistent with the Design Methodologies.

(1) Materials used for structures, systems, and components important to safety related to surface facility design are consistent either with the accepted design criteria, codes, standards, and specifications, or with those specifically developed by the U.S. Department of Energy;

(2) Materials are adequate, considering the material properties and allowable stresses and strains associated with the design;

(3) Materials and their properties are appropriate for the expected design loading conditions. In addition, anticipated stress limits for each material are based on maximum temperatures as established in the thermal analysis evaluation presented in the license application; and

(4) The potential for creep or brittle fracture and drop fracture resistance of materials is adequately assessed, to ensure that structures, systems, and components important to safety will perform their safety functions.

Acceptance Criterion 3 Design Analyses Use Appropriate Load Combinations for Normal and Categories 1 and 2 Event Sequence Conditions.

(1) The loads used in the U.S. Department of Energy design analyses are consistent with those normal and Categories 1 and 2 event sequence loadings of radiological importance;

(2) The load combinations used in the design analyses are consistent with those used and accepted by the U.S. Nuclear Regulatory Commission for the design of similar types of nuclear facilities and for steel and reinforced concrete structures; and

(3) The design analyses use appropriate techniques that are correctly applied to provide established design temperatures, mechanical loads, and pressures for the structures, systems, and components important to safety.

Acceptance Criterion 4 Design Analyses Are Properly Performed and Documented.

(1) The design analyses include relevant structural, thermal, shielding, criticality, confinement, and decommissioning factors;

(2) Values of material properties used for the design analyses have adequate technical bases and are consistent with site-specific data;

(3) Loads and load combinations used in the design analyses are consistent with defined normal operations and Categories 1 and 2 event sequences;

(4) Analytical methods, models, and codes used for the design analyses are appropriate for the conditions analyzed, and are properly benchmarked, if appropriate;

(5) Technical bases for the assumptions used in the design analyses are adequately defined, and are based on accepted engineering practice;

(6) The designs and design analyses for structures, systems, and components important to safety are performed correctly. These structures, systems, and components have sufficient capability to withstand normal and Categories 1 and 2 event sequence loadings; and

(7) Confirmatory checks indicate that the design analyses are adequate.

II. Designs and Design Analyses for Structures, Systems, and Components, Equipment, and Safety Controls That are Safety Related for Subsurface Facility

Acceptance Criterion 1 Design Assumptions, Codes, and Standards Used for the Design of Subsurface Facility Structures, Systems, and Components Important to Safety Are Acceptable.

(1) Applicable design codes, standards, or other detailed criteria used for the design of the subsurface facility are specified. Codes and standards are equivalent to, and consistent with, those accepted by the U.S. Nuclear Regulatory Commission for design of nuclear facilities with similar hazards and functions. If nonstandard approaches are used, the license application provides adequate technical bases to justify why they are used;

(2) Assumptions made for the design of the subsurface facility are technically defensible; and

(3) Designs for steel and concrete structures and components, air controlled systems, electrical power systems, and ventilation systems use applicable standards.

Acceptance Criterion 2 The Design of Subsurface Operating Systems Is Adequate.

(1) Methods, assumptions, and input data, used in the ventilation design, are consistent with proposed thermal loading performance goals. Confirmatory analyses verify the results in the license application. Analyses adequately address the thermal loads;

(2) Subsurface monitoring and control systems are appropriately designed for the safety functions of the structures, systems, and components during waste emplacement and monitoring;

(3) The design of the waste emplacement system is compatible with proposed waste emplacement procedures. Interfaces with other systems are identified and assessed, and continuity of operations and safety can be achieved;

(4) Emplacement drifts are located away from major faults, consistent with the seismic design, and the subsurface layout is appropriate for the quantity of waste to be emplaced and the design thermal load;

(5) The design of the geologic repository operations area accommodates implementation of the performance confirmation program, as specified in 10 CFR 63.111(d); and

(6) Standards and codes used for design of subsurface operating systems are properly applied.

Acceptance Criterion 3 Materials and Material Properties Used for the Subsurface Facility Design Are Appropriate.

(1) The selection of materials, and the properties of these materials, are appropriate for the anticipated subsurface environment;

(2) Materials and material properties are consistent with applicable design criteria, codes, standards, and specifications. If no standards are used, the technical bases provided are acceptable;

(3) The selection of ground support materials accounts for degradation of such materials under elevated temperature and thermal loading. Plausible mechanisms for material degradation are identified, and properly incorporated in assessments of subsystem structure, system, and component performance; and

(4) Fire-resistant materials are incorporated into the design of the subsurface ventilation systems. Ventilation equipment/components are designed to withstand prolonged high temperature conditions, effects of blast cooling, and wet and corrosive environments.

Acceptance Criterion 4 Design Analyses Use Appropriate Load Combinations for Normal and Categories 1 and 2 Event Sequence Conditions.

(1) The arrangement of waste packages within the subsurface facility satisfies the thermal load design criteria;

(2) The magnitude and time history of the applied thermal loading are consistent with the anticipated characteristics of the proposed waste, repository design configurations, and design areal mass loading;

(3) Thermal analyses have an appropriate technical basis, use site-specific thermal property data, consider temperature dependency and uncertainties of thermal property data, and

use thermal models and analyses that are properly documented. If credit is taken for use of ventilation, assessments of the effects of ventilation are adequate;

(4) Design analyses consider appropriate *in situ* stresses, potential running ground conditions, and hydrologic changes to the rock mass during the heating period; and

(5) The dynamic loads used in design analyses are consistent with seismic-design ground-motion parameters consider faulting effects, and are consistent with accepted methodologies for assessing faulting hazards.

Acceptance Criterion 5 Design Analyses Use Appropriate Models and Site-Specific Properties of the Host Rock, and Consider Spatial and Temporal Variation and Uncertainties in Such Properties.

(1) Appropriate combinations of continuum and discontinuum modeling, as well as two- and three-dimensional modeling, are conducted, to assess the behavior of a fractured rock mass under prolonged heated conditions and identified Categories 1 and 2 event sequences. The bases for the choice of specific models and model combinations are adequate. Appropriate bases for the assumptions and limitations of the modeling approach are provided;

(2) Principles for the design analyses, the underlying assumptions, and the anticipated limitations are documented, are consistent with modeling objectives, and are technically sound;

(3) Values for the rock-mass thermal-expansion coefficient are consistent with properly interpreted site-specific data, and such interpretation accounts for likely scale effects and temperature dependency. The uncertainty in the thermal-expansion coefficient is adequately assessed, and considered in the thermal-stress calculation;

(4) For continuum rock-mass modeling, the values for rock-mass elastic parameters (Young's modulus and Poisson's ratio) and strength parameters (friction angle and cohesion) are consistent with properly interpreted site-specific data. If the parameter values are obtained through empirical correlations with a rock-quality index, the empirical equations used are appropriate for the site and are applied correctly, and the values of the index are consistent with site-specific data. If intact-rock-scale values are used, the bases for application of the values to the rock-mass scale are adequate;

(5) For discontinuum rock-mass modeling, the selection of fracture patterns for numerical modeling is appropriate for the objectives of the design and analyses. The interpretation of modeling results adequately considers effects of the representation of the characteristics of the modeled fracture network, compared with those of the *in situ* fracture network;

(6) For discontinuum modeling, the selection of stiffness and strength parameters for rock blocks between any fractures that are explicitly represented in the model are appropriate, and account for fractures that are not explicitly represented;

(7) For discontinuum modeling, the values for fracture stiffness and strength parameters are consistent with properly interpreted site-specific data;

(8) If applicable, time-dependent mechanical degradation of the rock mass, fractures, and ground support that may occur after the emplacement of waste is adequately accounted for in thermal-mechanical analyses. The bases for the magnitude and rate of mechanical degradation applied in the analyses are appropriately established, and are technically defensible;

(9) Uncertainties in rock-mass and fracture-mechanical properties are adequately estimated, and considered in both continuum and discontinuum modeling;

(10) Models adequately address the stability of openings around drift intersections, considering the rock mass and its degraded properties and thermal loading; and

(11) Confirmatory checks indicate that the design analyses are adequate.

Acceptance Criterion 6 The Design of Ground Support Systems Is Based on Appropriate Design Methodologies and Interpretations of Modeling Results.

(1) Design methodologies, or combinations of design methodologies, are properly applied to the design of ground support systems. When used, the empirical design approach is consistent with accepted technology in the underground tunneling and mining industry. The evaluation and selection of ground support systems are supported by analyses that satisfy the previous two acceptance criteria, and that provide mechanical evaluation of ground support systems under thermal and dynamic loads;

(2) The ground support system responses are adequately evaluated, based on the results of model analyses. If the ground support system is explicitly modeled, the ground support responses include an adequate assessment of deformation and potential failure of the ground support systems. The interaction between the ground support system and the host rock units is adequately considered in the analysis. If the ground support system is not explicitly modeled, the anticipated ground support system responses from the modeling results are reasonably estimated, and the technical bases for these estimates are adequate;

(3) The geometrical, thermal, and mechanical characteristics of the support system used in the thermal-mechanical analyses are consistent with design and construction specifications. The time-dependent mechanical degradation of the support system, under heated conditions, is adequately accounted for in the analyses; and

(4) Stability of drifts, shafts, and ventilation tunnels is adequately assessed both with and without ground support. Such assessment includes identification of rock blocks that have potential to fall in the drifts; the potential for cave-in, collapse, or closure of the emplacement drifts; and the extent and severity of rock-mass disturbance near the excavations. The selection of a ground support system is consistent with the anticipated

rock-mass responses and potential failure mechanisms of the rock mass near the excavations.

Acceptance Criterion 7 The Subsurface Ventilation Systems Are Adequately Designed.

(1) The design of subsurface ventilation systems important to safety is consistent with the quality assurance classification and with appropriate codes and standards;

(2) The subsurface ventilation systems (including their power sources) important to safety are designed to continue functioning under normal subsurface operating conditions and under Categories 1 and 2 event sequences;

(3) Subsurface ventilation systems important to safety are designed to continue operating under Categories 1 and 2 event sequences and in case of a main subsurface power outage, if necessary;

(4) For subsurface ventilation systems important to safety, there is an adequate periodic inspection, testing, and maintenance program, to assure that concentrations of radioactive materials meet the limits specified in 10 CFR Parts 20 and 63, as practicable;

(5) The subsurface ventilation design for systems important to safety is adequate to seal off, or isolate, airborne radiation, within areas that could have a potential release;

(6) The ventilation design analysis is based on accepted industry codes or methods, incorporates site-specific data, and is based on an accurate representation of the subsurface drift structure. The ventilation design analysis shows that subsurface ventilation systems important to safety flow from the least likely contaminated areas to the most likely contaminated areas, and meets all other specified design criteria; and

(7) Confirmatory checks indicate that the design analyses for subsurface ventilation systems important to safety are adequate.

Acceptance Criterion 8 The Design of Subsurface Power and Power Distribution Systems for Structures, Systems, and Components and Operations Important to Safety Is Adequate.

(1) The design of subsurface electric power supplies and power distribution systems for structures, systems, and components important to safety is consistent with accepted design criteria, codes, standards, and specifications for underground usage, and is suitable for the normal operating environment and Categories 1 and 2 event sequences;

(2) The design incorporates proper grounding of electrical power sources/equipment;

(3) The design has sufficient emergency backup power capability for structures, systems, and components important to safety; and

(4) The design of electric power systems important to safety permits appropriate periodic inspection and testing.

Acceptance Criterion 9 An Adequate Maintenance Plan Exists for Subsurface Facility Structures, Systems, and Components, Equipment, and Controls Important to Safety.

(1) The maintenance plan developed to maintain drift stability, before permanent closure of the repository, is adequate. This maintenance plan considers the likely effects of uncertainties caused by high temperature and high radiation levels, and is based on an appropriate interpretation of modeling results that assesses the possibility of degradation of both the rock mass and the ground support system under sustained thermal load; and

(2) Adequate maintenance plans for other subsurface facility structures, systems, and components, equipment, and controls important to safety are in place, and they account for drift stability and accessibility during the period before permanent closure. The consideration of drift stability effects in the maintenance plan is based on an appropriate interpretation of modeling results.

III. Designs for Structures, Systems, and Components and Safety Controls That Are Safety-Related for Waste Package/Engineered Barrier System

Acceptance Criterion 1 Waste Package and Engineered Barrier System Structures, Systems, and Components and Their Controls Are Adequately Designed.

(1) The waste package/engineered barrier system design adequately incorporates containment, criticality control, shielding, structural strength of waste packages, thermal control, waste form degradation, drip shield, waste package support/invert, backfill, and sorption barrier, as appropriate;

(2) The description and assessment of the components for the various types of waste packages include containers and internal structures, such as structural guides, baskets, fuel baskets, fuel basket plates with neutron absorbers, neutron absorber rods, canisters, fillers, and fill gas, in addition to specific components of the engineered barrier system, such as drip shield, backfill, and sorption barrier. The design analyses for these components are adequate;

(3) The materials, methods, and processes used in the fabrication of containers, internal waste package components, and engineered barrier system components are consistent with accepted design criteria, codes, standards, and specifications. Processes specified for fabrication, assembly, closure, and inspection are based on accepted industry technology. The license application documents any significant uncertainties related to the corrosion and mechanical resistance of container materials and relevant engineered barrier system components, such as the drip shield. If the U.S. Department of Energy chooses to use design criteria, codes, standards, specifications, and industry technology different from those normally used, the technical bases provided are adequate;

(4) The specifications for container and internal waste package materials are in agreement with those established in the final design. The specifications for closure welding, preparation for welding, materials to be used in welds, and inspection of welding comply with applicable American Society of Mechanical Engineers codes;

(5) Appropriate methods for nondestructive examination of fabricated containers and other structural components of waste packages are identified to detect and evaluate fabrication defects and any other defects;

(6) Criticality design criteria are consistent with those used in model calculations that support the design. Isotopic enrichment of waste is properly characterized for these models. Model configurations are appropriate for the various postulated repository environments, and appropriate computer models are used in design calculations;

(7) The assessment of shielding provided by the containers is sufficient. The assessment includes estimates of dose rates, a description of the source of data for the evaluation, and the methods for estimating dose rate, including the use of computational codes;

(8) The components of the waste package and internals are designed to sustain loads from normal operation, drop events, and Categories 1 and 2 event sequences;

(9) Thermal control is such that the fuel cladding temperature will be sufficiently low to prevent cladding failure;

(10) The materials used in construction of the internal components of the waste package are compatible with the waste form, and interactions among these materials will not be detrimental to the stability of the waste form. No pyrophoric, explosive, or chemically reactive materials will be introduced in the waste package;

(11) The design of any drip shield, including materials of construction, configuration, and method of emplacement, is sufficient. The safety aspects of the engineered barrier system design and waste package handling are not impaired by the drip shield;

(12) The design of any backfill, including materials and physical characteristics, configuration, and methods of emplacement and compaction, is adequate; and

(13) The design of any sorption barrier is adequate.

2.1.1.7.4 Evaluation Findings

If the license application provides sufficient information and the regulatory acceptance criteria in Section 2.1.1.7.3 are appropriately satisfied, the staff concludes that this portion of the staff evaluation is acceptable. The reviewer writes material suitable for inclusion in the Safety Evaluation Report prepared for the entire application. The report includes a summary statement of what was reviewed and why the reviewer finds the submittal acceptable. The staff can document the review as follows.

U.S. Nuclear Regulatory Commission staff has reviewed the Safety Analysis Report and other information submitted in support of the license application, and has found, with reasonable assurance, that the requirements of 10 CFR 63.111(d) and 63.112(f) are satisfied. An adequate description of the geologic repository operations area design that satisfactorily defines the relationship between design criteria and the performance objectives, and that identifies the relationship between the design bases and the design criteria has been provided.

2.1.1.7.5 References

Kotra, J.P., et al. NUREG–1563, "Branch Technical Position on the Use of Expert Elicitation in the High-Level Radioactive Waste Program." Washington, DC: U.S. Nuclear Regulatory Commission. 1996.

U.S. Nuclear Regulatory Commission. Regulatory Guide 1.120, "Fire Protection for Operating Nuclear Power Plants." Washington, DC: U.S. Nuclear Regulatory Commission. 2001.

——. NUREG–1567, "Standard Review Plan for Spent Fuel Storage Facilities, Final Report." Washington, DC: U.S. Nuclear Regulatory Commission. 2000.

——. Regulatory Guide 3.71, "Nuclear Criticality Safety Standards for Fuels and Material Facilities." Washington, DC: U.S. Nuclear Regulatory Commission. 1998.

2.1.1.8 Meeting the 10 CFR Part 20 As Low As Is Reasonably Achievable Requirements for Normal Operations and Category 1 Event Sequences

Review Responsibilities—High-Level Waste Branch and Environmental and Performance Assessment Branch

2.1.1.8.1 Areas of Review

This section provides guidance on the review of meeting the 10 CFR Part 20 as low as is reasonably achievable requirements for normal operations and Category 1 event sequences. Reviewers will also evaluate the information required by 10 CFR 63.21(c)(5) and (c)(6).

The staff will evaluate the following parts of meeting the 10 CFR Part 20 as low as is reasonably achievable requirements for normal operations and Category 1 event sequences, using the review methods and acceptance criteria in Sections 2.1.1.8.2 and 2.1.1.8.3:

(1) Policy Considerations;
(2) Design Considerations; and
(3) Operational Considerations.

2.1.1.8.2 Review Methods

Review Method 1 Management Commitment to Maintain Exposures As Low As Is Reasonably Achievable

Confirm that the management commitment includes provisions and guidance for ensuring that:

(1) Supervisors will integrate appropriate radiation protection controls into work activities;

(2) Personnel are aware of the management commitment to as low as is reasonably achievable principles;

(3) Workers will receive sufficient and appropriate initial and periodic training related to as low as is reasonably achievable principles, considering the review of training and certification of personnel conducted, using Section 2.5.3 ("Training Certification of Personnel") of the Yucca Mountain Review Plan; and

(4) An operations program to control radiation exposure will be implemented. This program will ensure that individual and collective doses are as low as is reasonably achievable, considering the review of plans for conduct of normal operations conducted, using Section 2.5.6 ("Plans for Conduct of Normal Activities Including Maintenance, Surveillance, and Periodic Testing") of the Yucca Mountain Review Plan.

Review Method 2 Consideration of As Low As Is Reasonably Achievable Principles in the Geologic Repository Operations Area Design

Verify that the design of the geologic repository operations area has considered the as low as is reasonably achievable philosophy, as stated in Regulatory Guide 8.8, "Information Relevant to Ensuring that Occupational Radiation Exposure at Nuclear Power Stations Will Be As Low As Is Reasonably Achievable" (U.S. Nuclear Regulatory Commission, 1978). Note that Regulatory Guide 8.8 is for nuclear power plants, where radiation hazards are more severe than the radiation hazards at the geologic repository operations area; consider this aspect when using this guidance.

Confirm that as low as is reasonably achievable principles are adopted in the design considerations, to the extent practical, to ensure the following:

(1) Engineered design features minimize the time workers must stay in radiation areas;

(2) Remotely operated or robotic equipment, such as welders, wrenches, or radiation monitors, are used to minimize worker dose;

(3) Suitable methods are used to monitor for possible blockage of air cooling passages, or to perform inspection of materials;

(4) Design permits placement of equipment and temporary shielding by remote control to reduce doses, if appropriate;

(5) Materials and design features minimize the potential for accumulation of radioactive materials or surface contamination, to facilitate decontamination, or decontamination and dismantlement, of surface facilities;

(6) Radioactive material handling and storage facilities are located sufficiently far from the site boundary and from other on-site work stations. The controlled area of the facility is sufficient to maintain doses at locations accessible to members of the public at acceptable levels;

(7) Transfer routes for high-level radioactive waste will maintain the desired distance from the site perimeter; and

(8) Restricted areas, that is, high radiation and very high radiation areas, within the controlled area provide control of access to areas with radiation levels that would pose unacceptable risk to workers within those areas, if appropriate.

Confirm that modifications to the design of the geologic repository operations area to maintain doses as low as is reasonably achievable have been incorporated in the preclosure safety analysis, to ensure they do not adversely influence other components of the design.

Review Method 3 Incorporation of As Low As Is Reasonably Achievable Principles into Proposed Operations at the Geologic Repository Operations Area

Verify that operational procedures follow the as low as is reasonably achievable philosophy in Regulatory Guides 8.8 and 8.10 (U.S. Nuclear Regulatory Commission, 1978, 1977). Plans for conduct of normal activities, including maintenance, surveillance, and testing, should be reviewed, using Section 2.5.6 ("Plans for Conduct of Normal Activities Including Maintenance, Surveillance, and Periodic Testing") of the Yucca Mountain Review Plan.

Confirm that geologic repository operations area operational procedures will ensure that the doses to workers and members of the public will be as low as is reasonably achievable, and include:

(1) An operations program designed to control radiation exposure will be implemented, to ensure that both individual and collective doses are as low as is reasonably achievable. Plans for conduct of normal operations, and are reviewed, using Section 2.5.6 ("Plans for Conduct of Normal Activities, Including Maintenance, Surveillance, and Periodic Testing") of the Yucca Mountain Review Plan);

(2) Tradeoffs between requirements for increased monitoring or maintenance activities (and the increased exposures that would result), and the potential hazards associated with reduced frequency of these activities are assessed;

(3) Dry runs to develop proficiency in procedures involving radiation exposures, to determine exposures likely to be associated with specific procedures, and to consider alternative procedures, to minimize exposures are planed and conducted;

(4) Contingency procedures for potential off-normal occurrences are developed and tested; and

(5) As low as is reasonably achievable operational alternatives are developed based on relevant experience with independent spent nuclear fuel storage installations, pool facilities, and waste management facilities.

Confirm that modifications to proposed operations of the geologic repository operations area to maintain doses as low as is reasonably achievable have been incorporated in the preclosure safety analysis, to ensure that they do not adversely influence other aspects of geologic repository operations area operations.

2.1.1.8.3 Acceptance Criteria

The following acceptance criteria are based on meeting the requirements of 10 CFR 63.111(a)(1) and (c)(1), relating to meeting the 10 CFR Part 20 as low as is reasonably achievable requirements for normal operations and Category 1 event sequences.

Acceptance Criterion 1 An Adequate Statement of Management Commitment to Maintain Exposures to Workers and the Public as Low as Is Reasonably Achievable Is Provided.

(1) The management commitment includes provisions and guidance for ensuring that:

(a) Supervisors will integrate appropriate radiation protection controls into work activities;

(b) Personnel are aware of the management commitment to as low as is reasonably achievable principles;

(c) Workers will receive sufficient and appropriate initial and periodic training related to as low as is reasonably achievable principles; and

(d) An operations program to control radiation exposure will be implemented. This program will ensure that individual and collective doses are as low as is reasonably achievable.

Acceptance Criterion 2 As Low as Is Reasonably Achievable Principles Are Adequately Considered in Geologic Repository Operations Area Design.

(1) The design of the geologic repository operations area adequately considers the as low as is reasonably achievable philosophy; and

(2) As low as is reasonably achievable principles are adopted in the design considerations, to the extent practical, to ensure the following:

(a) Engineered design features minimize the time workers must stay in radiation areas;

(b) Remotely operated or robotic equipment such as welders, wrenches, or radiation monitors are used to minimize worker dose;

(c) Suitable methods are used to perform inspection of materials;

(d) Design permits placement of equipment and temporary shielding by remote control, to reduce doses, if appropriate;

(e) Materials and design features minimize the potential for accumulation of radioactive materials or surface contamination, to facilitate decontamination, or decontamination and dismantlement, of surface facilities;

(f) Radioactive material handling and storage facilities are located sufficiently far from the site boundary and from other on-site work stations. The controlled area of the facility is sufficient to maintain doses at locations accessible to members of the public at acceptable levels;

(g) Transfer routes for high-level radioactive waste will maintain the desired distance from the site perimeter; and

(h) Restricted areas, that is, high radiation and very high radiation areas, within the controlled area, provide control of access to areas with radiation levels that would pose unacceptable risk to workers within those areas.

(3) Modifications to the design of the geologic repository operations area to maintain doses as low as is reasonably achievable have been incorporated in the preclosure safety analysis, to ensure they do not adversely influence other components of the design.

Acceptance Criterion 3 Proposed Operations at the Geologic Repository Operations Area Adequately Incorporate as Low as Is Reasonably Achievable Principles.

(1) Operational procedures follow the as low as is reasonably achievable philosophy;

(2) Geologic repository operations area operational procedures will ensure that the doses to workers and members of the public will be as low as is reasonably achievable, including the consideration of items such as:

(a) An operations program designed to control radiation exposure will be implemented, to ensure both individual and collective doses are as low as is reasonably achievable;

(b) Tradeoffs between requirements for increased monitoring or maintenance activities (and the increased exposures that would result) and the potential hazards associated with reduced frequency of these activities;

(c) Dry runs to develop proficiency in procedures involving radiation exposures, to determine exposures likely to be associated with specific procedures, and to consider alternative procedures to minimize exposures;

(d) Development of tested contingency procedures for potential off-normal occurrences; and

(e) As low as is reasonably achievable operational alternatives, based on relevant experience with independent spent nuclear fuel storage installations, pool facilities, and waste management facilities.

(3) Modifications to proposed operations of the geologic repository operations area to maintain doses as low as is reasonably achievable have been incorporated in the preclosure safety analysis, to ensure that they do not adversely influence other aspects of geologic repository operations area operations;

2.1.1.8.4 Evaluation Findings

If the license application provides sufficient information and the regulatory acceptance criteria in Section 2.1.1.8.3 are appropriately satisfied, the staff concludes that this portion of the staff evaluation is acceptable. The reviewer writes material suitable for inclusion in the safety evaluation report prepared for the entire application. The report includes a summary statement of what was reviewed and why the reviewer finds the submittal acceptable. The staff can document the review as follows.

U.S. Nuclear Regulatory Commission staff has reviewed the Safety Analysis Report and other information submitted in support of the license application, and has found, with reasonable assurance, that the requirements of 10 CFR 63.111(a)(1) are satisfied. The operations at the geologic repository operations area, through permanent closure, will comply with the as low as is reasonably achievable requirements in 10 CFR Part 20.

U.S. Nuclear Regulatory Commission staff has reviewed the Safety Analysis Report and other information submitted in support of the license application, and has found, with reasonable assurance, that the performance objective at 10 CFR 63.111(c)(1) are satisfied. The requirements of 10 CFR 63.111(a) for as low as is reasonably achievable will be met.

2.1.1.8.5 References

U.S. Nuclear Regulatory Commission. Regulatory Guide 8.8, “Information Relevant to Ensuring that Occupational Radiation Exposures at Nuclear Power Stations Will Be As Low As Is Reasonably Achievable.” Revision 3. Washington, DC: U.S. Nuclear Regulatory Commission, Office of Standards Development. 1978.

———. Regulatory Guide 8.10, “Operating Philosophy for Maintaining Occupational Radiation Exposures As Low As Is Reasonably Achievable.” Revision 1. Washington, DC: U.S. Nuclear Regulatory Commission, Office of Standards Development. 1977.

2.1.2 Plans for Retrieval and Alternate Storage of Radioactive Wastes

Review Responsibilities—High-Level Waste Branch and Environmental and Performance Assessment Branch

2.1.2.1 Areas of Review

This section provides guidance on the review of plans for retrieval and alternate storage of radioactive wastes. Reviewers will also evaluate the information specified in 10 CFR 63.21(c)(7).

The staff will evaluate the following parts of plans for retrieval and alternate storage of radioactive wastes, using the review methods and acceptance criteria in Sections 2.1.2.2 and 2.1.2.3.

(1) Plans meeting performance objectives in 10 CFR 63.111(a) and (b);
(2) Adequate alternate storage for retrieved wastes; and
(3) Reasonable retrieval schedule.

2.1.2.2 Review Methods

Review Method 1 Waste Retrieval Plans

Confirm that waste retrieval plans include a discussion of: (i) retrieval operations processes; (ii) equipment to be used; and (iii) compliance with 10 CFR 63.111(a) and (b) preclosure performance objectives, during retrieval of waste.

Verify that the U.S. Department of Energy has developed scenarios under which retrieval operations will take place. Confirm that development of the scenarios has considered the protection of health and safety and keeping radiation exposures as low as is reasonably achievable. Assess the reasonableness of the scenarios developed throughout the potential waste retrieval period in compliance with 10 CFR 63.111(e).

Confirm that adequate methodologies have been established for identifying and analyzing potential problems for the various retrieval operations scenarios. Evaluate whether the solutions proposed for the problems identified are feasible, and are based on sound engineering principles. Verify that the extent of degradation of the emplacement drifts, during the period of retrieval operations, has been appropriately considered in the retrieval plans. Verify that retrieval plans contain acceptable maintenance plans to support the completion of retrieval, in a manner that would protect health and safety as well as keeping radiation exposures as low as is reasonably achievable.

If the backfilling option is used in emplacement drifts before the end of the period of design for retrievability, determine whether the retrieval plans adequately address the requirements of 10 CFR 63.111(e).

Verify that the U.S. Department of Energy has provided a discussion of the potential effect of the duration of the planned performance confirmation program on the time frame required to maintain the option of waste retrieval. Assess whether there is a need for a different time frame for the period of design for retrievability so it will be consistent with the duration proposed by the U.S. Department of Energy for conducting the performance confirmation program.

Review Method 2 Compliance with Preclosure Performance Objectives

Verify the U.S. Department of Energy has demonstrated that preclosure performance objectives in 10 CFR 63.111(a) and (b) can be met during waste retrieval. Assess if the as low as is reasonably achievable requirements are met during retrieval operation using the review methods and acceptance criteria in Section 2.1.1.8 ("Meeting the 10 CFR Part 20 As Low As Is Reasonably Achievable Requirements for Normal Operations and Category 1 Event Sequences") of the Yucca Mountain Review Plan.

Review Method 3 Proposed Alternate Storage

Confirm that the physical location and boundary of the proposed alternate storage area are adequately defined.

Verify that the proposed alternate storage area is sufficient to hold the waste to be retrieved.

Assess whether the plans are adequate for protection of workers and the public, while transporting the retrieved wastes to the alternate storage area.

Review Method 4 Retrieval Operations Schedule

Verify that plans for retrieval meet the 10 CFR 63.111(e)(3) requirement that retrieval can be completed under a reasonable schedule, that is, the same time as that required to construct the geologic repository operations area and emplace waste.

2.1.2.3 Acceptance Criteria

The following acceptance criteria are based on meeting the requirements of 10 CFR 63.111(e), relating to plans for retrieval and alternate storage of radioactive wastes.

Acceptance Criterion 1 Plans for Retrieval of Waste Packages, Based on a Reasonable Schedule Are Provided and Can Be Implemented, if Necessary.

(1) Waste retrieval plans include an adequate discussion of: (i) retrieval operations processes; (ii) equipment to be used; and (iii) compliance with 10 CFR 63.111(a) and (b) preclosure performance objectives, during retrieval of waste;

(2) The U.S. Department of Energy has prepared reasonable scenarios under which retrieval operations will take place. The scenarios consider the projected duration required to complete retrieval operations;

(3) Adequate methodologies are established for identifying and analyzing potential problems for the various retrieval operations scenarios. The solutions proposed for the problems identified are feasible, and are based on sound engineering principles. The extent of degradation of emplacement drifts, during the period of retrieval operations, is appropriately considered in the retrieval plans. The retrieval plans contain acceptable maintenance plans to support the completion of retrieval, in a manner that would protect health and safety as well as keeping radiation exposures as low as is reasonably achievable;

(4) Should the backfilling option be used in emplacement drifts, before the end of the period of design for retrievability, the retrieval plans adequately address the requirements of 10 CFR 63.111(e); and

(5) The U.S. Department of Energy provides a discussion of the potential effect of the duration of the planned performance confirmation program on the time frame required, to maintain the option of waste retrieval. If there is a need for a different time frame for the period of design for retrievability, the time frame is consistent with the duration proposed by the U.S. Department of Energy for conducting the performance confirmation program.

Acceptance Criterion 2 The Proposed Retrieval Operations Comply with the Requirements of the Preclosure Performance Objectives.

(1) The retrieval plan is adequate to meet preclosure performance objectives of 10 CFR 63.111(a) and (b) and incorporates the as low as is reasonably achievable requirements.

Acceptance Criterion 3 The Proposed Alternate Storage of Retrieved Radioactive Wastes Is Reasonable.

(1) The physical location and boundary of the proposed alternate storage area are adequately defined;

(2) The proposed alternate storage area is sufficient to hold the waste to be retrieved; and

(3) Plans are adequate for protection of workers and the public, while transporting the retrieved wastes to the alternate storage area.

Acceptance Criterion 4 A Reasonable Schedule for Potential Retrieval Operations Is Provided.

(1) Plans for retrieval meet the 10 CFR 63.111(e)(3) requirement that retrieval can be completed in about the same time as that required to construct the geologic repository operations area and emplace waste.

2.1.2.4 Evaluation Findings

If the license application provides sufficient information and the regulatory acceptance criteria in Section 2.1.2.3 are appropriately satisfied, the staff concludes that this portion of the staff evaluation is acceptable. The reviewer writes material suitable for inclusion in the safety evaluation report prepared for the entire application. The report includes a summary statement of what was reviewed and why the reviewer finds the submittal acceptable. The staff can document the review as follows.

U.S. Nuclear Regulatory Commission staff has reviewed the Safety Analysis Report and other information submitted in support of the license application, and has found, with reasonable assurance, that the requirements of 10 CFR 63.111(e) are satisfied. The geologic repository operations area has been designed to allow for retrievability of wastes. The option of waste retrieval has been preserved and allows a U.S. Nuclear Regulatory Commission review of that program. The design allows for retrieval on a reasonable schedule.

2.1.2.5 References

None.

2.1.3 Plans for Permanent Closure and Decontamination, or Decontamination and Dismantlement of Surface Facilities

Review Responsibilities—High-Level Waste Branch and Environmental and Performance Assessment Branch

2.1.3.1 Areas of Review

This section provides guidance on the review of plans for permanent closure and decontamination, or decontamination and dismantlement of surface facilities. Reviewers will evaluate the information required by 10 CFR 63.21(c)(8) and (c)(22)(vi).

In determining the acceptability of these plans, the reviewer should consider that plans submitted at the time of initial licensing will be prospective in nature, and will not reflect knowledge gained over the course of facility operation (e.g., detailed knowledge of the types, extent, and precise locations of contamination). Therefore, it is not reasonable to expect plans submitted with the initial license application to have the same level of detail as final plans, especially with respect to elements, such as planned decontamination activities and the final radiation survey. The U.S. Department of Energy will be required to submit final plans; these will be reviewed and approved before license termination.

In preparing for the review of the proposed plans for permanent closure, decontamination, and dismantlement, the reviewer should consult the general review procedures contained in any Office of Nuclear Material Safety and Safeguards decommissioning standard review plan. However, the reviewer should keep in mind that these documents are for use with final plans that are prepared at the time of license termination.

The staff will review the following parts of plans for permanent closure and decontamination, or decontamination and dismantlement of surface facilities, using the review methods and acceptance criteria in Sections 2.1.3.2 and 2.1.3.3.

(1) The description of design considerations that are intended to facilitate permanent closure and decontamination, or decontamination and dismantlement of surface facilities; and

(2) Plans for permanent closure and decontamination, or decontamination and dismantlement.

2.1.3.2 Review Methods

Review Method 1 Design Considerations That Will Facilitate Permanent Closure and Decontamination, or Decontamination and Dismantlement

Verify that the license application describes the functions of design features as they relate to permanent closure and decontamination, or decontamination and dismantlement.

Confirm that the repository design is compatible with the objectives of permanent closure and decontamination, or decontamination and dismantlement. Note that the design could be considered to meet this requirement if design provisions included, where feasible and economical, design choices that support closure and decontamination, or decontamination and dismantlement over competing alternatives. If such features were not chosen, an acceptable rationale for not adopting the more favorable alternatives should be provided. Examples of favorable design features include:

(1) Selection of materials and processes to minimize waste production;

(2) Minimization of the mass of shielding materials, subject to neutron activation;

(3) Use of modular design and inclusion of lifting points, to facilitate removal and dismantlement;

(4) Selection of materials for compatibility with projected closure and decontamination, or decontamination and dismantlement, or waste processing procedures;

(5) Use of minimum surface roughness finishes on structures, systems, and components that have potential for contamination;

(6) Use of coatings that preclude penetration into porous materials by radioactive gas, condensate, deposited aerosols, or spills, to permit decontamination by surface treatment;

(7) Incorporation of features to contain leaks and spills;

(8) Incorporation of waste minimization techniques; and

(9) Incorporation of features that would maintain occupational and public radiation exposures as low as is reasonably achievable during decommissioning.

Coordinate with reviewers of the design of waste management systems for Section 2.1.1.7 ("Design of Structures, Systems, and Components Important to Safety and Safety Controls") of the Yucca Mountain Review Plan, to confirm that these designs will facilitate closure and decontamination, or decontamination and dismantlement.

Review Method 2 Plans for Permanent Closure and Decontamination, or Decontamination and Dismantlement

Confirm that the license application presents adequate preliminary plans for permanent closure and decontamination, or decontamination and dismantlement of the surface facilities, as appropriate. Use any Nuclear Material Safety and Safeguards decommissioning standard review plan as guidance for evaluating the adequacy of the preliminary plans. In conducting the review, consider that permanent closure and decommissioning and dismantlement would not begin for many years after the submittal of the license application. However, the preliminary plans that the U.S. Department of Energy does submit with the license application should have detail sufficient to indicate that the U.S. Department of Energy has considered what the requirements, process, and impact of permanent closure and decommissioning and dismantlement may be in the future.

Evaluate whether the U.S. Department of Energy, in its preliminary plans for permanent closure and decommissioning and dismantlement, has addressed the content areas in any decommissioning standard review plan. For each section of such standard review plan, evaluate whether the preliminary plans provided by the U.S. Department of Energy indicate that the U.S. Department of Energy has evaluated the requirements, process, and impacts of permanent closure, and decommissioning and dismantlement. Specifically, evaluate the following:

Facility History: The U.S. Department of Energy should describe the type of information that will be required to facilitate decommissioning, with respect to the facility's operating history. This will include information such as records documenting the radionuclides received and processed at the facility and the locations of the processing activities. The U.S. Department of Energy should also indicate how it would document the routine and nonroutine contamination of areas, within the facility, to facilitate future decommissioning activities. The reviewer should refer to any decommissioning standard review plan for a description of the types of information related to the facility's operating history that the U.S. Department of Energy will be required to provide at permanent closure and decommissioning. The U.S. Department of Energy should indicate how it will ensure that the necessary information will be available and defensible at the time of permanent closure and decommissioning.

Facility Description: The U.S. Department of Energy should describe the type of information related to the facility and its environs that will be required to evaluate estimation of doses to on-site and off-site populations during, and at the completion of, permanent closure and decommissioning. Refer to any decommissioning standard review plan for a description of the types of information related to the facility and its environs that the U.S. Department of Energy will be required to provide at the time of permanent closure and decommissioning. The U.S. Department of Energy should indicate how it will ensure the necessary information will be available and defensible at permanent closure and decommissioning.

Radiological Status of the Facility: The U.S. Department of Energy should describe the type of information that will be required to facilitate decommissioning with respect to the facility's radiological status at permanent closure and decommissioning. This will include information such as the types and extent of radionuclide contamination in media at the facility, including buildings, systems and equipment, surface and subsurface soil, and surface and ground water. The U.S. Department of Energy should provide a preliminary description of the anticipated magnitude of decommissioning activities, with respect to these and any other media. Refer to any decommissioning standard review plan for a description of the types of information related to the facility's radiological status that the U.S. Department of Energy will be required to provide at permanent closure and decommissioning. The U.S. Department of Energy should indicate how it will ensure that the necessary information will be available and defensible at permanent closure and decommissioning.

Dose Modeling Evaluations: The U.S. Department of Energy should describe the general type of information that will be required to facilitate decommissioning, with respect to the dose modeling at the time of permanent closure and decommissioning. The U.S. Department of Energy should indicate how it will ensure the necessary information will be available and defensible at permanent closure and decommissioning.

Alternatives for Decommissioning: The U.S. Department of Energy should describe the general type of information that will be required to facilitate decommissioning, with respect to evaluating alternative decommissioning strategies. The U.S. Department of Energy should indicate how it will ensure the necessary information will be available and defensible at permanent closure and decommissioning.

As Low as Is Reasonably Achievable Analysis: The U.S. Department of Energy should describe the general type of information that will be required to facilitate decommissioning, with respect to as low as is reasonably achievable analyses. The U.S. Department of Energy should indicate how it will ensure the necessary information will be available and defensible at permanent closure and decommissioning.

Planned Decommissioning Activities: The U.S. Department of Energy should describe the type of information that will be required to facilitate decommissioning, with respect to the planned closure and decommissioning activities. The U.S. Department of Energy should provide preliminary information to allow the reviewer to understand the general approach to decommissioning activities. The U.S. Department of Energy should also provide a preliminary schedule for completing the activities. Refer to any decommissioning standard review plan for a description of the types of information related to planned decommissioning activities that the U.S. Department of Energy will be required to provide at permanent closure and decommissioning. The U.S. Department of Energy should indicate how it will ensure the necessary information will be available and defensible at permanent closure and decommissioning.

Project Management and Organization: The U.S. Department of Energy should describe the type of information that will be required to facilitate decommissioning, with respect to project management and organization. The U.S. Department of Energy should provide preliminary information to allow the reviewer to understand the general approach to managing closure and decommissioning activities. Refer to any decommissioning standard review plan for a description of the types of information related to the management of closure and decommissioning activities that the U.S. Department of Energy will be required to provide at permanent closure and decommissioning.

Health and Safety Program During Decommissioning: The U.S. Department of Energy should describe the type of information that will be required to facilitate decommissioning, with respect to health and safety program. The U.S. Department of Energy should indicate how the program would be developed and integrated with the preclosure health and safety program.

Environmental Monitoring and Control Program: The U.S. Department of Energy should describe the type of information that will be required to facilitate decommissioning, with respect to environmental monitoring and control. The U.S. Department of Energy should indicate how the program would be developed and integrated with the preclosure environmental and control program.

Radioactive Waste Management Program: The U.S. Department of Energy should describe the type of information that will be required to facilitate decommissioning, with respect to the management of radioactive waste generated through planned closure and decommissioning activities. The U.S. Department of Energy should provide preliminary estimates of the types and quantities of radioactive waste that may be generated through closure and decommissioning activities. The U.S. Department of Energy should provide preliminary plans for minimizing the quantities of radioactive waste, and discuss preliminary plans for disposing of the radioactive waste. Refer to any decommissioning standard review plan for a description of the types of information related to radioactive waste management that the U.S. Department of Energy will be required to provide at permanent closure and decommissioning. The U.S. Department of Energy should indicate how it will ensure the necessary information will be available and defensible at permanent closure and decommissioning.

Quality Assurance Program: The U.S. Department of Energy should describe the type of information that will be required to facilitate decommissioning, with respect to quality assurance. The U.S. Department of Energy should indicate how the program would be developed and integrated with the preclosure quality assurance program. The U.S. Department of Energy quality assurance program is reviewed using Section 2.5.1 of the Yucca Mountain Review Plan.

Facility Radiation Surveys: The U.S. Department of Energy should describe the general type of information that will be required to facilitate decommissioning, with respect to radiation surveys to support closure and decommissioning activities.

Financial Assurance: The U.S. Department of Energy is not required to provide a financial assurance plan in support of closure or decommissioning.

2.1.3.3 Acceptance Criteria

The following acceptance criteria are based on meeting the requirements of 10 CFR 63.21(c)(8) and (c)(22)(vi), relating to plans for permanent closure and decontamination, or decontamination and dismantlement of surface facilities.

Acceptance Criterion 1 The License Application Describes and Provides Bases for Features of the Geologic Repository Operations Area Design That Will Facilitate Permanent Closure and Decontamination, or Decontamination and Dismantlement of Surface Facilities.

(1) The license application describes the functions of design features as they relate to permanent closure and decontamination, or decontamination and dismantlement;

(2) The repository design is compatible with the objectives of permanent closure and decontamination, or decontamination and dismantlement. Design provisions are included, where feasible and economical, and those design choices that support closure and decontamination, or decontamination and dismantlement are selected over competing alternatives. An acceptable rationale for not adopting the more favorable alternatives is provided; and

(3) Designs will facilitate closure and decontamination, or decontamination and dismantlement.

Acceptance Criterion 2 The License Application Includes Adequate Preliminary Plans for Permanent Closure and Decontamination, or Decontamination and Dismantlement of Surface Facilities.

(1) The license application demonstrates that the U.S. Department of Energy is cognizant of the information, analyses, and programs that will be required at permanent closure, decommissioning, and dismantlement;

(2) The license application demonstrates that the U.S. Department of Energy will ensure that the necessary information to support closure and decommissioning—related to operating history, facility description and radiological status, dose evaluations, alternatives for decommissioning, and as low as is reasonably achievable requirements—will be available at the time of permanent closure and decommissioning;

(3) The license application demonstrates that the U.S. Department of Energy has an estimate of the scope of closure and decommissioning activities, has preliminary plans for conducting and managing the activities, and has preliminary estimates and plans for managing radioactive waste generated through closure and decommissioning activities; and

(4) The license application demonstrates that the U.S. Department of Energy has considered the requirements of the health and safety, environmental monitoring, and quality assurance programs required during closure and decommissioning, and has

considered how these programs will be developed and integrated with the comparable preclosure programs.

2.1.3.4 Evaluation Findings

If the license application provides sufficient information and the regulatory acceptance criteria in Section 2.1.3.3 are appropriately satisfied, the staff concludes that this portion of the staff evaluation is acceptable. The reviewer writes material suitable for inclusion in the safety evaluation report prepared for the entire application. The report includes a summary statement of what was reviewed and why the reviewer finds the submittal acceptable. The staff can document the review as follows.

The U.S. Nuclear Regulatory Commission staff has reviewed the Safety Analysis Report and other information submitted in support of the license application, and has found, with reasonable assurance, that the requirements of 10 CFR 63.21(c)(8) are satisfied. Requirements for the content of the license application have been met in that the U.S. Department of Energy has provided an adequate description of design considerations that are intended to facilitate permanent closure and decontamination, or decontamination and dismantlement of surface facilities.

The U.S. Nuclear Regulatory Commission staff has reviewed the Safety Analysis Report and other information submitted in support of the license application, and has found, with reasonable assurance, that the requirements of 10 CFR 63.21(c)(22)(vi) are satisfied. The U.S. Department of Energy has provided adequate plans for permanent closure and decontamination, or decontamination and dismantlement of surface facilities.

2.1.3.5 References

None.

2.2 Repository Safety After Permanent Closure

2.2.1 Performance Assessment

Risk-Informed Review Process for Performance Assessment—The performance assessment quantifies repository performance, as a means of demonstrating compliance with the postclosure performance objectives at 10 CFR 63.113. The U.S. Department of Energy performance assessment is a systematic analysis that answers the triplet risk questions: what can happen; how likely is it to happen; and what are the consequences. The Yucca Mountain performance assessment is a sophisticated analysis that involves various complex considerations and evaluations. Examples include evolution of the natural environment, degradation of engineered barriers over a 10,000-year period, and disruptive events, such as seismicity and igneous activity. The staff needs to consider the technical support for models and parameters of the performance assessment, based on detailed process models, laboratory and field experiments, and natural analogs. In their evaluation of the technical support for models and parameter distributions, the staff will consider the implications for the repository system and the effects on the calculated dose. Because the performance assessment encompasses such a broad range of issues, the staff needs to use risk information throughout the review process. Using risk information will ensure the review focuses on those items most important to waste isolation.

Section 2.2.1 requires the staff to apply risk information throughout the review of the performance assessment. First, the staff reviews the barriers important to waste isolation in Section 2.2.1.1. The U.S. Department of Energy must identify the important barriers (engineered and natural) of the performance assessment, describe each barrier's capability, and provide the technical basis for that capability. This risk information describes the U.S. Department of Energy understanding of each barrier's capability to prevent or substantially delay the movement of water or radioactive materials. Staff review of the U.S. Department of Energy performance assessment—first the barrier analysis and later the rest of the performance assessment—considers risk insights from previous performance assessments conducted for the Yucca Mountain site, detailed process modeling efforts, laboratory and field experiments, and natural analog studies. The result of the initial multiple barrier review is a staff understanding of each barrier's importance to waste isolation, which will influence the emphasis placed on the reviews conducted in Sections 2.2.1.2, "Scenario Analysis and Event Probability" and 2.2.1.3, "Model Abstraction." The emphasis placed on particular parts of the staff review will change based on changes to the risk insights or in response to preliminary review results.

Scenario analysis and model abstraction are the key attributes of the performance assessment. The risk information, drawn from the review of the multiple barriers section, will direct the staff review to those topics within scenario analysis and model abstraction that are important to waste isolation. Section 2.2.1.2 provides the review methods and acceptance criteria for scenarios for both nominal and disruptive events. An acceptable scenario selection method includes identification and classification, screening, and construction of scenarios from the features, events, and processes considered at the Yucca Mountain site. Then, it is necessary to review abstracted models used in the performance assessment for the retained scenarios. The performance assessment review focuses on the 14 model abstractions in Section 2.2.1.3 and the implementation of the model abstractions in the total system performance assessment model. These model abstractions stemmed from those aspects of the engineered, geosphere, and biosphere subsystems shown to be most important to waste isolation, based on prior performance assessments and knowledge of site characteristics and repository design. The staff developed each of the fourteen model abstraction sections in substantial detail, to allow for a detailed review. However, it is unlikely that each of the abstractions will have the same risk significance. The staff will review the abstractions according to the risk significance determined in the multiple barrier review, using Section 2.2.1.1. Nevertheless, until the U.S. Department of Energy completes its license application, the review plan sections dealing with model abstractions must remain flexible and in enough detail, so that the U.S. Department of Energy will understand how the U.S. Nuclear Regulatory Commission will conduct the reviews .

The review of the model abstraction process begins with the review of the repository design and the data characterizing the geology and the performance of the design and proceeds through the development of models used in the performance assessment. The model abstraction review process ends with a review of how the abstracted models are implemented in the total system performance assessment model (e.g., parameter ranges and distributions, integration with model abstractions for other parts of the repository system, representation of spatial and temporal scales, and whether the performance assessment model appropriately implements the abstracted model). Reviews conducted on the early stages of the model abstraction process will be influenced by the final application of the information. For example, the review of parameter distributions will consider the relevant data, the corresponding uncertainty, and effects on the performance of the repository (i.e., the dose to the reasonably maximally exposed individual). The potential for risk dilution—the lowering of the risk, or dose, from an unsupported parameter range and distribution—will also be part of this review of model abstraction.

An unwanted risk dilution can easily result, if care is not exercised in selecting parameter ranges. For example, the parameter range for the retardation factor of a particular radionuclide could be expanded beyond that found in the supporting data in an effort to represent uncertainty. This expanded range could increase the spread in calculated arrival time for the radionuclide and, consequently, result in a smaller expected annual dose. The staff will review parameter ranges and distributions to evaluate whether they are technically defensible, whether they appropriately represent uncertainty, and the potential for risk dilution.

In many regulatory applications, a conservative approach can be used to decrease the need to collect additional information or to justify a simplified modeling approach. Conservative estimates for the dose to the reasonably maximally exposed individual may be used to demonstrate that the proposed repository meets U.S. Nuclear Regulatory Commission regulations and provides adequate protection of public health and safety. Approaches designed to overestimate a specific aspect of repository performance (e.g., higher temperatures within the drifts) may be conservative with respect to temperature but could lead to nonconservative results with respect to dose. The total system performance assessment is a complex analysis with many parameters, and the U.S. Department of Energy may use conservative assumptions to simplify its approaches and data collection needs. However, a technical basis that supports the selection of models and parameter ranges or distributions must be provided. The staff evaluation of the adequacy of technical bases supporting models and parameter ranges or distributions will consider whether the approach results in calculated doses that would overestimate, rather than underestimate, the dose to the reasonably maximally exposed individual. In particular, the claim of conservatism as a basis for simplifying models and parameters should be carefully evaluated to verify that any simplifications are justified and do not unintentionally result in nonconservative results.

The intentional use of conservatism to manage uncertainty also has implications for the staff's efforts to risk-inform its review. The staff will evaluate assertions that a given model or parameter distribution is conservative from the perspective of overall system performance (i.e., the dose to the reasonably maximally exposed individual). The staff will use any available information to risk-inform its review. For example, if the U.S. Department of Energy were to use an approach that overestimates a specific aspect of repository performance, then the staff would consider the effects of this approach on other parts of the total system performance assessment model, overall repository performance, and the representation or sensitivity of important phenomena.

2.2.1.1 System Description and Demonstration of Multiple Barriers

Review Responsibilities—High-Level Waste Branch and Environmental and Performance Assessment Branch

2.2.1.1.1 Areas of Review

This section addresses review of the system description and demonstration of multiple barriers. Reviewers will evaluate the information required by 10 CFR 63.21(c)(1), (9), (10), (14), and (15).

The staff will evaluate the following parts of the system description and demonstration of multiple barriers, using the review methods and acceptance criteria in Sections 2.2.1.1.2 and 2.2.1.1.3.

(1) Identification of barriers relied on for postclosure performance; (including at least one barrier from the engineered system and one from the natural system);

(2) Description of the capability of identified barriers to prevent or substantially reduce the rate of movement of water or radionuclides from the Yucca Mountain repository to the accessible environment, or prevent the release or substantially reduce the release of radionuclides from the waste including the uncertainty associated with this capacity and the consistency with approaches used in the total system performance assessment; and

(3) Discussion of the technical bases for assertions of barrier capability commensurate with the importance of a particular barrier in the performance assessment and with the associated uncertainties.

2.2.1.1.2 Review Methods

Review Method 1 Identification of Barriers

Verify that the U.S. Department of Energy has described the repository system in terms of the engineered components and attributes of the geologic setting, which are barriers contributing to the postclosure performance of the repository. Confirm that the U.S. Department of Energy has clearly linked identified barriers to a capability to reduce the rate of movement of water or radioactive materials. Verify that, among the materials, structures, and features and processes identified as barriers, at least one is engineered and one is part of the geologic setting.

Review Method 2 Description of Barrier Capability

Verify that the U.S. Department of Energy description of barrier capability is explained in terms of a capability to prevent or substantially reduce the rate of movement of water or radionuclides from the Yucca Mountain repository to the accessible environment, or prevent the release or substantially reduce the release rate of radionuclides from the waste; and includes a characterization of the related uncertainty. There are no quantitative limits placed on the capability of individual barriers. The intent of the review is to understand the capability of each barrier to perform its intended function and the relationship of that barrier's role in limiting radiological exposure in the context of the overall performance assessment.

Confirm that information is provided on the time period over which each barrier performs its intended function, including any changes during the compliance period.

Confirm that the U.S. Department of Energy adequately describes the capability of each barrier, including uncertainties, consistent with the quantitative analyses in the U.S. Department of Energy total system performance assessment (e.g., sensitivity and uncertainty analyses, and intermediate results for individual barriers).

To the extent possible, use information gained from alternative total system performance assessment code audit calculations and/or other appropriate quantitative analyses to confirm each barrier's capabilities.

Review Method 3 Technical Basis for Barrier Capability

Use information gained from the review conducted, using Review Method 2, to focus review of the adequacy of the technical bases. Verify Department of Energy has provided technical bases to support the descriptions of barrier capability commensurate with the significance of each barrier's capability and the associated uncertainties. Confirm the technical bases are based on and consistent with the technical bases for the performance assessment. Based on the reviews conducted using Sections 2.2.1.2 ("Scenario Analysis and Event Probability") and 2.2.1.3 ("Model Abstraction"), confirm the quality and completeness of the technical bases for the barrier capabilities.

2.2.1.1.3 Acceptance Criteria

The following acceptance criteria are based on meeting the requirements at 10 CFR 63.113(a) and 63.115(a)–(c).

Acceptance Criterion 1 Identification of Barriers Is Adequate.

Barriers relied on to achieve compliance with 10 CFR 63.113(b), as demonstrated in the total system performance assessment, are adequately identified, and are clearly linked to their capability. The barriers identified include at least one from the engineered system and one from the natural system.

Acceptance Criterion 2 Description of Barrier Capability to Isolate Waste Is Acceptable.

The capability of the identified barriers to prevent or substantially reduce the rate of movement of water or radionuclides from the Yucca Mountain repository to the accessible environment, or prevent the release or substantially reduce the release rate of radionuclides from the waste is adequately identified and described:

(1) The information on the time period over which each barrier performs its intended function, including any changes during the compliance period, is provided;

(2) The uncertainty associated with barrier capabilities is adequately described;

(3) The described capabilities are consistent with the results from the total system performance assessment; and

(4) The described capabilities are consistent with the definition of a barrier at 10 CFR 63.2.

Acceptance Criterion 3 Technical Basis for Barrier Capability Is Adequately Presented.

The technical bases are consistent with the technical basis for the performance assessment. The technical basis for assertions of barrier capability is commensurate with the importance of each barrier's capability and the associated uncertainties.

2.2.1.1.4 Evaluation Findings

If the license application provides sufficient information and the regulatory acceptance criteria in Section 2.2.1.1.3 are appropriately satisfied, the staff concludes that this portion of the staff evaluation is acceptable. The reviewer writes material suitable for inclusion in the safety evaluation report prepared for the entire application. The report includes a summary statement of what was reviewed and why the reviewer finds the submittal acceptable. The staff can document the review as follows.

U.S. Nuclear Regulatory Commission staff has reviewed the Safety Analysis Report and other information submitted in support of the license application, and has found, with reasonable expectation, that the requirements of 10 CFR 63.113(a) are satisfied. An engineered barrier system has been designed that, working in combination with natural barriers, satisfies the requirement for a system of multiple barriers, in compliance with the postclosure performance objectives.

U.S. Nuclear Regulatory Commission staff has reviewed the Safety Analysis Report and other information submitted in support of the license application, and has found, with reasonable expectation, that the requirements at 10 CFR 63.115(a)–(c) are satisfied. Those design features of the engineered barrier system and natural features of the geologic setting that are considered barriers important to waste isolation have been identified. A description has been provided of the capability of barriers identified as important to waste isolation to isolate waste, taking into account uncertainties in characterizing and modeling the barriers, and the technical basis for this description has been provided that is based on and consistent with the technical basis for the performance assessment.

2.2.1.1.5 References

None.

2.2.1.2 Scenario Analysis and Event Probability

2.2.1.2.1 Scenario Analysis

Review Responsibilities—High-Level Waste Branch and Environmental and Performance Assessment Branch

2.2.1.2.1.1 Areas of Review

This section reviews identification of features, events, and processes affecting compliance with the overall performance objective. Reviewers will also evaluate the information required by 10 CFR 63.21(c)(1) and (9).

Review the U.S. Department of Energy methodology for inclusion or exclusion of features, events, and processes in the total system performance assessment. The U.S. Department of Energy is not required to use steps provided here that involve categorization and screening of the initial comprehensive features, events, and processes list for an acceptable license application. However, many steps can be used in accordance with the requirements in 10 CFR Part 63 to reduce the burden of the analysis and to focus the representation of the system on those features, events, and processes that most affect compliance with the overall performance objective. All included features, events, and processes must be appropriately incorporated into the total system performance assessment, and will be reviewed as part of the model abstraction review conducted, using Section 2.2.1.3 of the Yucca Mountain Review Plan. To evaluate repository postclosure safety, verify that the U.S. Department of Energy has conducted analyses that consider potential future conditions a repository may be subjected to, during the period of regulatory concern. These analyses should address those features, events, and processes necessary to describe the future evolution of the repository system.

The staff will review the following parts of the identification of features, events, and processes affecting compliance with the overall performance objective, using the review methods and acceptance criteria in Sections 2.2.1.2.1.2 and 2.2.1.2.1.3:

(1) Identification of an initial list of features, events, and processes;

(2) Screening of the initial list of features, events, and processes;

(3) Formation of scenario classes using the reduced set of features, events and processes; and

(4) Screening of scenario classes.

2.2.1.2.1.2 Review Methods

Review Method 1 Identification of a List of Features, Events, and Processes

Verify that the U.S. Department of Energy list of features, events, and processes includes all features, events, and processes having a potential to influence repository performance. Use knowledge gained reviewing the Yucca Mountain site and regional characterization data and the description of the modes of degradation, deterioration, and alteration of the engineered barriers to assess the completeness of the features, events, and processes list. The staff should use, as appropriate, available generic lists of features, events, and processes (e.g., Nuclear Energy Agency, 1997), as a reference to determine the completeness of the U.S. Department of Energy list of features, events, and processes.

Review Method 2 Screening of the List of Features, Events, and Processes

Examine the excluded features and processes. Evaluate the adequacy of the rationale for excluding each feature and process, based on the description of the site, the design specifications, and the waste characteristics. Consider information from site and regional characterization, natural analog studies, and the repository design, during this evaluation.

Examine the U.S. Department of Energy event-screening rationale, to determine whether an event is appropriately defined. Use the results of the review, conducted using Section 2.2.1.2.2 of the Yucca Mountain Review Plan, for this purpose. Assess the U.S. Department of Energy justification (i.e., whether the probability of occurrence can be technically supported) for those events that fall below the regulatory probability criterion, to evaluate whether the U.S. Department of Energy defined these events too narrowly, and they were inappropriately excluded.

Review the criteria used to screen features, events, and processes related to the geologic setting, and the degradation, deterioration, or alteration of engineered barriers from the performance assessment, based on having no significant change on the magnitude and time of the resulting radiological exposures to the reasonably maximally exposed individual, or radionuclide releases to the accessible environment. Evaluate the U.S. Department of Energy analyses or calculations supporting this screening and the use of bounding or representative estimates for the consequences. Independently assess, using tools such as an alternative total system performance assessment code, the potential consequences to confirm the U.S. Department of Energy screening of features, events, and processes.

Review Method 3 Formation of Scenario Classes Using the Reduced Set of Events

Evaluate the U.S. Department of Energy description of the approach and technical bases, to determine whether the resulting scenario classes are mutually exclusive and include all events that have not been screened from the performance assessment.

Review Method 4 Screening of Scenario Classes

Review the criteria used by the U.S. Department of Energy to screen scenario classes from the performance assessment on the basis that their omission would not significantly change the magnitude and time of the resulting radiological exposures to the reasonably maximally exposed individual, or radionuclide releases to the accessible environment. Examine the U.S. Department of Energy analyses or calculations supporting this screening and the use of bounding or representative estimates for the consequences. Independently assess, using tools such as an alternative total system performance assessment code, as needed, the potential consequences to confirm the U.S. Department of Energy screening of scenario classes.

Evaluate whether the U.S. Department of Energy has adequately considered coupling of processes in estimates of consequences used to screen scenario classes. For each screened scenario class, assess related scenario classes to evaluate whether a narrow definition resulted in the premature exclusion of the scenario class.

Examine those scenario classes excluded for the Yucca Mountain repository and the supporting technical bases. Consider the site description, design specifications, and waste characteristics in this examination. Also, consider information from site and regional characterization, natural analog studies, and repository design, in this evaluation.

Use the results of the review, conducted using Section 2.2.1.2.2 of the Yucca Mountain Review Plan, to examine the U.S. Department of Energy technical justification for screening scenario classes from the performance assessment, based on their probability of being below the regulatory criterion.

2.2.1.2.1.3 Acceptance Criteria

The following acceptance criteria are based on meeting the requirements at 10 CFR 63.114(e) and (f).

Acceptance Criterion 1 The Identification of a List of Features, Events, and Processes Is Adequate.

(1) The Safety Analysis Report contains a complete list of features, events, and processes, related to the geologic setting or the degradation, deterioration, or alteration of engineered barriers (including those processes that would affect the performance of natural barriers), that have the potential to influence repository performance. The list is consistent with the site characterization data. Moreover, the comprehensive features, events, and processes list includes, but is not limited to, potentially disruptive events related to igneous activity (extrusive and intrusive); seismic shaking (high-frequency-low magnitude, and rare large-magnitude events); tectonic evolution (slip on existing faults and formation of new faults); climatic change (change to pluvial conditions); and criticality.

Acceptance Criterion 2 Screening of the List of Features, Events, and Processes Is Appropriate.

(1) The U.S. Department of Energy has identified all features, events, and processes related to either the geologic setting or to the degradation, deterioration, or alteration of engineered barriers (including those processes that would affect the performance of natural barriers) that have been excluded;

(2) The U.S. Department of Energy has provided justification for those features, events, and processes that have been excluded. An acceptable justification for excluding features, events, and processes is that either the feature, event, and process is specifically excluded by regulation; probability of the feature, event, and process (generally an event) falls below the regulatory criterion; or omission of the feature, event, and process does not significantly change the magnitude and time of the resulting radiological exposures to the reasonably maximally exposed individual, or radionuclide releases to the accessible environment; and

(3) The U.S. Department of Energy has provided an adequate technical basis for each feature, event, and process, excluded from the performance assessment, to support the conclusion that either the feature, event, or process is specifically excluded by regulation; the probability of the feature, event, and process falls below the regulatory criterion; or omission of the feature, event, and process does not significantly change the magnitude and time of the resulting radiological exposures to the reasonably maximally exposed individual, or radionuclide releases to the accessible environment.

Acceptance Criterion 3 Formation of Scenario Classes Using the Reduced Set of Events Is Adequate.

(1) Scenario classes are mutually exclusive and complete, clearly documented, and technically acceptable.

Acceptance Criterion 4 Screening of Scenario Classes Is Appropriate.

(1) Screening of scenario classes is comprehensive, clearly documented, and technically acceptable;

(2) The U.S. Department of Energy has adequately considered coupling of processes in estimates of consequences used to screen scenario classes. Scenario classes were not prematurely excluded by a narrow definition;

(3) Scenario classes that are screened from the performance assessment, on the basis that they are specifically ruled out by regulation or are contrary to stated regulatory assumptions are identified, and sufficient justifications are provided;

(4) Scenario classes that are screened from the performance assessment, on the basis that their probabilities fall below the regulatory criterion, are identified, and sufficient justifications are provided; and

(5) Scenario classes that are screened from the performance assessment, on the basis that their omission would not significantly change the magnitude and time of the resulting radiological exposure to the reasonably maximally exposed individual, or radionuclide releases to the accessible environment, are identified, and sufficient justifications are provided.

2.2.1.2.1.4 Evaluation Findings

If the license application provides sufficient information and the regulatory acceptance criteria in Section 2.2.1.2.1.3 are appropriately satisfied, the staff concludes that this portion of the staff evaluation is acceptable. The reviewer writes material suitable for inclusion in the safety evaluation report prepared for the entire application. The report includes a summary statement of what was reviewed and why the reviewer finds the submittal acceptable. The staff can document the review as follows.

U.S. Nuclear Regulatory Commission staff has reviewed the Safety Analysis Report and other information submitted in support of the license application, and has found, with reasonable expectation, that the requirements of 10 CFR 63.114(e) and (f) are satisfied in that:

(1) The Safety Analysis Report provides an adequate initial list of features, events, and processes related to the geologic setting or the degradation, deterioration, or alteration of engineered barriers (including those processes that would affect the performance of natural barriers) that have the potential to influence repository performance;

(2) The list of initial features, events, and processes has been appropriately screened;

(3) Scenario classes formed from the screened list of features, events, and processes are adequate; and

(4) Scenario classes have been appropriately screened.

2.2.1.2.1.5 Reference

Nuclear Energy Agency. "An International Database of Features, Events, and Processes [Draft]." Nuclear Energy Agency Working Group on the "Development of a Database of Features, Events, and Processes Relevant to the Assessment of Post-Closure Safety of Radioactive Waste Repositories, Safety Assessment of Radioactive Waste Repositories Series." United Kingdom: Safety Assessment Management Limited. June 24, 1997.

2.2.1.2.2 Identification of Events with Probabilities Greater Than 10^{-8} Per Year

Review Responsibilities—High-Level Waste Branch and Environmental and Performance Assessment Branch

2.2.1.2.2.1 Areas of Review

This section reviews identification of events with probabilities greater than 10^{-8} per year, that may affect compliance with the postclosure performance standards. Reviewers will also evaluate information required by 10 CFR 63.21(c)(1) and (9).

The staff will evaluate the following parts of the identification of events with probabilities greater than 10^{-8} per year, using the review methods and acceptance criteria in Sections 2.2.1.2.2.2 and 2.2.1.2.2.3:

(1) Definitions of events that may affect compliance with the postclosure performance standards, such as faulting, seismicity, igneous activity, and criticality;

(2) The probability assigned to each event, and the technical bases used to support this assignment;

(3) Conceptual models evaluated or considered in determining the probabilities of events;

(4) Parameters used to calculate the probabilities of events; and

(5) Uncertainty in models and parameters used to calculate the probabilities of events.

2.2.1.2.2.2 Review Methods

Review Method 1 Event Definition

Evaluate whether the definitions for events (potentially beneficial or disruptive), applicable to the Yucca Mountain repository, are unambiguous; probabilities are estimated for the specific event; and event definitions are used consistently and appropriately in probability models.

Confirm that probabilities of intrusive and extrusive igneous events are estimated separately. Verify that definitions of faulting and earthquakes are derived from the historical record, paleoseismic studies, or geological analyses. Confirm that criticality events, for the purpose of initial screening of the features, events, and processes list, are calculated separately, only by location of the criticality event (e.g., in-package, near-field, and far-field).

Review Method 2 Probability Estimates

Evaluate whether the probability estimates for events applicable to Yucca Mountain consider past patterns of natural events in the Yucca Mountain region or natural analogs. Evaluate whether probability estimates are consistent with the design of the proposed repository system. Evaluate whether the U.S. Department of Energy interpretations of the likelihood of future occurrence of the events are compatible with current understandings of present and likely future conditions of the natural and engineered repository systems.

Verify that probability estimates for future igneous events have considered past patterns of igneous events in the Yucca Mountain region. Evaluate the adequacy and sufficiency of the U.S. Department of Energy characterization and documentation of past igneous activity. This should include uncertainties about the distribution, timing, and characteristics of past igneous activity. Confirm that, at a minimum, documentation of past igneous activity, since about 12 million years ago, encompasses the area within about 50 kilometers [30 miles] of the proposed repository site. Give particular attention to the documentation of the locations, ages, volumes, geochemistry, and geologic settings of less than 6-million-year-old basaltic igneous features, such as cinder cones, lava flows, igneous dikes, and sills. Verify that the

U.S. Department of Energy used geological and geophysical information relevant to past igneous activity contained in the literature.

Verify that probability estimates for future faulting and seismic events have considered past patterns of these events in the Yucca Mountain region. Examine the adequacy and sufficiency of characterization and documentation of past faulting and seismicity in the Yucca Mountain region, since 2 million years ago. This should include characterization of uncertainties in the age, timing, magnitude (i.e., displacements), distribution, size, location, and style of faulting and seismicity. Evaluate whether interpretations of faulting and seismicity from surficial and underground mapping, interpretations of geophysical data, or analog investigations are internally consistent and geologically feasible, so reasonable projections can be made about the probability of future faulting and earthquake-induced ground vibrations at the site.

Evaluate whether probability estimates for future criticality events have considered design characteristics and natural features of the proposed Yucca Mountain repository system. Verify that the U.S. Department of Energy has included various fuel types to be disposed at the proposed Yucca Mountain repository in calculating probability of future criticality events. Confirm that the estimate of probability of criticality is determined using methodology outlined in the "U.S. Department of Energy Topical Report on Disposal Criticality" (U.S. Department of Energy, 1998), as amended by responses to the U.S. Nuclear Regulatory Commission request for additional information,[1] and subject to conditions and limitations in the U.S. Nuclear Regulatory Commission Safety Evaluation Report (U.S. Nuclear Regulatory Commission, 2000).

Review Method 3 Probability Model Support

Confirm that a technical justification is provided for models used to estimate the probability for events applicable to the Yucca Mountain repository. Verify that justifications include comparison with results from detailed process models, or comparison with empirical observations, such as reasonably analogous natural systems or appropriate laboratory tests. Confirm that alternative modeling approaches, consistent with available data and current scientific understanding, are investigated, and results and limitations are appropriately factored into the probability models.

Examine whether the U.S. Department of Energy probability models are consistent with known less than 12-million-year-old basaltic igneous events in the Yucca Mountain magmatic system. Confirm that the U.S. Department of Energy probability models are consistent with patterns of igneous activity in other, comparable volcanic fields outside the Yucca Mountain region. Independent models may be used, as appropriate, to review the probabilities of igneous activity, based on geologic information from the Yucca Mountain region. Evaluate the potential risk significance of differences between independent probability models and U.S. Department of Energy probability models. Verify that the U.S. Department of Energy considered alternative interpretations of probability for igneous events. Assess whether igneous activity probability models are consistent with the range of tectonic models used to assess other geological

---

[1]U.S. Department of Energy. "U.S. Department of Energy Response to U.S. Nuclear Regulatory Commission Request for Additional Information on the DOE Topical Report on Disposal Criticality Analysis Methodology." Letter (November 19) to C.W. Reamer, U.S. Nuclear Regulatory Commission. Washington, DC: U.S. Department of Energy. 1999.

processes, such as seismic source characterization, site geological models, and patterns of ground-water flow.

Confirm that results of the U.S. Department of Energy probabilistic and total system performance assessment models compare reasonably to results from seismotectonic process models, and/or empirical observations from appropriate analogs. Verify that the U.S. Department of Energy appropriately adopted acceptable and documented procedures, to construct and test empirical and physical models used to estimate the seismic and fault-displacement hazards. For faulting, ascertain whether the U.S. Department of Energy models, used to describe primary and secondary (or distributed) faulting, are justified technically, and are adequate to predict the effects of faulting on repository performance. For seismicity, verify whether the U.S. Department of Energy considered credible alternative modeling approaches for determining tectonic ground motions that relate to repository performance. Assess whether faulting models are consistent with fault-slip rates, fault displacements, or earthquake data used in the seismic hazard analysis; and evaluate whether the timing and magnitude of future seismic events are consistent with the results of the fault-hazard analysis.

Confirm that models, used to estimate the probability of future criticality events, are validated, using methodology outlined in the "U.S. Department of Energy Topical Report on Disposal Criticality" (U.S. Department of Energy, 1998), as amended by responses to the U.S. Nuclear Regulatory Commission request for additional information,[2] and subject to conditions and limitations contained in the U.S. Nuclear Regulatory Commission Safety Evaluation Report (U.S. Nuclear Regulatory Commission, 2000).

Probability model support for infrequent events should include data from analog systems that contain significantly more events than the Yucca Mountain system. This support should also include justification that the models produce results consistent with the timing and characteristics of past events in the Yucca Mountain system. Confirm that probability models for natural events use geologic bases that are consistent with other relevant features, events, and processes.

Review Method 4 Probability Model Parameters

Verify whether the parameters used to calculate the probability of events, applicable to the Yucca Mountain repository, are reasonable, based on data from the Yucca Mountain region or analogous natural systems, and/or design and engineering characteristics of the proposed Yucca Mountain repository system.

Assess whether the parameters used in probabilistic volcanic hazard assessments are reasonable, based on data from the Yucca Mountain region. If appropriate data are not readily

---

[2]U.S. Department of Energy. "U.S. Department of Energy Response to U.S. Nuclear Regulatory Commission Request for Additional Information on the DOE Topical Report on Disposal Criticality Analysis Methodology." Letter (November 19) to C.W. Reamer, U.S. Nuclear Regulatory Commission. Washington, DC: U.S. Department of Energy. 1999.

obtainable from the Yucca mountain region, confirm that comparable volcanic systems outside the Yucca Mountain region were considered in developing such parameters.

Verify whether parameter values used in probabilistic seismic and fault-displacement hazard assessments are adequately supported by Yucca Mountain region faulting and earthquake data or appropriate analogs, so the effects of faulting and seismicity are appropriately factored into repository performance. Verify that parameters are consistent with the range of faulting characteristics and seismicity observed in the Yucca Mountain region, or with parameters derived from representative analogs, and ascertain that the parameters account for variability in data precision and accuracy. For example, confirm that the U.S. Department of Energy adequately evaluated uncertainties in faulting or earthquake activity (i.e., recurrence). Confirm that the U.S. Department of Energy has established reasonable and consistent correlations between parameters, where appropriate.

Where sufficient data do not exist, confirm that parameter values and conceptual models are based on appropriate use of other sources, such as expert elicitation, using NUREG–1563 (Kotra, et al., 1996).

Review Method 5 Uncertainty in Event Probability

For events applicable to the Yucca Mountain repository, verify whether the U.S. Department of Energy has adequately identified and propagated uncertainties in estimating probabilities. Confirm that an adequate technical basis, that includes treatment of uncertainty, is provided for the probability value. For probability distributions or ranges, confirm that a technical basis for the analysis is provided, and that the distribution or range accounts for the uncertainty in the probability estimates. [Note: Although probability distributions or ranges can include probabilities less than 10^{-8} per year, the mean of the distribution range is to be used to screen an event from the performance assessment.]

Assess the probability values used for igneous events by considering the range of values available in the literature for the Yucca Mountain region and comparable volcanic fields outside the Yucca Mountain region. To confirm that probability models are sufficiently robust to reasonably approximate the distribution of Yucca Mountain region igneous features, evaluate probability models by testing their sensitivity to uncertainties about the past distribution of volcanic vents, recurrence rates of volcanism, and relationships between igneous activity and tectonism.

Verify that probabilities used in the evaluation of faulting and seismicity effects on repository performance include both infrequent seismic and faulting events with relatively large-magnitude ground motions and fault displacements, and the cumulative effects of repeated ground motions or fault displacements from more frequent and lower-magnitude seismic or faulting events.

2.2.1.2.2.3 Acceptance Criteria

The following acceptance criteria are based on meeting the requirements at 10 CFR 63.114(d).

Acceptance Criterion 1 Events Are Adequately Defined.

(1) Events or event classes are defined without ambiguity and used consistently in probability models, such that probabilities for each event or event class are estimated separately; and

(2) Probabilities of intrusive and extrusive igneous events are calculated separately. Definitions of faulting and earthquakes are derived from the historical record, paleoseismic studies, or geological analyses. Criticality events are calculated separately by location.

Acceptance Criterion 2 Probability Estimates for Future Events Are Supported by Appropriate Technical Bases.

(1) Probabilities for future natural events have considered past patterns of the natural events in the Yucca Mountain region, considering the likely future conditions and interactions of the natural and engineered repository system. These probability estimates have specifically included igneous events, faulting and seismic events, and criticality events.

Acceptance Criterion 3 Probability Model Support Is Adequate.

(1) Probability models are justified through comparison with output from detailed process-level models and/or empirical observations (e.g., laboratory testing, field measurements, or natural analogs, including Yucca Mountain site data). Specifically:

   (a) For infrequent events, the U.S. Department of Energy justifies, to the extent appropriate, proposed probability models with data from reasonably analogous systems. Analog systems should contain significantly more events than the Yucca Mountain system, to provide reasonable evaluations of probability model performance;

   (b) The U.S. Department of Energy justifies, to the extent appropriate, the ability of probability models to produce results consistent with the timing and characteristics (e.g., location and magnitude) of successive past events in the Yucca Mountain system; and

   (c) The U.S. Department of Energy probability models for natural events use underlying geologic bases (e.g., tectonic models) that are consistent with other relevant features, events, and processes evaluated, using Section 2.2.1.2.1.

Acceptance Criterion 4 Probability Model Parameters Have Been Adequately Established.

(1) Parameters used in probability models are technically justified and documented by the U.S. Department of Energy. Specifically:

   (a) Parameters for probability models are constrained by data from the Yucca Mountain region and engineered repository system to the extent practical;

(b) The U.S. Department of Energy appropriately establishes reasonable and consistent correlations between parameters; and

(c) Where sufficient data do not exist, the definition of parameter values and conceptual models is based on appropriate use of other sources, such as expert elicitation conducted in accordance with appropriate guidance.

Acceptance Criterion 5 Uncertainty in Event Probability Is Adequately Evaluated.

(1) Probability values appropriately reflect uncertainties. Specifically:

(a) The U.S. Department of Energy provides a technical basis for probability values used, and the values account for the uncertainty in the probability estimates; and

(b) The uncertainty for reported probability values adequately reflects the influence of parameter uncertainty on the range of model results (i.e., precision) and the model uncertainty, as it affects the timing and magnitude of past events (i.e., accuracy).

2.2.1.2.2.4 Evaluation Findings

If the license application provides sufficient information and the regulatory acceptance criteria in Section 2.2.1.2.2.3 are appropriately satisfied, the staff concludes that this portion of the staff evaluation is acceptable. The reviewer writes material suitable for inclusion in the safety evaluation report prepared for the entire application. The report includes a summary statement of what was reviewed and why the reviewer finds the submittal acceptable. The staff can document the review as follows.

U.S. Nuclear Regulatory Commission staff has reviewed the Safety Analysis Report and other information submitted in support of the license application, and has found, with reasonable expectation, that the requirements of 10 CFR 63.114(d) are satisfied. The license application considers those events that have at least one chance in 10,000 of occurring over 10,000 years.

2.2.1.2.2.5 References

Kotra, et al. NUREG–1563, "Branch Technical Position on the Use of Expert Elicitation in the High-Level Radioactive Waste Program." Washington, DC: U.S. Nuclear Regulatory Commission. 1996.

U.S. Department of Energy. "Disposal Criticality Analysis Methodology Topical Report." YMP/TR–004Q. Revision 0. Las Vegas, Nevada: U.S. Department of Energy, Office of Civilian Radioactive Waste Management. November 1998.

U.S. Nuclear Regulatory Commission. "Draft Safety Evaluation Report on Disposal Criticality Analysis Methodology Topical Report." Revision 0. Washington, DC: U.S. Nuclear Regulatory Commission. 2000.

2.2.1.3 Model Abstraction

There are 14 model abstraction sections the staff will use to determine compliance with 10 CFR 63.114. The abstractions consider the engineered, geosphere, and biosphere subsystems that may be important to waste isolation. Important to waste isolation means important to meeting the performance objectives specified in 10 CFR 63.113. The staff will decide which abstractions are important to waste isolation, by using risk insights gained from performance assessments, knowledge of site characteristics and repository design, and review of the U.S. Department of Energy license application. Each section provides enough review methods and acceptance criteria to allow for a detailed review. However, it is unlikely that each of the 14 abstraction topics will have the same risk significance and need the same review level. Nevertheless, until the U.S. Department of Energy completes its license application, the sections about model abstractions need to be flexible and in enough detail that the staff clearly understands how to conduct the review of abstraction information provided by the licensee. The staff will focus its review to understand the importance to performance of the various assumptions, models, and data in the performance assessment. The staff will also focus its review to verify that the degree of technical support for models and data abstractions is equal to their contribution to risk. This means the staff will review each model abstraction to a detail level suitable to the degree the U.S. Department of Energy relies on it to prove its license application. The staff will be familiar with the U.S. Department of Energy license application, because of the multiple barrier review (refer to Section 2.2.1.1). In the multiple barrier review, the staff will evaluate the capability of the barriers. For example, if the U.S. Department of Energy relies on the unsaturated zone to provide significant delay in the transport of radionuclides to the reasonably maximally exposed individual, then the staff will perform a detailed review of the abstraction of radionuclide transport in the unsaturated zone. However, if the U.S. Department of Energy shows that this abstraction has a minor impact on the delay in the transport of radionuclides to the reasonably maximally exposed individual, then the staff will conduct a simplified review focusing on the bounding assumptions. The staff will use the review methods and acceptance criteria in these sections to decide whether the U.S. Department of Energy properly characterized the features, events, and processes and properly factored them into the performance assessment. This is necessary to decide whether the U.S. Department of Energy performance assessment is acceptable and complies with 10 CFR 63.114 and 63.115. The review methods and acceptance criteria the staff will use to evaluate compliance with the performance objectives (numerical standard) are in Section 2.2.1.4 of the Yucca Mountain Review Plan.

2.2.1.3.1 Degradation of Engineered Barriers

To review this model abstraction, consider the degree to which the U.S. Department of Energy relies on degradation of engineered barriers to demonstrate compliance. Review this model abstraction, considering the risk information evaluated in the “Multiple Barriers” Section (2.2.1.1). For example, if the U.S. Department of Energy relies on the engineered barriers to provide significant delay in the transport of radionuclides to the reasonably maximally exposed individual, then perform a detailed review of this abstraction. If, on the other hand, the U.S. Department of Energy demonstrates this abstraction to have a minor impact on the dose to the reasonably maximally exposed individual, then conduct a simplified review focusing on the bounding assumptions. The review methods and acceptance criteria provided here are for a detailed review. Some of the review methods and acceptance criteria may not be necessary in a simplified review for those abstractions that have a minor impact on performance. The demonstration of compliance with the performances objective is evaluated using Section 2.2.1.4 of the Yucca Mountain Review Plan.

Review Responsibilities—High-Level Waste Branch and Environmental and Performance Assessment Branch

2.2.1.3.1.1 Areas of Review

This section reviews degradation of engineered barriers within the emplacement drift. Reviewers will also evaluate information, required by 10 CFR 63.21(c)(3), (9), (10), (15) and (19), that is relevant to the abstraction of degradation of engineered barriers. It is important to note that the scope of this review includes various parts of the engineered barrier system, as specified in 10 CFR 63.2.

The staff will evaluate the following parts of the abstraction of degradation of engineered barriers, using review methods and acceptance criteria in Sections 2.2.1.3.1.2 and 2.2.1.3.1.3:

(1) Description of the engineered barrier system, hydrology, geochemistry, and thermal effects related to the degradation of the engineered barrier system and the technical basis the U.S. Department of Energy provides to support model integration, across the total system performance assessment abstractions;

(2) Sufficiency of the data and parameters used to justify the total system performance assessment model abstraction;

(3) Methods the U.S. Department of Energy uses to characterize data uncertainty, and propagate the effects of this uncertainty through the total system performance assessment model abstraction;

(4) Methods the U.S. Department of Energy uses to characterize model uncertainty, and propagate the effects of this uncertainty through the total system performance assessment model abstraction;

(5) Approaches the U.S. Department of Energy uses to compare the total system performance assessment output to process-level model outputs and empirical studies; and

(6) Use of expert elicitation.

2.2.1.3.1.2 Review Methods

To review the abstraction of degradation of engineered barriers, recognize that models used in the total system performance assessments may range from highly complex process-level models to simplified models, such as response surfaces or look-up tables. Evaluate model adequacy, regardless of the level of complexity.

Review Method 1 Model Integration

Examine the U.S. Department of Energy license application description of design features, physical phenomena, and couplings, as well as the description of the waste package, and features of the engineered barrier system that contribute to high-level radioactive waste isolation. Assess the adequacy of the technical bases for these descriptions and for incorporating them in the total system performance assessment abstraction for the degradation of engineered barriers.

Examine assumptions, technical bases, data, and models used by the U.S. Department of Energy in the total system performance assessment abstraction of degradation process models for consistency with other related U.S. Department of Energy abstractions. Evaluate whether the descriptions and technical bases provide transparent and traceable support for the abstraction of the degradation of the engineered barriers.

Evaluate whether the U.S. Department of Energy description of aspects of environmental conditions, within the waste package emplacement drifts, design features, physical phenomena, and couplings that may affect the degradation of the engineered barriers, is adequate. Verify that conditions and assumptions, used in the total system performance assessment abstraction of the degradation of the engineered barriers, are consistent with the body of data presented in the abstraction.

Confirm that the U.S. Department of Energy has propagated boundary and initial conditions, used in the total system performance assessment abstraction of the degradation of engineered barriers, throughout its abstraction approaches.

Examine how the features, events, and processes, related to the degradation of the engineered barriers have been included in the total system performance assessment abstraction.

Evaluate the technical bases that the U.S. Department of Energy used for selecting the design criteria, that mitigate any potential impact of in-package criticality on repository performance, including all features, events, and processes that may increase the reactivity of the system inside the waste package; all the configuration classes and configurations that have potential for nuclear criticality; and changes in radionuclide inventory and thermal conditions, in the abstraction of the degradation of engineered barriers.

Verify that the U.S. Department of Energy reviews follow guidance such as NUREG–1297 and NUREG–1298 (Altman, et al., 1988a,b), or other acceptable approaches.

Review Method 2 Data and Model Justification

Evaluate the sufficiency of the experimental and site characterization data used to support parameters used in conceptual models, process-level models, and alternative conceptual models, considered in the total system performance assessment abstraction of degradation of engineered barriers.

Verify whether sufficient data have been collected to adequately model degradation processes, as well as characteristics of the geochemistry, hydrology, design features, and thermal effects, to establish initial and boundary conditions for the total system performance assessment abstraction of degradation of engineered barriers. For example, mechanical property data should cover the range of anticipated temperatures and microstructural conditions. The corrosion data should consider the appropriate range of environmental conditions, such as chloride concentration.

Evaluate and confirm that data used to support the U.S. Department of Energy total system performance assessment abstraction of the degradation of engineered barriers are based on appropriate techniques, and are adequate for the accompanying sensitivity/uncertainty analyses. Evaluate the need for additional data, based on the sensitivity analyses.

Verify that the U.S. Department of Energy demonstrates the adequacy of the degradation of engineered barriers models used to assess the range of possible degradation processes.

Review Method 3 Data Uncertainty

Evaluate the technical bases for parameter values, assumed ranges, probability distributions, and bounding values used in conceptual models, process models, and alternative conceptual models, considered in the total system performance assessment abstraction of degradation of engineered barriers. The reviewer should verify that the technical bases supports the treatment of uncertainty and variability of these parameters in the performance assessment. If conservative values are used as a method for addressing uncertainty and variability, the reviewer should verify that the conservative values result in conservative estimates of risk and do not cause unintended results (i.e., conservative representation of one aspect of the repository behavior that leads to an overall reduction in risk; inappropriate dilution of the risk estimate by assuming an approach is conservative when a parameter range is increased beyond the supporting data).

Examine the abstraction for those degradation processes that the U.S. Department of Energy assumes are not important to waste isolation and confirm that the parameters, used in these abstractions, are assigned values consistent with the abstractions of other degradation processes, determined to be significant to performance of the engineered barriers, as well as the initial and boundary conditions used in other abstractions for the total system performance assessment.

Confirm that the U.S. Department of Energy has used parameters, in the abstraction of the degradation of engineered barriers, that are based on laboratory experiments, field measurements, natural analog or industrial analog research, and process-level modeling studies, conducted under conditions relevant to the range of environmental conditions in the emplacement drifts located in the unsaturated zone at Yucca Mountain. Examine the results of the U.S. Department of Energy engineered barrier degradation tests, and confirm that the U.S. Department of Energy has provided adequate models.

Evaluate the methods used by the U.S. Department of Energy for nondestructive examination of fabricated engineered barriers, including the type, size, and location of fabrication defects, that may lead to premature failure, as a result of rapidly initiated engineered barrier degradation.

Examine the justification for the allowable distribution of fabrication defects in the engineered barriers, and evaluate how the U.S. Department of Energy assesses the effect on engineered barrier performance of defects that cannot be detected.

Verify that the U.S. Department of Energy appropriately established possible statistical correlations between parameters. Verify that an adequate technical basis or bounding argument is provided for neglected correlations.

Evaluate the methods used by the U.S. Department of Energy in conducting expert elicitation to define parameter values.

Review Method 4 Model Uncertainty

Evaluate the U.S. Department of Energy alternative conceptual models used in developing the total system performance assessment abstraction for degradation of engineered barriers. Examine the model parameters in the context of available site characterization data, laboratory corrosion tests, field measurements, and process-level modeling studies.

Where appropriate, use an alternative total system performance assessment model to evaluate selected parts of the U.S. Department of Energy abstraction of the degradation of engineered barriers, including waste package corrosion.

Evaluate the treatment of conceptual model uncertainty in light of the available site characterization data, laboratory experiments, field measurements, natural analog information and process-level modeling studies. If adoption of a conservative model is used as an approach for addressing conceptual model uncertainty, the reviewer should verify that the selected conceptual model: (i) is conservative relative to alternative conceptual models that are consistent with the available data and current scientific understanding; and (ii) results in conservative estimates of risk and not cause unintended results (i.e., conservative representation of one aspect of the repository behavior that leads to an overall reduction in the risk estimate).

Examine the mathematical models used in the analyses of degradation of engineered barriers. Examine and evaluate the bases for excluding alternative conceptual models and the limitations and uncertainties of the chosen model.

Review Method 5 Model Support

Evaluate the output from the abstraction of the degradation of engineered barriers, and compare the results with a combination of data from laboratory corrosion testing and field measurements, as well as results obtained through process-level modeling. Evaluate the sensitivity analyses used to support the abstraction of the degradation of engineered barriers in the total system performance assessment.

Use detailed models of degradation processes to evaluate the total system performance assessment abstractions of the degradation of engineered barriers. If practical, use an alternative to the total system performance assessment model to evaluate selected parts of the U.S. Department of Energy abstraction of the degradation of the engineered barriers, and assess the effects on repository performance. Compare results of the U.S. Department of Energy abstraction to approximations shown to be appropriate for closely analogous systems, industrial experience, and experimental results.

Evaluate evidence to show that models used to evaluate performance are not likely to underestimate the actual degradation and failure of engineered barriers, as a result of corrosion or other degradation processes.

In developing supporting evidence for the models, verify that mathematical models for the degradation of engineered barriers are based on the same environmental parameters, material factors, assumptions, and approximations shown to be appropriate for closely analogous engineering or industrial applications and experimental investigations.

Examine the procedures used by the U.S. Department of Energy to construct and test its mathematical and numerical models.

As appropriate, use an alternative total system performance assessment model to evaluate the U.S. Department of Energy sensitivity or bounding analyses, and confirm that the U.S. Department of Energy has used ranges consistent with available site characterization data, field and laboratory tests, and industrial and natural analog research.

2.2.1.3.1.3 Acceptance Criteria

The following acceptance criteria are based on meeting the requirements of 10 CFR 63.114(a)–(c) and (e)–(g), relating to the degradation of engineered barriers model abstraction. U.S. Nuclear Regulatory Commission staff should apply the following acceptance criteria, according to the level of importance established in the U.S. Department of Energy risk-informed license application.

Acceptance Criterion 1 System Description and Model Integration Are Adequate.

(1) The total system performance assessment adequately incorporates important design features, physical phenomena, and couplings, and uses consistent and appropriate assumptions throughout the degradation of engineered barriers abstraction process;

(2) Assessment abstraction of the degradation of engineered barriers uses assumptions, technical bases, data, and models that are appropriate and consistent with other related U.S. Department of Energy abstractions. For example, the assumptions used for degradation of engineered barriers should be consistent with the abstractions of the quantity and chemistry of water contacting waste packages and waste forms (Section 2.2.1.3.3); climate and infiltration (Section 2.2.1.3.5); and mechanical disruption of waste packages (Section 2.2.1.3.2). The descriptions and technical bases provide transparent and traceable support for the abstraction of the degradation of engineered barriers;

(3) The descriptions of engineered barriers, design features, degradation processes, physical phenomena, and couplings that may affect the degradation of the engineered barriers are adequate. For example, materials and methods used to construct the

engineered barriers are included, and degradation processes, such as uniform corrosion, pitting corrosion, crevice corrosion, stress corrosion cracking, intergranular corrosion, microbially influenced corrosion, dry-air oxidation, hydrogen embrittlement, and the effects of wet and dry cycles, material aging and phase stability, welding, and initial defects on the degradation modes for the engineered barriers are considered;

(4) Boundary and initial conditions used in the total system performance assessment abstractions are propagated consistently throughout the abstraction approaches. For example, the conditions and assumptions used in the degradation of engineered barriers abstraction are consistent with those used to model the quantity and chemistry of water contacting waste packages and waste forms (Section 2.2.1.3.3); climate and infiltration (Section 2.2.1.3.5); and mechanical disruption of waste packages (Section 2.2.1.3.2);

(5) Sufficient technical bases for the inclusion of features, events, and processes related to degradation of engineered barriers in the total system performance assessment abstractions are provided;

(6) Adequate technical bases are provided, for selecting the design criteria, that mitigate any potential impact of in-package criticality on repository performance, including considering all features, events, and processes that may increase the reactivity of the system inside the waste package. For example, the technical bases for the abstraction of the degradation of engineered barriers include configuration classes and configurations that have potential for nuclear criticality, changes in radionuclide inventory, and changes in thermal conditions; and

(7) Guidance in NUREG–1297 and NUREG–1298 (Altman, et al., 1988a,b), or other acceptable approaches, is followed.

Acceptance Criterion 2 Data Are Sufficient for Model Justification.

(1) Parameters used to evaluate the degradation of engineered barriers in the license application are adequately justified (e.g., laboratory corrosion tests, site-specific data such as data from drift-scale tests, in-service experience in pertinent industrial applications, and test results not specifically performed for the Yucca Mountain site, etc.). The U.S. Department of Energy describes how the data were used, interpreted, and appropriately synthesized into the parameters;

(2) Sufficient data have been collected on the characteristics of the engineered components, design features, and the natural system to establish initial and boundary conditions for abstraction of degradation of engineered barriers;

(3) Data on the degradation of the engineered barriers (e.g., general and localized corrosion, microbially influenced corrosion, galvanic interactions, hydrogen embrittlement, and phase stability), used in the abstraction, are based on laboratory measurements, site-specific field measurements, industrial analog and/or natural analog research, and tests designed to replicate the range of conditions that may occur at the Yucca Mountain site. As appropriate, sensitivity or uncertainty analyses, used to support

the U.S. Department of Energy total system performance assessment abstraction, are adequate to determine the possible need for additional data; and

(4) Degradation models for the processes that may be significant to the performance of the engineered barriers are adequate. For example, the U.S. Department of Energy models consider the possible degradation of the engineered barriers, as a result of uniform and localized corrosion processes, stress-corrosion cracking, microbially influenced corrosion, hydrogen embrittlement, and incorporate the effects of fabrication processes, thermal aging, and phase stability.

Acceptance Criterion 3 Data Uncertainty Is Characterized and Propagated Through the Model Abstraction.

(1) Models use parameter values, assumed ranges, probability distributions, and/or bounding assumptions that are technically defensible, reasonably account for uncertainties and variabilities, and do not result in an under-representation of the risk estimate;

(2) For those degradation processes that are significant to the performance of the engineered barriers, the U.S. Department of Energy provides appropriate parameters, based on techniques that may include laboratory experiments, field measurements, industrial analogs, and process-level modeling studies conducted under conditions relevant to the range of environmental conditions within the waste package emplacement drifts. The U.S. Department of Energy also demonstrates the capability to predict the degradation of the engineered barriers in laboratory and field tests;

(3) For the selection of parameters used in conceptual and process-level models of engineered barrier degradation that can be expected under repository conditions, assumed range of values and probability distributions are not likely to underestimate the actual degradation and failure of engineered barriers as a result of corrosion;

(4) The U.S. Department of Energy uses appropriate methods for nondestructive examination of fabricated engineered barriers to assess the type, size, and location of fabrication defects that may lead to premature failure as a result of rapidly initiated engineered barrier degradation. The U.S. Department of Energy specifies and justifies the allowable distribution of fabrication defects in the engineered barriers, and assesses the effects of defects that cannot be detected on the performance of the engineered barriers; and

(5) Where sufficient data do not exist, the definition of parameter values and conceptual models, used by the U.S. Department of Energy, is based on appropriate use of other sources, such as expert elicitation conducted in accordance with NUREG–1563 (Kotra, et al., 1996). If other approaches are used, the U.S. Department of Energy adequately justifies their use.

Acceptance Criterion 4 Model Uncertainty Is Characterized and Propagated Through the Model Abstraction.

(1) Alternative modeling approaches of features, events, and processes are considered and are consistent with available data and current scientific understanding, and the results and limitations are appropriately considered in the abstraction;

(2) Consideration of conceptual model uncertainty is consistent with available site characterization data, laboratory experiments, field measurements, natural analog information and process-level modeling studies; and the treatment of conceptual model uncertainty does not result in an under-representation of the risk estimate; and

(3) The U.S. Department of Energy uses alternative modeling approaches, consistent with available data and current scientific understanding, and evaluates the model results and limitations, using tests and analyses that are sensitive to the processes modeled. For example, for processes such as uniform corrosion, localized corrosion, and stress corrosion cracking of the engineered barriers, the U.S. Department of Energy considers alternative modeling approaches, to develop its understanding of environmental conditions and material factors significant to these degradation processes.

Acceptance Criterion 5 Model Abstraction Output Is Supported by Objective Comparisons.

(1) Models implemented in this total system performance assessment abstraction provide results consistent with output from detailed process-level models and/or empirical observations (laboratory and field testings and/or natural analogs);

(2) Numerical corrosion models used to calculate the lifetimes of the engineered barriers are adequate representations, considering the associated uncertainties in the expected long-term behaviors, the range of conditions (including residual stresses), and the variability in engineered barrier fabrication processes (including welding);

(3) Evidence is sufficient to show that models used to evaluate performance are not likely to underestimate the actual degradation and failure of engineered barriers, as a result of corrosion or other degradation processes;

(4) Mathematical models for the degradation of engineered barriers are based on the same environmental parameters, material factors, assumptions, and approximations shown to be appropriate for closely analogous engineering or industrial applications and experimental investigations;

(5) Accepted and well-documented procedures are used to construct and test the numerical models that simulate the engineered barrier chemical environment and degradation of engineered barriers; and

(6) Sensitivity analyses or bounding analyses are provided to support the abstraction of degradation of engineered barriers that cover ranges consistent with the site data, field or laboratory experiments and tests, and industrial analogs.

2.2.1.3.1.4 Evaluation Findings

If the license application provides sufficient information and the regulatory acceptance criteria in Section 2.2.1.3.1.3 are appropriately satisfied, the staff concludes that this portion of the staff evaluation is acceptable. The reviewer writes material suitable for inclusion in the safety evaluation report prepared for the entire application. The report includes a summary statement of what was reviewed and why the reviewer finds the submittal acceptable. The staff can document the review as follows.

U.S. Nuclear Regulatory Commission staff has reviewed the Safety Analysis Report and other information submitted in support of the license application, and has found, with reasonable expectation, that the requirements of 10 CFR 63.114 are satisfied, regarding the abstraction of degradation of engineered barriers in the performance assessment. In particular, the U.S. Nuclear Regulatory Commission staff found that:

(1) Appropriate data from the site and surrounding region, uncertainties and variabilities in parameter values, and alternative conceptual models have been used in the analyses, in compliance with 10 CFR 63.114(a)–(c);

(2) Specific features, events, and processes have been included in the analyses, and appropriate technical bases have been provided for inclusion or exclusion, in compliance with 10 CFR 63.114(e);

(3) Specific degradation, deterioration, and alteration processes have been included in the analyses, taking into consideration their effects on annual dose, and appropriate technical bases have been provided for inclusion or exclusion, in compliance with 10 CFR 63.114(f); and

(4) Adequate technical bases have been provided for models used in the performance assessment, as required by 10 CFR 63.114(g).

2.2.1.3.1.5 References

Altman, W.D., J.P. Donnelly, and J.E. Kennedy. NUREG–1297, "Generic Technical Position on Peer-Review for High-Level Nuclear Waste Repositories." Washington, DC: U.S. Nuclear Regulatory Commission. 1988a.

———. NUREG–1298, "Generic Technical Position on Qualification of Existing Data for High-Level Nuclear Waste Repositories." Washington, DC: U.S. Nuclear Regulatory Commission. 1988b.

Kotra, J.P., et al. NUREG–1563, "Branch Technical Position on the Use of Expert Elicitation in the High-Level Radioactive Waste Program." Washington, DC: U.S. Nuclear Regulatory Commission. 1996.

2.2.1.3.2 Mechanical Disruption of Engineered Barriers

Mechanical disruption of an engineered barrier is defined as partial or total mechanical failure of the barrier resulting from external events (man-made and/or natural), which immediately or eventually reduces its design life and intended performance, and, consequently, causes release of radionuclides. For example, a rock fall may cause a container to rupture or may cause a dent in its structure, which could lead to an accelerated rate of corrosion and failure sooner than under normal conditions.

To review this model abstraction, consider the degree to which the U.S. Department of Energy relies on mechanical disruption of engineered barriers to demonstrate compliance. Review this model abstraction, considering the risk information evaluated in the "Multiple Barriers" Section (2.2.1.1). For example, if the U.S. Department of Energy relies on the engineered barriers to provide significant delay in the transport of radionuclides to the reasonably maximally exposed individual, then perform a more detailed review of this abstraction. If, on the other hand, the U.S. Department of Energy demonstrates this abstraction to have a minor impact on the delay in the transport of radionuclides to the reasonably maximally exposed individual, then conduct a simplified review focusing on the bounding assumptions. The review methods and acceptance criteria provided here are for a detailed review. Some of the review methods and acceptance criteria may not be necessary in a simplified review for those abstractions that have a minor impact on performance. The demonstration of compliance with the performance objectives is evaluated using Section 2.2.1.4 of the Yucca Mountain Review Plan.

Review Responsibilities—High-Level Waste Branch and Environmental and Performance Assessment Branch

2.2.1.3.2.1 Areas of Review

This section reviews mechanical disruption of engineered barriers. Reviewers will also evaluate information, required by 10 CFR 63.21(c)(1)–(3), (9), (10), (15), and (19), that is relevant to the abstraction of mechanical disruption of engineered barriers.

The staff will evaluate the following parts of the abstraction of mechanical disruption of engineered barriers, using the review methods and acceptance criteria in Sections 2.2.1.3.2.2 and 2.2.1.3.2.3:

(1) Description of the geological and engineering aspects of mechanical disruption of engineered barriers and the technical bases the U.S. Department of Energy provides to support model integration across the total system performance assessment abstractions;

(2) Sufficiency of the data and parameters used to justify the model abstraction;

(3) Methods the U.S. Department of Energy uses to characterize data uncertainty, and propagate the effects of this uncertainty through the total system performance assessment;

(4) Methods the U.S. Department of Energy uses to characterize model uncertainty, and propagate the effects of this uncertainty through the total system performance assessment;

(5) Approaches the U.S. Department of Energy uses to compare total system performance assessment output to process-level model outputs and empirical studies; and

(6) Use of expert elicitation.

2.2.1.3.2.2 Review Methods

To review the abstraction of mechanical disruption of engineered barriers, recognize that models used in the total system performance assessment may range from highly complex process-level models to simplified models, such as response surfaces or look-up tables. Evaluate model adequacy, regardless of the level of complexity.

Review Method 1 Model Integration

Examine the description of design features, physical phenomena, and couplings included in the mechanical disruption of engineered barriers abstraction. Assess the adequacy of the technical bases for these descriptions and for incorporating them in the total system performance assessment abstraction of mechanical disruption of engineered barriers.

Evaluate whether the description of design features, physical phenomena, and couplings that may affect mechanical disruption of engineered barriers is adequate. Verify that conditions and assumptions, used in the total system performance assessment abstraction of mechanical disruption of engineered barriers, are consistent with the body of data presented in the description.

Examine assumptions, technical bases, data, and models, used by the U.S. Department of Energy in the total system performance assessment abstraction of mechanical disruption of engineered barriers, for consistency with other related U.S. Department of Energy abstractions. Evaluate whether the descriptions and technical bases provide transparent and traceable support for the abstraction of mechanical disruption of engineered barriers.

Confirm that the U.S. Department of Energy has propagated boundary and initial conditions, used in the total system performance assessment abstraction of mechanical disruption of engineered barriers, throughout its abstraction approaches.

Examine how the features, events, and processes, related to mechanical disruption of engineered barriers, have been included in the total system performance assessment abstraction.

Evaluate the U.S. Department of Energy conclusion with respect to the impact of transient criticality on the integrity of the engineered barriers.

Verify that the U.S. Department of Energy reviews follow guidance, such as NUREG–1297 and NUREG–1298 (Altman, et al., 1988a,b), or make an acceptable case for using alternative approaches.

Review Method 2 Data and Model Justification

Evaluate the sufficiency of the geological and engineering data used to support parameters for conceptual models, process-level models, and alternative conceptual models considered in the abstraction of mechanical disruption of engineered barriers. Evaluate the basis for the data on physical phenomena, couplings, geology, and engineering used in the abstraction of mechanical disruption of engineered barriers. This basis may include a combination of techniques, such as laboratory experiments, site-specific field measurements, natural analog research, process-level modeling studies, and expert elicitation.

Verify that sufficient data have been collected to adequately characterize the geology of the natural system, engineering materials, and initial manufacturing defects to establish initial and boundary conditions for the abstraction of mechanical disruption of engineered barriers.

Evaluate and confirm that data used to support the U.S. Department of Energy abstraction of mechanical disruption of engineered barriers are based on appropriate techniques, and are adequate for the accompanying sensitivity/uncertainty analyses. Evaluate the need for additional data based on sensitivity analyses.

Verify that the U.S. Department of Energy demonstrates the adequacy of engineered barrier mechanical failure models for disruption events.

Review Method 3 Data Uncertainty

Evaluate the technical bases for parameter values, assumed ranges, probability distributions, and bounding values, used in conceptual models, process-level models, and alternative conceptual models, considered in the abstraction of mechanical disruption of engineered barriers. The reviewer should verify that the technical bases supports the treatment of uncertainty and variability of these parameters in the performance assessment. If conservative values are used as a method for addressing uncertainty and variability, the reviewer should verify that the conservative values result in conservative estimates of risk and do not cause unintended results (i.e., conservative representation of one aspect of the repository behavior that leads to an overall reduction in risk; inappropriate dilution of the risk estimate by assuming an approach is conservative when a parameter range is increased beyond the supporting data).

Evaluate the U.S. Department of Energy justification of process-level models used to represent mechanically disruptive events within the emplacement drifts at the proposed Yucca Mountain repository. Verify that the U.S. Department of Energy parameter values are adequately constrained by Yucca Mountain site data, such that the effects of mechanically disruptive events on engineered barrier integrity are not underestimated. Confirm that the U.S. Department of Energy identifies parameters within conceptual models for mechanically disruptive events that are consistent with the range of characteristics observed at Yucca Mountain.

Verify that the U.S. Department of Energy appropriately establishes possible statistical correlations between parameters. Verify that an adequate technical basis or bounding argument is provided for neglected correlations.

Evaluate the methods used by the U.S. Department of Energy in conducting expert elicitation to define parameter values.

Review Method 4 Model Uncertainty

Evaluate the U.S. Department of Energy alternative conceptual models used in developing the abstraction for mechanical disruption of engineered barriers. Examine the model parameters, considering available site characterization data, laboratory experiments, field measurements, natural analog research, and process-level modeling studies and evaluate their consistency.

Where appropriate, use an alternative total system performance assessment model to evaluate selected parts of the U.S. Department of Energy abstraction of mechanical disruption of engineered barriers.

Evaluate the treatment of conceptual model uncertainty in light of the available site characterization data, laboratory experiments, field measurements, natural analog information and process-level modeling studies. If adoption of a conservative model is used as an approach for addressing conceptual model uncertainty, the reviewer should verify that the selected conceptual model: (i) is conservative relative to alternative conceptual models that are consistent with the available data and current scientific understanding; and (ii) results in conservative estimates of risk and not cause unintended results (i.e., conservative representation of one aspect of the repository behavior that leads to an overall reduction in the risk estimate).

Examine the mathematical models included in the analyses of mechanical disruption of engineered barriers. Also, examine and evaluate the bases for excluding alternative conceptual models, and the limitations and uncertainties of the chosen model.

Review Method 5 Model Support

Evaluate the output from the abstraction of mechanical disruption of engineered barriers, and compare the results with an appropriate combination of site characterization data, process-level modeling, laboratory testing, field measurements, and natural analog research.

Use detailed models of geological and engineering processes to evaluate the total system performance assessment abstractions of mechanical disruption of engineered barriers. If practical, use an alternative total system performance assessment model to evaluate selected parts of the U.S. Department of Energy abstraction of mechanical disruption of engineered barriers, and evaluate the effects on repository performance. Compare results of the U.S. Department of Energy abstraction to approximations shown to be appropriate for closely analogous natural systems or experimental systems.

Examine the procedures used by the U.S. Department of Energy to develop and test its mathematical and numerical models.

As appropriate, use an alternative total system performance assessment model to evaluate the U.S. Department of Energy sensitivity or bounding analyses, and confirm that the U.S. Department of Energy has used ranges consistent with available site characterization data, field and laboratory tests, and natural analog research.

2.2.1.3.2.3 Acceptance Criteria

The following acceptance criteria are based on meeting the requirements of 10 CFR 63.114(a)–(c) and (e)–(g), relating to the mechanical disruption of engineered barriers model abstraction. U.S. Nuclear Regulatory Commission staff should apply the following acceptance criteria, according to the level of importance established in the U.S. Department of Energy risk-informed license application.

Acceptance Criterion 1 System Description and Model Integration Are Adequate.

(1) Total system performance assessment adequately incorporates important design features, physical phenomena, and couplings, and uses consistent and appropriate assumptions throughout the mechanical disruption of engineered barrier abstraction process;

(2) The description of geological and engineering aspects of design features, physical phenomena, and couplings, that may affect mechanical disruption of engineered barriers, is adequate. For example, the description may include materials used in the construction of engineered barrier components, environmental effects (e.g., temperature, water chemistry, humidity, radiation, etc.) on these materials, and mechanical-failure processes and concomitant failure criteria used to assess the performance capabilities of these materials. Conditions and assumptions in the abstraction of mechanical disruption of engineered barriers are readily identified and consistent with the body of data presented in the description;

(3) The abstraction of mechanical disruption of engineered barriers uses assumptions, technical bases, data, and models that are appropriate and consistent with other related U.S. Department of Energy abstractions. For example, assumptions used for mechanical disruption of engineered barriers are consistent with the abstraction of degradation of engineered barriers (Section 2.2.1.3.1 of the Yucca Mountain Review Plan). The descriptions and technical bases provide transparent and traceable support for the abstraction of mechanical disruption of engineered barriers;

(4) Boundary and initial conditions used in the total system performance assessment abstraction of mechanical disruption of engineered barriers are propagated throughout its abstraction approaches;

(5) Sufficient data and technical bases to assess the degree to which features, events, and processes have been included in this abstraction are provided;

(6) The conclusion, with respect to the impact of transient criticality on the integrity of the engineered barriers, is defensible; and

(7) Guidance in NUREG–1297 and NUREG–1298 (Altman, et al., 1988a,b), or other acceptable approaches, is followed.

Acceptance Criterion 2 Data Are Sufficient for Model Justification.

(1) Geological and engineering values, used in the license application to evaluate mechanical disruption of engineered barriers, are adequately justified. Adequate descriptions of how the data were used, interpreted, and appropriately synthesized into the parameters are provided;

(2) Sufficient data have been collected on the geology of the natural system, engineering materials, and initial manufacturing defects, to establish initial and boundary conditions for the total system performance assessment abstraction of mechanical disruption of engineered barriers;

(3) Data on geology of the natural system, engineering materials, and initial manufacturing defects, used in the total system performance assessment abstraction, are based on appropriate techniques. These techniques may include laboratory experiments, site-specific field measurements, natural analog research, and process-level modeling studies. As appropriate, sensitivity or uncertainty analyses used to support the U.S. Department of Energy total system performance assessment abstraction are adequate to determine the possible need for additional data; and

(4) Engineered barrier mechanical failure models for disruption events are adequate. For example, these models may consider effects of prolonged exposure to the expected emplacement drift environment, material test results not specifically designed or performed for the Yucca Mountain site, and engineered barrier component fabrication flaws.

Acceptance Criterion 3 Data Uncertainty Is Characterized and Propagated Through the Model Abstraction.

(1) Models use parameter values, assumed ranges, probability distributions, and bounding assumptions that are technically defensible, reasonably account for uncertainties and variabilities, and do not result in an under-representation of the risk estimate;

(2) Process-level models used to represent mechanically disruptive events, within the emplacement drifts at the proposed Yucca Mountain repository, are adequate. Parameter values are adequately constrained by Yucca Mountain site data, such that the effects of mechanically disruptive events on engineered barrier integrity are not underestimated. Parameters within conceptual models for mechanically disruptive events are consistent with the range of characteristics observed at Yucca Mountain;

(3) Uncertainty is adequately represented in parameter development for conceptual models, process-level models, and alternative conceptual models considered in developing the assessment abstraction of mechanical disruption of engineered barriers. This may be done either through sensitivity analyses or use of conservative limits; and

(4) Where sufficient data do not exist, the definition of parameter values and conceptual models is based on appropriate use of expert elicitation, conducted in accordance with NUREG–1563 (Kotra, et al., 1996). If other approaches are used, the U.S. Department of Energy adequately justifies their use.

Acceptance Criterion 4 Model Uncertainty Is Characterized and Propagated Through the Model Abstraction.

(1) Alternative modeling approaches of features, events, and processes are considered and are consistent with available data and current scientific understanding, and the results and limitations are appropriately considered in the abstraction;

(2) Consideration of conceptual model uncertainty is consistent with available site characterization data, laboratory experiments, field measurements, natural analog information and process-level modeling studies; and the treatment of conceptual model uncertainty does not result in an under-representation of the risk estimate; and

(3) Appropriate alternative modeling approaches are investigated that are consistent with available data and current scientific knowledge, and appropriately consider their results and limitations using tests and analyses that are sensitive to the processes modeled.

Acceptance Criterion 5 Model Abstraction Output Is Supported by Objective Comparisons.

(1) Models implemented in this total system performance assessment abstraction provide results consistent with output from detailed process-level models and/or empirical observations (laboratory and field testings and/or natural analogs);

(2) Outputs of mechanical disruption of engineered barrier abstractions reasonably produce or bound the results of corresponding process-level models, empirical observations, or both;

(3) Well-documented procedures, that have been accepted by the scientific community to construct and test the mathematical and numerical models, are used to simulate mechanical disruption of engineered barriers; and

(4) Sensitivity analyses or bounding analyses are provided to support the total system performance assessment abstraction of mechanical disruption of engineered barriers that cover ranges consistent with site data, field or laboratory experiments and tests, and natural analog research.

2.2.1.3.2.4 Evaluation Findings

If the license application provides sufficient information and the regulatory acceptance criteria in Section 2.2.1.3.2.3 are appropriately satisfied, the staff concludes that this portion of the staff evaluation is acceptable. The reviewer writes material suitable for inclusion in the safety evaluation report prepared for the entire application. The report includes a summary statement

of what was reviewed and why the reviewer finds the submittal acceptable. The staff can document the review as follows.

U.S. Nuclear Regulatory Commission staff has reviewed the Safety Analysis Report and other information submitted in support of the license application, and has found, with reasonable expectation, that the requirements of 10 CFR 63.114 are satisfied, regarding the abstraction of mechanical disruption of engineered barriers in the performance assessment. In particular, the U.S. Nuclear Regulatory Commission staff found that:

(1) Appropriate data from the site and surrounding region, uncertainties and variabilities in parameter values, and alternative conceptual models have been used in the analyses, in compliance with 10 CFR 63.114(a)–(c);

(2) Specific features, events, and processes have been included in the analyses, and appropriate technical bases have been provided for inclusion or exclusion, in compliance with 10 CFR 63.114(e);

(3) Specific degradation, deterioration, and alteration processes have been included in the analyses, taking into consideration their effects on annual dose, and appropriate technical bases have been provided for inclusion or exclusion, in compliance with 10 CFR 63.114(f); and

(4) Adequate technical bases have been provided for models used in the performance assessment, as required by 10 CFR 63.114(g).

2.2.1.3.2.5 References

Altman, W.D., J.P. Donnelly, and J.E. Kennedy. NUREG–1297, "Generic Technical Position on Peer-Review for High-Level Nuclear Waste Repositories." Washington, DC: U.S. Nuclear Regulatory Commission. 1988a.

———. NUREG–1298, "Generic Technical Position on Qualification of Existing Data for High-Level Nuclear Waste Repositories." Washington, DC: U.S. Nuclear Regulatory Commission. 1988b.

Kotra, J.P., et al. NUREG–1563, "Branch Technical Position on the Use of Expert Elicitation in the High-Level Radioactive Waste Program." Washington, DC: U.S. Nuclear Regulatory Commission. 1996.

2.2.1.3.3 Quantity and Chemistry of Water Contacting Engineered Barriers and Waste Forms

To review this model abstraction, consider the degree to which the U.S. Department of Energy relies on quantity and chemistry of water contacting engineered barrier and waste forms to demonstrate compliance. Review this model abstraction, considering the risk information evaluated in the "Multiple Barriers" Section (2.2.1.1). For example, if the U.S. Department of Energy relies on the processes affecting the quantity and chemistry of water contacting engineered barriers and waste forms to significantly reduce dose to the reasonably maximally exposed individual, then a detailed review of this abstraction will be performed. If, on the other hand, the U.S. Department of Energy demonstrates that this abstraction has a minor impact on the dose to the reasonably maximally exposed individual, then a simplified review will be conducted focusing on the bounding assumptions. The review methods and acceptance criteria provided here are for a detailed review. Some of the review methods and acceptance criteria may not be necessary, in a simplified review, for those abstractions that have a minor impact on performance. The demonstration of the performance objectives is evaluated in Section 2.2.1.4 of the Yucca Mountain Review Plan.

Review Responsibilities—High-Level Waste Branch and Environmental and Performance Assessment Branch

2.2.1.3.3.1 Areas of Review

This section reviews quantity and chemistry of water contacting engineered barriers and waste forms. Reviewers will also evaluate information, required by 10 CFR 63.21(c)(1)–(4), (9), (10), (15), and (19), that is relevant to the abstraction of quantity and chemistry of water contacting engineered barriers and waste forms.

The staff will evaluate the following parts of the abstraction of quantity and chemistry of water contacting engineered barriers and waste forms, using the review methods and acceptance criteria in Sections 2.2.1.3.3.2 and 2.2.1.3.3.3:

(1) Description of the geological, hydrological, and geochemical aspects of quantity and chemistry of water contacting engineered barriers and waste forms, and the technical bases the U.S. Department of Energy provides to support model integration across the total system performance assessment abstractions;

(2) Sufficiency of the data and parameters used to justify the model abstraction;

(3) Methods the U.S. Department of Energy uses to characterize data uncertainty, and propagate the effects of this uncertainty through the total system performance assessment model abstraction;

(4) Methods the U.S. Department of Energy uses to characterize model uncertainty, and propagate the effects of this uncertainty through the total system performance assessment model abstraction;

(5) Approaches the U.S. Department of Energy uses to compare total system performance assessment output to process-level model outputs and empirical studies; and

(6) Use of expert elicitation.

2.2.1.3.3.2 Review Methods

To review the abstraction of quantity and chemistry of water contacting engineered barriers and waste forms, recognize that models used in the total system performance assessments may range from highly complex process-level models to simplified models, such as response surfaces or look-up tables. Evaluate model adequacy regardless of the level of complexity.

Review Method 1 Model Integration

Examine the descriptions of design features (including drip shield, backfill, waste packages, drift design and support, thermal loading, and other engineered barrier components); relevant physical features; physical phenomena; and couplings, as well as the description of the geological, hydrological, geochemical, and geomechanical aspects of the unsaturated zone, included in the abstraction of quantity and chemistry of water contacting engineered barriers and waste forms. Assess the adequacy of the technical bases for these descriptions, and for incorporating them in the total system performance assessment to represent quantity and chemistry of water contacting engineered barriers and waste forms.

Evaluate whether the description of hydrology, geology, geochemistry, design features, physical phenomena, and couplings, that may affect the quantity and chemistry of water contacting engineered barriers and waste forms, is adequate. Verify that conditions, assumptions, and the technical bases, used in the abstraction of quantity and chemistry of water contacting engineered barriers and waste forms, are consistent with other related U.S. Department of Energy abstractions.

Verify that important design features, such as waste package design and material selection, backfill, drip shield, ground support, thermal loading strategy, and degradation processes, are included in determining the initial and boundary conditions for calculations of the quantity and chemistry of water contacting engineered barriers and waste forms.

Examine the spatial and temporal abstractions to verify whether they appropriately address the physical couplings (thermal-hydrologic-mechanical-chemical).

Assess the technical bases for the geological, hydrological, geochemical, and geomechanical descriptions, and for incorporating them in the total system performance assessment abstraction for coupled thermal-hydrologic-mechanical-chemical effects. Confirm that the technical bases used for modeling assumptions and approximations have been documented, and are adequate. Evaluate whether the descriptions provide transparent and traceable support to the abstraction, and are consistent with other model abstractions.

Evaluate the model abstraction for quantity and chemistry of water contacting engineered barriers and waste forms, to confirm that it reasonably bounds the expected ranges of environmental conditions within the waste package emplacement drifts, inside of breached engineered barriers and contacting the waste forms.

Evaluate the consistency of the model abstraction for quantity and chemistry of water contacting engineered barriers and waste forms with detailed information on waste package design and other engineered features.

Examine how the features, events, and processes, related to the quantity and chemistry of water contacting engineered barriers and waste forms have been included in the total system performance assessment abstraction.

Verify that processes that have been observed in thermal-hydrologic tests and experiments and that are significant to performance are included in the total system performance assessment model abstraction.

Verify that the U.S. Department of Energy includes likely modes for container corrosion (Section 2.2.1.3.1 of the Yucca Mountain Review Plan) in determining the quantity and chemistry of water entering the engineered barriers and contacting waste forms. Evaluate the treatment of parameters such as pH and carbonate concentration, and the effect of waste package corrosion on the quantity and chemistry of water contacting engineered barriers and waste forms.

Evaluate the abstraction of in-package criticality or external-to-package criticality within the emplacement drift, and the associated technical basis for screening these events.

Confirm that if either event is included in the total system performance assessment, the U.S. Department of Energy uses acceptable technical bases for selecting the design criteria that mitigate the potential impact of in-package criticality on repository performance; identifies the features, events, and processes that may increase the reactivity of the system inside the waste package; identifies the configuration classes and configurations that have potential for nuclear criticality; and includes changes in thermal conditions and degradation of engineered barriers in the abstraction of the quantity and chemistry of water contacting engineered barriers packages and waste forms.

Verify that the U.S. Department of Energy reviews follow the guidance in NUREG–1297 and NUREG–1298 (Altman, et al., 1988a,b), or make an acceptable case for using alternative approaches.

Review Method 2 Data and Model Justification

Evaluate the sufficiency of the geological, hydrological, and geochemical data used to support parameters used in conceptual models, process-level models, and alternative conceptual models (if any) considered in the abstraction of quantity and chemistry of water contacting engineered barriers and waste forms. Evaluate whether the basis for the data includes a combination of techniques, such as laboratory experiments, site-specific field measurements, natural analog research, process-level modeling studies, and expert elicitation. Assess how the data were used, interpreted, and synthesized into the parameters. Examine and confirm the sufficiency, transparency, and traceability of the data that support the technical bases for features, events, and processes, related to the quantity and chemistry of water contacting engineered barriers and waste forms, that have been included in the total system performance assessment abstraction.

Verify that sufficient data were collected on the characteristics of the natural system and engineered materials to establish initial and boundary conditions for conceptual models of thermal-hydrologic-mechanical-chemical coupled processes that affect seepage and flow and the engineered barrier chemical environment, and the chemical environment for radionuclide release.

Verify that the U.S. Department of Energy has used results from thermal-hydrologic tests to identify important processes and establish temperature ranges for repository conditions in developing its mathematical models. Verify that the data are sufficient to support thermal-hydrologic conceptual models.

Evaluate the sufficiency of data used to support the conceptual approaches for water contact with the drip shield, waste package, and waste forms.

Examine the sufficiency of data used to support the analysis of the potential for microbial activity affecting the engineered barrier chemical environment and the chemical environment for radionuclide release. Confirm that the data are sufficient to support determination of the probability for microbially influenced corrosion and microbially enhanced dissolution of the high-level radioactive waste glass form.

Review Method 3 Data Uncertainty

Evaluate the technical bases for parameter values and assumed ranges, probability distributions, and bounding values used in conceptual models, process models, and alternative conceptual models considered in the total system performance assessment for this abstraction. The reviewer should verify that the technical bases supports the treatment of uncertainty and variability of these parameters in the performance assessment. If conservative values are used as a method for addressing uncertainty and variability, the reviewer should verify that the conservative values result in conservative estimates of risk and do not cause unintended results (i.e., conservative representation of one aspect of the repository behavior that leads to an overall reduction in risk; inappropriate dilution of the risk estimate by assuming an approach is conservative when a parameter range is increased beyond the supporting data).

Confirm that the parameter values are based on site-specific data obtained from techniques such as laboratory and field experiments. As necessary, evaluate whether the parameter values and ranges derived from natural analog research or process-level models are correctly incorporated in the model abstraction of quantity and chemistry of water contacting engineered barriers and waste forms.

Evaluate the initial and boundary conditions used to evaluate coupled thermal-hydrologic-mechanical-chemical effects on the quantity and chemistry of water contacting engineered barriers and waste forms for consistency with available data. As necessary, confirm that correlations between input values have been appropriately established in the U.S. Department of Energy total system performance assessment.

Evaluate the U.S. Department of Energy assessment of uncertainty and variability in parameters. Confirm that the U.S. Department of Energy incorporates data uncertainty and temporal and spatial variability in conditions affecting coupled thermal-hydrologic-mechanical-chemical effects into parameter ranges.

If in-package criticality or external-to-package criticality is included in the total system performance assessment, examine the methods and parameters used by the U.S. Department of Energy to calculate the effective neutron multiplication factor.

Verify that the U.S. Department of Energy appropriately established possible statistical correlations between parameters. Verify that an adequate technical basis or bounding argument is provided for neglected correlations.

If expert elicitations were used as a basis for data uncertainty for this abstraction, confirm they were conducted in accordance with appropriate guidance (Kotra, et al., 1996).

Review Method 4 Model Uncertainty

Verify that the U.S. Department of Energy has considered appropriate alternative conceptual models. Examine the bases for alternative conceptual models, considered in the model abstraction of quantity and chemistry of water contacting engineered barriers and waste forms, and the limitations and uncertainties of the chosen model. Evaluate the discussion of alternative modeling approaches not considered in the final analysis, and the limitations and uncertainties of the chosen model. Evaluate the selected model for consistency with available data.

Evaluate the treatment of conceptual model uncertainty in light of the available site characterization data, laboratory experiments, field measurements, natural analog information and process-level modeling studies. If adoption of a conservative model is used as an approach for addressing conceptual model uncertainty, the reviewer should verify that the selected conceptual model: (i) is conservative relative to alternative conceptual models that are consistent with the available data and current scientific understanding; and (ii) results in conservative estimates of risk and not cause unintended results (i.e., conservative representation of one aspect of the repository behavior that leads to an overall reduction in the risk estimate).

Evaluate the U.S. Department of Energy assessment of the effects of model uncertainty on conclusions regarding performance.

Review the methods used by the U.S. Department of Energy in considering the effects of thermal-hydrologic-mechanical-chemical coupled processes in different alternative conceptual models.

Confirm that the U.S. Department of Energy has provided an adequate demonstration of the effects on radiological exposures to the reasonably maximally exposed individual and releases of radionuclides into the accessible environment in its assessment of alternative conceptual models of coupled thermal-hydrologic-mechanical-chemical processes.

Review Method 5 Model Support

Evaluate the output from the abstraction of the quantity and chemistry of water contacting engineered barriers and waste forms, and compare the results with an appropriate combination of site characterization and design data, process-level modeling, laboratory testing, field measurements, and natural analog data.

Examine the analytical and numerical models used in the thermal-mechanical analyses for consistency with site-specific or natural analog data. Evaluate predicted changes in hydrologic properties and the magnitudes and distributions of changes resulting from effects of thermal-mechanical processes, for consistency with results of thermal-mechanical analyses of the underground facility.

Examine the output from the mathematical models for abstractions of coupled-process effects on the quantity and chemistry of water contacting engineered barriers and waste forms for consistency with conceptual models, based on inferences about the near-field environment, field data, and natural alteration observed at the site, and expected engineered materials properties. Examine the use of abstracted model results, and compare mathematical models to judge the robustness of results. Evaluate the acceptability of the sensitivity analyses used to support the abstraction of the quantity and chemistry of water contacting engineered barriers and waste forms in the total system performance assessment. To the extent practical, use an alternative total system performance assessment model to evaluate selected parts of the U.S. Department of Energy abstraction, and to evaluate the effects of the quantity and chemistry of water contacting engineered barriers and waste forms on repository performance.

2.2.1.3.3.3 Acceptance Criteria

The following acceptance criteria are based on meeting the requirements of 10 CFR 63.114(a)–(c) and (e)–(g), relating to the quantity and chemistry of water contacting engineered barriers and waste forms model abstraction. U.S. Nuclear Regulatory Commission staff should apply the following acceptance criteria, according to the level of importance established in the U.S. Department of Energy risk-informed license application.

Acceptance Criterion 1 System Description and Model Integration are Adequate.

(1) Total system performance assessment adequately incorporates important design features, physical phenomena, and couplings, and uses consistent and appropriate assumptions throughout the quantity and chemistry of water contacting engineered barriers and waste forms abstraction process;

(2) The abstraction of the quantity and chemistry of water contacting engineered barriers and waste forms uses assumptions, technical bases, data, and models, that are appropriate and consistent with other related U.S. Department of Energy abstractions. For example, the assumptions used for the quantity and chemistry of water contacting engineered barriers and waste forms are consistent with the abstractions of "Degradation of Engineered Barriers" (Section 2.2.1.3.1); "Mechanical Disruption of Engineered Barriers (Section 2.2.1.3.2); "Radionuclide Release Rates and Solubility Limits" (Section 2.2.1.3.4); "Climate and Infiltration" (Section 2.2.1.3.5); and "Flow Paths in the Unsaturated Zone" (Section 2.2.1.3.6). The descriptions and technical bases provide transparent and traceable support for the abstraction of quantity and chemistry of water contacting engineered barriers and waste forms;

(3) Important design features, such as waste package design and material selection, backfill, drip shield, ground support, thermal loading strategy, and degradation processes, are adequate to determine the initial and boundary conditions for calculations of the quantity and chemistry of water contacting engineered barriers and waste forms;

(4) Spatial and temporal abstractions appropriately address physical couplings (thermal-hydrologic-mechanical-chemical). For example, the U.S. Department of Energy evaluates the potential for focusing of water flow into drifts, caused by coupled thermal-hydrologic-mechanical-chemical processes;

(5) Sufficient technical bases and justification are provided for total system performance assessment assumptions and approximations for modeling coupled thermal-hydrologic-mechanical-chemical effects on seepage and flow, the waste package chemical environment, and the chemical environment for radionuclide release. The effects of distribution of flow on the amount of water contacting the engineered barriers and waste forms are consistently addressed, in all relevant abstractions;

(6) The expected ranges of environmental conditions within the waste package emplacement drifts, inside of breached waste packages, and contacting the waste forms and their evolution with time are identified. These ranges may be developed to include: (i) the effects of the drip shield and backfill on the quantity and chemistry of water (e.g., the potential for condensate formation and dripping from the underside of the shield); (ii) conditions that promote corrosion of engineered barriers and degradation of waste forms; (iii) irregular wet and dry cycles; (iv) gamma-radiolysis; and (v) size and distribution of penetrations of engineered barriers;

(7) The model abstraction for quantity and chemistry of water contacting engineered barriers and waste forms is consistent with the detailed information on engineered barrier design and other engineered features. For example, consistency is demonstrated for: (i) dimensionality of the abstractions; (ii) various design features and site characteristics; and (iii) alternative conceptual approaches. Analyses are adequate to demonstrate that no deleterious effects are caused by design or site features that the U.S. Department of Energy does not take into account in this abstraction;

(8) Adequate technical bases are provided, including activities such as independent modeling, laboratory or field data, or sensitivity studies, for inclusion of any thermal-hydrologic-mechanical-chemical couplings and features, events, and processes;

(9) Performance-affecting processes that have been observed in thermal-hydrologic tests and experiments are included into the performance assessment. For example, the U.S. Department of Energy either demonstrates that liquid water will not reflux into the underground facility or incorporates refluxing water into the performance assessment calculation, and bounds the potential adverse effects of alteration of the hydraulic pathway that result from refluxing water;

(10) Likely modes for container corrosion (Section 2.2.1.3.1 of the Yucca Mountain Review Plan) are identified and considered in determining the quantity and chemistry of water entering the engineered barriers and contacting waste forms. For example, the model abstractions consistently address the role of parameters, such as pH, carbonate concentration, and the effect of corrosion on the quantity and chemistry of water contacting engineered barriers and waste forms;

(11) The abstraction of in-package criticality or external-to-package criticality, within the emplacement drift, provides an adequate technical basis for screening these events. If either event is included in the assessment, then the U.S. Department of Energy uses acceptable technical bases for selecting the design criteria that mitigate the potential impact of in-package criticality on repository performance; identifies the features, events, and processes that may increase the reactivity of the system inside the waste package; identifies the configuration classes and configurations that have potential for nuclear criticality; and includes changes in thermal conditions and degradation of engineered barriers in the abstraction of the quantity and chemistry of water contacting engineered barriers and waste forms; and

(12) Guidance in NUREG–1297 and NUREG–1298 (Altman, et al., 1988a,b), or other acceptable approaches, is followed.

Acceptance Criterion 2 Data are Sufficient for Model Justification.

(1) Geological, hydrological, and geochemical values used in the license application are adequately justified. Adequate description of how the data were used, interpreted, and appropriately synthesized into the parameters is provided;

(2) Sufficient data were collected on the characteristics of the natural system and engineered materials to establish initial and boundary conditions for conceptual models of thermal-hydrologic-mechanical-chemical coupled processes, that affect seepage and flow and the engineered barrier chemical environment;

(3) Thermo-hydrologic tests were designed and conducted with the explicit objectives of observing thermal-hydrologic processes for the temperature ranges expected for repository conditions and making measurements for mathematical models. Data are sufficient to verify that thermal-hydrologic conceptual models address important thermal-hydrologic phenomena;

(4) Sufficient information to formulate the conceptual approach(es) for analyzing water contact with the drip shield, engineered barriers, and waste forms is provided; and

(5) Sufficient data are provided to complete a nutrient- and energy-inventory calculation, if it has been used to justify the inclusion of the potential for microbial activity affecting the engineered barrier chemical environment and the chemical environment for radionuclide release. As necessary, data are adequate to support determination of the probability for microbially influenced corrosion and microbial effects, such as production of organic byproducts and microbially enhanced dissolution of the high-level radioactive waste glass form.

Acceptance Criterion 3 Data Uncertainty Is Characterized and Propagated Through the Model Abstraction.

(1) Models use parameter values, assumed ranges, probability distributions, and bounding assumptions that are technically defensible, reasonably account for uncertainties and variabilities, and do not result in an under-representation of the risk estimate;

(2) Parameter values, assumed ranges, probability distributions, and bounding assumptions used in the total system performance assessment calculations of quantity and chemistry of water contacting engineered barriers and waste forms are technically defensible and reasonable, based on data from the Yucca Mountain region (e.g., results from large block and drift-scale heater and niche tests), and a combination of techniques that may include laboratory experiments, field measurements, natural analog research, and process-level modeling studies;

(3) Input values used in the total system performance assessment calculations of quantity and chemistry of water contacting engineered barriers (e.g., drip shield and waste package) are consistent with the initial and boundary conditions and the assumptions of the conceptual models and design concepts for the Yucca Mountain site. Correlations between input values are appropriately established in the U.S. Department of Energy total system performance assessment. Parameters used to define initial conditions, boundary conditions, and computational domain in sensitivity analyses involving coupled thermal-hydrologic-mechanical-chemical effects on seepage and flow, the waste package chemical environment, and the chemical environment for radionuclide release, are consistent with available data. Reasonable or conservative ranges of parameters or functional relations are established;

(4) Adequate representation of uncertainties in the characteristics of the natural system and engineered materials is provided in parameter development for conceptual models, process-level models, and alternative conceptual models. The U.S. Department of Energy may constrain these uncertainties using sensitivity analyses or conservative limits. For example, the U.S. Department of Energy demonstrates how parameters used to describe flow through the engineered barrier system bound the effects of backfill and excavation-induced changes;

(5) If criticality is included in the total system performance assessment, then the U.S. Department of Energy uses an appropriate range of input parameters for calculating the effective neutron multiplication factor; and

(6) Where sufficient data do not exist, the definition of parameter values and conceptual models is based on other appropriate sources, such as expert elicitation conducted in accordance with NUREG–1563 (Kotra, et al., 1996).

Acceptance Criterion 4 Model Uncertainty Is Characterized and Propagated Through the Model Abstraction.

(1) Alternative modeling approaches of features, events, and processes are considered and are consistent with available data and current scientific understanding, and the results and limitations are appropriately considered in the abstraction;

(2) Alternative modeling approaches are considered and the selected modeling approach is consistent with available data and current scientific understanding. A description that includes a discussion of alternative modeling approaches not considered in the final analysis and the limitations and uncertainties of the chosen model is provided;

(3) Consideration of conceptual model uncertainty is consistent with available site characterization data, laboratory experiments, field measurements, natural analog information and process-level modeling studies; and the treatment of conceptual model uncertainty does not result in an under-representation of the risk estimate;

(4) Adequate consideration is given to effects of thermal-hydrologic-mechanical-chemical coupled processes in the assessment of alternative conceptual models. These effects may include: (i) thermal-hydrologic effects on gas, water, and mineral chemistry; (ii) effects of microbial processes on the engineered barrier chemical environment and the chemical environment for radionuclide release; (iii) changes in water chemistry that may result from the release of corrosion products from the engineered barriers and interactions between engineered materials and ground water; and (iv) changes in boundary conditions (e.g., drift shape and size) and hydrologic properties, relating to the response of the geomechanical system to thermal loading; and

(5) If the U.S. Department of Energy uses an equivalent continuum model for the total system performance assessment abstraction, the models produce conservative estimates of the effects of coupled thermal-hydrologic-mechanical-chemical processes on calculated compliance with the postclosure public health and environmental standards.

Acceptance Criterion 5 Model Abstraction Output is Supported by Objective Comparisons.

(1) The models implemented in this total system performance assessment abstraction provide results consistent with output from detailed process-level models and/or empirical observations (laboratory and field testings and/or natural analogs);

(2) Abstracted models for coupled thermal-hydrologic-mechanical-chemical effects on seepage and flow and the engineered barrier chemical environment, as well as on the chemical environment for radionuclide release, are based on the same assumptions and approximations demonstrated to be appropriate for process-level models or closely analogous natural or experimental systems. For example, abstractions of processes, such as thermally induced changes in hydrological properties, or estimated diversion of percolation away from the drifts, are adequately justified by comparison to results of process-level modeling, that are consistent with direct observations and field studies; and

(3) Accepted and well-documented procedures are used to construct and test the numerical models that simulate coupled thermal-hydrologic-mechanical-chemical effects on seepage and flow, engineered barrier chemical environment, and the chemical environment for radionuclide release. Analytical and numerical models are appropriately supported. Abstracted model results are compared with different mathematical models, to judge robustness of results.

2.2.1.3.3.4 Evaluation Findings

If the license application provides sufficient information and the regulatory acceptance criteria in Section 2.2.1.3.3.3 are appropriately satisfied, the staff concludes that this portion of the staff evaluation is acceptable. The reviewer writes material suitable for inclusion in the safety evaluation report prepared for the entire application. The report includes a summary statement of what was reviewed and why the reviewer finds the submittal acceptable. The staff can document the review as follows.

U.S. Nuclear Regulatory Commission staff has reviewed the Safety Analysis Report and other information submitted in support of the license application, relevant to the quantity and chemistry of water contacting engineered barriers and waste forms, and has found, with reasonable expectation, that the requirements of 10 CFR 63.114 are satisfied for this abstraction. Technical requirements for conducting a performance assessment in the area of quantity and chemistry of water contacting engineered barriers and waste forms have been met. In particular, the U.S. Nuclear Regulatory Commission staff that:

(1) Appropriate data from the site and surrounding region, uncertainties and variabilities in parameter values, and alternative conceptual models have been used in the analyses, in compliance with 10 CFR 63.114(a)–(c);

(2) Specific features, events, and processes have been included in the analyses, and appropriate technical bases have been provided for inclusion or exclusion, in compliance with 10 CFR 63.114(e);

(3) Specific degradation, deterioration, and alteration processes have been included in the analyses, taking into consideration their effects on annual dose, and appropriate technical bases have been provided for inclusion or exclusion, in compliance with 10 CFR 63.114(f); and

(4) Adequate technical bases have been provided for models used in the performance assessment, as required by 10 CFR 63.114(g).

2.2.1.3.3.5 References

Altman, W.D., J.P. Donnelly, and J.E. Kennedy. NUREG–1297, "Generic Technical Position on Peer Review for High-Level Nuclear Waste Repositories." Washington, DC: U.S. Nuclear Regulatory Commission. 1988a.

———. NUREG–1298, "Generic Technical Position on Qualification of Existing Data for High-Level Nuclear Waste Repositories." Washington, DC: U.S. Nuclear Regulatory Commission. 1988b.

Kotra, J.P., et al. NUREG–1563, "Branch Technical Position on the Use of Expert Elicitation in the High-Level Radioactive Waste Program." Washington, DC: U.S. Nuclear Regulatory Commission. 1996.

2.2.1.3.4 Radionuclide Release Rates and Solubility Limits

To review this model abstraction, consider the degree to which the U.S. Department of Energy relies on radionuclide release rates and solubility limits, to demonstrate compliance. Review this model abstraction considering the risk information evaluated in the "Multiple Barriers" Section (2.2.1.1). For example, if the U.S. Department of Energy license application relies on the release rates and solubility limits to significantly reduce dose to the reasonably maximally exposed individual, then perform a detailed review of this abstraction. If, on the other hand, the U.S. Department of Energy demonstrates that this abstraction has a minor impact on the dose to the reasonably maximally exposed individual, then conduct a simplified review focusing on the bounding assumptions. The review methods and acceptance criteria provided here are for a detailed review. Some of the review methods and acceptance criteria may not be necessary, in a simplified review, for those abstractions that have minor impacts on performance. The demonstration of the performance objectives is evaluated in Section 2.2.1.4 of the Yucca Mountain Review Plan.

Review Responsibilities—High-Level Waste Branch and Environmental and Performance Assessment Branch

2.2.1.3.4.1 Areas of Review

This section reviews radionuclide release rates and solubility limits. Reviewers will also evaluate information, required by 10 CFR 63.21(c)(1)–(4), (9), (10), (15), and (19), that is relevant to the abstraction of radionuclide release rates and solubility limits.

The staff will evaluate the following parts of the abstraction of radionuclide release rates and solubility limits using the review methods and acceptance criteria in Sections 2.2.1.3.4.2 and 2.2.1.3.4.3:

(1) Description of the geological, hydrological, and geochemical aspects of radionuclide release rates and solubility limits, and the technical bases the U.S. Department of Energy provides to support model integration across the total system performance assessment abstractions;

(2) Sufficiency of the data and parameters used to justify the total system performance assessment model abstraction;

(3) Methods the U.S. Department of Energy uses to characterize data uncertainty, and propagate the effects of this uncertainty through the total system performance assessment model abstraction;

(4) Methods the U.S. Department of Energy uses to characterize model uncertainty, and propagate the effects of this uncertainty through the total system performance assessment model abstraction;

(5) Approaches the U.S. Department of Energy uses to compare total system performance assessment output model abstraction to process-level model outputs and empirical studies; and

(6) Use of expert elicitation.

2.2.1.3.4.2 Review Methods

To review the abstraction of radionuclide release rates and solubility limits, recognize that models used in the total system performance assessments may range from highly complex process-level models to simplified models, such as response surfaces or look-up tables. Evaluate model adequacy, regardless of the level of complexity.

Review Method 1 Model Integration

Examine the descriptions of design features (including drip shield, backfill, waste packages, waste forms, thermal loading, and other engineered barrier components); relevant physical features; physical phenomena; and couplings, as well as the description of the geological, hydrological, and geochemical aspects of the unsaturated zone included in the abstraction of radionuclide release rates and solubility limits. Verify that the description is adequate, and that the conditions and assumptions in the total system performance assessment abstraction are consistent with the information presented in the description of barriers important to waste isolation, as reviewed using Section 2.2.1.1 of the Yucca Mountain Review Plan.

Assess the technical bases for these descriptions and for incorporating them in the total system performance assessment abstractions. Where simplifications for modeling coupled thermal-hydrologic-chemical effects on the chemical environment for radionuclide release rates and solubility limits were used in the total system performance assessment abstractions, confirm that the technical bases used for modeling assumptions and approximations have been documented and are adequate. Evaluate whether the descriptions provide transparent and traceable support to the abstractions, and are consistent with other model abstractions.

Evaluate the design information on waste packages and engineered barrier systems, provided in the abstraction of radionuclide release rates and solubility limits. Verify that the information is sufficient and consistent with design information in other model abstractions.

Examine the U.S. Department of Energy description of environmental conditions expected inside breached waste packages and in the engineered barrier environment surrounding the waste package. Verify that the ranges in conditions are described in sufficient detail.

Verify that the U.S. Department of Energy description of process-level conceptual and mathematical models is sufficiently complete, with respect to thermal-hydrologic processes affecting radionuclide release from the emplacement drifts.

Examine how the features, events, and processes related to radionuclide release rates and solubility limits have been included in the total system performance assessment abstraction of radionuclide release rates and solubility limits.

Evaluate the total system performance assessment abstraction of in-package criticality or external-to-package criticality, within the emplacement drift, and the associated technical basis for screening these events. Confirm that if either event is included in the total system performance assessment, the U.S. Department of Energy uses acceptable technical bases for selecting the design criteria that mitigate the potential impact of in-package criticality on the repository performance; identifies the features, events, and processes that may increase the reactivity of the system inside the waste package; identifies the configuration classes and configurations that have potential for nuclear criticality; and includes changes in thermal conditions and degradation of engineered barriers in the abstraction of radionuclide release rates and solubility limits.

Verify that the U.S. Department of Energy reviews follow the guidance in NUREG–1297 and NUREG–1298 (Altman, et al., 1988a,b), or make an acceptable case for using alternative approaches.

Review Method 2 Data and Model Justification

Evaluate the sufficiency of the geological, hydrological, and geochemical data used to support conceptual models, process-level models, and alternative conceptual models considered in the abstraction of radionuclide release rates and solubility limits. Evaluate the basis for the data on design features (including drip shield, backfill, waste packages, waste forms, and other engineered barrier components) used in the abstraction of radionuclide release rates and solubility limits.

Examine and confirm that the U.S. Department of Energy has provided sufficient data on the characteristics of the natural system, and engineered materials to establish initial and boundary conditions for conceptual models and simulations of thermal-hydrologic-chemical coupled processes.

Examine and evaluate the models used to support abstraction of solubility limits, and verify that they are consistent with guidance in "Determination of Radionuclide Solubility in Ground Water for Assessment of High-Level Waste Isolation, Technical Position" (U.S. Nuclear Regulatory Commission, 1984).

Evaluate the U.S. Department of Energy corrosion and radionuclide release testing program for high-level radioactive waste forms intended for disposal. Verify that it provides consistent, sufficient, and suitable data for the in-package and in-drift chemistry, used in the abstraction of radionuclide release rates and solubility limits. Evaluate the justification for the use of test results not specifically collected from the Yucca Mountain site.

Review Method 3 Data Uncertainty

Evaluate the technical bases for parameter values and assumed ranges, probability distributions, and bounding values used in conceptual models, process models, and alternative conceptual models considered in the total system performance assessment radionuclide release rates and solubility. The reviewer should verify that the technical bases supports the treatment of uncertainty and variability of these parameters in the performance assessment. If conservative values are used as a method for addressing uncertainty and variability, the reviewer should verify that the conservative values result in conservative estimates of risk and do not cause unintended results (i.e., conservative representation of one aspect of the repository behavior that leads to an overall reduction in risk; inappropriate dilution of the risk estimate by assuming an approach is conservative when a parameter range is increased beyond the supporting data).

Evaluate the technical bases for parameter ranges, probability distributions, or bounding values. The reviewer should verify that the parameter values are derived from site-specific data, or an analysis is included to show that the assumed parameter values lead to a conservative assessment of performance. Examine the technical bases for parameter values and ranges in conceptual models, process-level models, and alternative conceptual models considered in the abstraction.

Examine the initial conditions, boundary conditions, and computational domain used in sensitivity analyses, involving coupled thermal-hydrologic-chemical effects on radionuclide release, for consistency with available data.

Evaluate the U.S. Department of Energy assessment of uncertainty and variability in parameters used in model abstractions. Confirm that uncertainty in data from both temporal and spatial variations in conditions affecting radionuclide release, was incorporated into the parameter ranges.

Evaluate the parameters used to describe flow through and out of the engineered barrier, and confirm that they are sufficient to bound the effects of backfill, excavation-induced changes, and thermally induced mechanical changes that affect flow.

If in-package criticality or external-to-package criticality is included in the total system performance assessment, examine the methods and parameters used by the U.S. Department of Energy to calculate the effective neutron multiplication factor.

Verify that the U.S. Department of Energy uses an appropriate range of time-history of temperature, humidity, and dripping to constrain the probability for microbial effects.

Verify that the U.S. Department of Energy adequately considers the uncertainties in the characteristics of the natural system and engineered materials, such as the type, quantity, and reactivity of material, in establishing initial and boundary conditions for conceptual models and simulations of thermal-hydrologic-chemical coupled processes that affect radionuclide release.

Verify that the U.S. Department of Energy appropriately establishes possible statistical correlations between parameters. Verify that an adequate technical basis or bounding argument is provided for neglected correlations.

Determine whether expert elicitations were used as a basis for data uncertainty for this abstraction, and whether they were conducted in accordance with appropriate guidance.

Review Method 4 Model Uncertainty

Evaluate the U.S. Department of Energy alternative conceptual models used in developing the total system performance assessment abstraction for radionuclide release rates and solubility limits. Examine the model parameters in the context of available site characterization data; design data (engineered barrier system, waste packages, and waste forms); laboratory experiments; field measurements; natural analog research; and process-level modeling studies. When practical, use an alternative total system performance assessment model to evaluate the effect of alternative conceptual models on the assessment of repository performance.

Confirm that the U.S. Department of Energy uses appropriate models, tests, and analyses that are sensitive to the processes modeled for both natural and engineering systems. Verify that conceptual model uncertainties are adequately defined and documented, and effects on conclusions regarding performance are properly assessed.

Examine the mathematical models included in the analyses of coupled thermal-hydrologic-chemical effects on the chemical environment for radionuclide release. Evaluate the bases for excluding alternative conceptual models, and the limitations and uncertainties of the chosen model.

Evaluate the treatment of conceptual model uncertainty in light of the available site characterization data, laboratory experiments, field measurements, natural analog information and process-level modeling studies. If adoption of a conservative model is used as an approach for addressing conceptual model uncertainty, the reviewer should verify that the selected conceptual model: (i) is conservative relative to alternative conceptual models that are consistent with the available data and current scientific understanding; and (ii) results in conservative estimates of risk and not cause unintended results (i.e., conservative representation of one aspect of the repository behavior that leads to an overall reduction in the risk estimate).

Review Method 5 Model Support

Evaluate the output from the abstraction of radionuclide release rates and solubility limits, and verify that the U.S. Department of Energy has compared the results with an appropriate combination of site characterization and design data, process-level modeling, laboratory testing, field measurements, and natural analog data.

Examine the analytical and numerical models used in the thermal-mechanical analyses for consistency with site-specific or natural analog data. Evaluate predicted changes in hydrologic properties and the magnitudes and distributions of changes resulting from effects of thermal-mechanical processes for consistency with results of thermal-mechanical analyses of the underground facility. To the extent practical, use an alternative total system performance assessment model to evaluate selected parts of the U.S. Department of Energy abstraction, and to evaluate the effects of the quantity and chemistry of water contacting the waste packages and waste forms on repository performance.

Examine the output from the mathematical models for abstractions of coupled-process effects on radionuclide release for consistency with conceptual models. Compare the output from the abstractions with inferences about the near-field environment, field data, and natural alteration observed at the site, and expected engineered materials properties.

Evaluate where the U.S. Department of Energy will rely on performance confirmation for this model abstraction, and whether specific plans for monitoring radionuclide release are adequate for further testing, to acquire additional necessary information, as part of the performance confirmation program, using Section 2.4 of the Yucca Mountain Review Plan.

2.2.1.3.4.3 Acceptance Criteria

The following acceptance criteria are based on meeting the relevant requirements of 10 CFR 63.114(a)–(c) and (e)–(g), as they relate to the radionuclide release rates and solubility limits model abstraction. U.S. Nuclear Regulatory Commission staff should apply the following acceptance criteria, according to the level of importance established in the U.S. Department of Energy risk-informed license application.

Acceptance Criterion 1 System Description and Model Integration Are Adequate.

(1) Total system performance assessment adequately incorporates important design features, physical phenomena, and couplings, and uses consistent and appropriate assumptions throughout the radionuclide release rates and solubility limits abstraction process;

(2) The abstraction of radionuclide release rates and solubility limits uses assumptions, technical bases, data, and models that are appropriate and consistent with other related U.S. Department of Energy abstractions. For example, the assumptions used for this model abstraction are consistent with the abstractions of "Degradation of Engineered Barriers" (Section 2.2.1.3.1); "Mechanical Disruption of Waste Packages" (Section 2.2.1.3.2); "Quantity and Chemistry of Water Contacting Engineered Barriers and Waste Forms" (Section 2.2.1.3.3); "Climate and Infiltration" (Section 2.2.1.3.5); and "Flow Paths in the Unsaturated Zone" (Section 2.2.1.3.6). The descriptions and technical bases provide transparent and traceable support for the abstraction of radionuclide release rates and solubility limits;

(3) The abstraction of radionuclide release rates and solubility limits provides sufficient, consistent design information on waste packages and engineered barrier systems. For example, inventory calculations and selected radionuclides are based on the detailed information provided on the distribution (both spatially and by compositional phase) of the radionuclide inventory, within the various types of high-level radioactive waste;

(4) The U.S. Department of Energy reasonably accounts for the range of environmental conditions expected inside breached waste packages and in the engineered barrier environment surrounding the waste package. For example, the U.S. Department of Energy should provide a description and sufficient technical bases for its abstraction of changes in hydrologic properties in the near field, caused by coupled thermal-hydrologic-mechanical-chemical processes;

(5) The description of process-level conceptual and mathematical models is sufficiently complete, with respect to thermal-hydrologic processes affecting radionuclide release from the emplacement drifts. For example, if the U.S. Department of Energy uncouples coupled processes, the demonstration that uncoupled model results bound predictions of fully coupled results is adequate;

(6) Technical bases for inclusion of any thermal-hydrologic-mechanical-chemical couplings and features, events, and processes in the radionuclide release rates and solubility limits model abstraction are adequate. For example, technical bases may include activities, such as independent modeling, laboratory or field data, or sensitivity studies;

(7) The abstraction of in-package criticality or external-to-package criticality, within the emplacement drift, provides an adequate technical basis for screening these events. If either event is included in the total system performance assessment, then the U.S. Department of Energy uses acceptable technical bases for selecting the design criteria that mitigate the potential impact of in-package criticality on the repository performance; identifies the features, events, and processes that may increase the reactivity of the system inside the waste package; identifies the configuration classes and configurations that have potential for nuclear criticality; and includes changes in thermal conditions and degradation of engineered barriers in the abstraction of radionuclide release rates and solubility limits; and

(8) Guidance in NUREG–1297 and NUREG–1298 (Altman, et al., 1988a,b), or other acceptable approaches for peer reviews and data qualification, is followed.

Acceptance Criterion 2 Data Are Sufficient for Model Justification.

(1) Geological, hydrological, and geochemical values used in the license application are adequately justified. Adequate description of how the data were used, interpreted, and appropriately synthesized into the parameters is provided;

(2) Sufficient data have been collected on the characteristics of the natural system and engineered materials to establish initial and boundary conditions for conceptual models and simulations of thermal-hydrologic-chemical coupled processes. For example, sufficient data should be provided on design features, such as the type, quantity, and reactivity of materials, that may affect radionuclide release for this abstraction;

(3) Where the U.S. Department of Energy uses data supplemented by models to support abstraction of solubility limits, the anticipated range of proportions and compositions of phases under the various physicochemical conditions expected are supported by experimental data (U.S. Nuclear Regulatory Commission, 1984); and

(4) The corrosion and radionuclide release testing program for high-level radioactive waste forms intended for disposal provides consistent, sufficient, and suitable data for the in-package and in-drift chemistry used in the abstraction of radionuclide release rates and solubility limits. For expected environmental conditions, the U.S. Department of Energy provides sufficient justification for the use of test results, not specifically collected from the Yucca Mountain site, for engineered barrier components, such as high-level radioactive waste forms, drip shield, and backfill.

Acceptance Criterion 3 Data Uncertainty Is Characterized and Propagated Through the Model Abstraction.

(1) Models use parameter values, assumed ranges, probability distributions, and bounding assumptions that are technically defensible, reasonably account for uncertainties and variabilities, and do not result in an under-representation of the risk estimate;

(2) Parameter values, assumed ranges, probability distributions, and bounding assumptions used in the abstractions of radionuclide release rates and solubility limits in the total system performance assessment are technically defensible and reasonable based on data from the Yucca Mountain region, laboratory tests, and natural analogs. For example, parameter values, assumed ranges, probability distributions, and bounding assumptions adequately reflect the range of environmental conditions expected inside breached waste packages;

(3) The U.S. Department of Energy uses reasonable or conservative ranges of parameters or functional relations to determine effects of coupled thermal-hydrologic-chemical processes on radionuclide release. These values are consistent with the initial and boundary conditions and the assumptions for the conceptual models and design concepts for natural and engineered barriers at the Yucca Mountain site. If any correlations between the input values exist, they are adequately established in the total system performance assessment. For example, estimations are based on a thermal loading and ventilation strategy; engineered barrier system design (including drift liner, backfill, and drip-shield); and natural system masses and fluxes that are consistent with those used in other abstractions;

(4) Uncertainty is adequately represented in parameter development for conceptual models, process models, and alternative conceptual models considered in developing the abstraction of radionuclide release rates and solubility limits, either through sensitivity analyses or use of bounding analyses;

(5) Parameters used to describe flow through and out of the engineered barrier, sufficiently bound the effects of backfill, excavation-induced changes, and thermally induced mechanical changes that affect flow;

(6) If criticality cannot be excluded from total system performance assessment, then the U.S. Department of Energy provides an appropriate range of input parameters for calculating the effective neutron multiplication factor;

(7) The U.S. Department of Energy uses an appropriate range of time-history of temperature, humidity, and dripping to constrain the probability for microbial effects, such as production of organic by-products that act as complexing ligands for actinides and microbially enhanced dissolution of the high-level radioactive waste glass form;

(8) The U.S. Department of Energy adequately considers the uncertainties, in the characteristics of the natural system and engineered materials, such as the type, quantity, and reactivity of material, in establishing initial and boundary conditions for

conceptual models and simulations of thermal-hydrologic-chemical coupled processes that affect radionuclide release; and

(9) Where sufficient data do not exist, the definition of parameter values and conceptual models is based on appropriate other sources, such as expert elicitation conducted in accordance with NUREG–1563 (Kotra, et al., 1996).

Acceptance Criterion 4 Model Uncertainty Is Characterized and Propagated Through the Model Abstraction.

(1) Alternative modeling approaches of features, events, and processes are considered and are consistent with available data and current scientific understanding, and the results and limitations are appropriately considered in the abstraction;

(2) In considering alternative conceptual models for radionuclide release rates and solubility limits, the U.S. Department of Energy uses appropriate models, tests, and analyses that are sensitive to the processes modeled for both natural and engineering systems. Conceptual model uncertainties are adequately defined and documented, and effects on conclusions regarding performance are properly assessed. For example, in modeling flow and radionuclide release from the drifts, the U.S. Department of Energy represents significant discrete features, such as fault zones, separately, or demonstrates that their inclusion in the equivalent continuum model produces a conservative effect on calculated performance; and

(3) Consideration of conceptual model uncertainty is consistent with available site characterization data, laboratory experiments, field measurements, natural analog information and process-level modeling studies; and the treatment of conceptual model uncertainty does not result in an under-representation of the risk estimate; and

(4) The effects of thermal-hydrologic-chemical coupled processes that may occur in the natural setting, or from interactions with engineered materials, or their alteration products, on radionuclide release, are appropriately considered.

Acceptance Criterion 5 Model Abstraction Output Is Supported by Objective Comparisons.

(1) The models implemented in this total system performance assessment abstraction provide results consistent with output from detailed process-level models and/or empirical observations (laboratory and field testings and/or natural analogs);

(2) Results of thermal-hydrologic process-level models are verified by demonstrating consistency with observations and results from laboratory and field-scale thermal-hydrologic tests. In particular, the U.S. Department of Energy demonstrates that sufficient physical evidence exists, to support conceptual models used to predict thermally driven flow in the near field;

(3) The U.S. Department of Energy adopts well-documented procedures that have been accepted by the scientific community to construct and test the numerical models, used to

simulate coupled thermal-hydrologic-chemical effects on radionuclide release. For example, the U.S. Department of Energy demonstrates that the numerical models used for high-level radioactive waste degradation and dissolution, and radionuclide release from the engineered barrier system, are adequate representations; include consideration of uncertainties; and are not likely to underestimate radiological exposures to the reasonably maximally exposed individual and releases of radionuclides into the accessible environment; and

(4) If the U.S. Department of Energy will rely on the performance confirmation program to assess whether the natural system and engineered materials are functioning as intended, an adequate program for monitoring radionuclide release from the waste packages, during the performance confirmation period, is established, using assumptions and calculations of radionuclide release from the waste packages that are appropriately substantiated (the acceptability of the performance confirmation program is reviewed using Section 2.4 of the Yucca Mountain Review Plan).

2.2.1.3.4.4 Evaluation Findings

If the license application provides sufficient information and the regulatory acceptance criteria in Section 2.2.1.3.4.3 are appropriately satisfied, the staff concludes that this portion of the staff evaluation is acceptable. The reviewer writes material suitable for inclusion in the safety evaluation report prepared for the entire application. The report includes a summary statement of what was reviewed and why the reviewer finds the submittal acceptable. The staff can document the review as follows.

U.S. Nuclear Regulatory Commission staff has reviewed the Safety Analysis Report and other information submitted in support of the license application, relevant to radionuclide release rates and solubility limits, and has found, with reasonable expectation, that the requirements of 10 CFR 63.114 are satisfied for model abstraction in this section. Technical requirements for conducting a performance assessment in the area of radionuclide release rates and solubility limits have been met. In particular, the U.S. Nuclear Regulatory Commission staff found that:

(1) Appropriate data from the site and surrounding region, uncertainties and variabilities in parameter values, and alternative conceptual models have been used in the analyses, in compliance with 10 CFR 63.114(a)–(c);

(2) Specific features, events, and processes have been included in the analyses, and appropriate technical bases have been provided for inclusion or exclusion, in compliance with 10 CFR 63.114(e);

(3) Specific degradation, deterioration, and alteration processes have been included in the analyses, taking into consideration their effects on annual dose, and appropriate technical bases have been provided for inclusion or exclusion, in compliance with 10 CFR 63.114(f); and

(4) Adequate technical bases have been provided for models used in the performance assessment, as required by 10 CFR 63.114(g).

2.2.1.3.4.5 References

Altman, W.D., J.P. Donnelly, and J.E. Kennedy. NUREG–1297, "Generic Technical Position on Peer Review for High-Level Nuclear Waste Repositories." Washington, DC: U.S. Nuclear Regulatory Commission. 1988a.

———. NUREG–1298, "Generic Technical Position on Qualification of Existing Data for High-Level Nuclear Waste Repositories." Washington, DC: U.S. Nuclear Regulatory Commission. 1988b.

Kotra, J.P., et al. NUREG–1563, "Branch Technical Position on the Use of Expert Elicitation in the High-Level Radioactive Waste Program." Washington, DC: U.S. Nuclear Regulatory Commission. 1996.

U.S. Nuclear Regulatory Commission. "Determination of Radionuclide Solubility in Ground Water for Assessment of High-Level Waste Isolation, Technical Position." Washington, DC: U.S. Nuclear Regulatory Commission. 1984.

2.2.1.3.5 Climate and Infiltration

To review this model abstraction, consider the degree to which the U.S. Department of Energy relies on climate and infiltration to demonstrate compliance. Review this model abstraction, considering the risk information evaluated in the "Multiple Barriers" Section (2.2.1.1). For example, if the U.S. Department of Energy relies on climate and infiltration to provide significant delay in the transport of radionuclides or a significant dilution in dose to the reasonably maximally exposed individual, then perform a detailed review of this abstraction. If, on the other hand, the U.S. Department of Energy demonstrates this abstraction to have a minor impact on the delay in the transport of radionuclides to the reasonably maximally exposed individual, or insignificant dilution in dose, then conduct a simplified review focusing on the bounding assumptions. The review methods and acceptance criteria provided here are for a detailed review. Some of the review methods and acceptance criteria may not be necessary in a simplified review for those abstractions that have a minor impact on performance. The demonstration of compliance with the performance objectives is evaluated using Section 2.2.1.4 of the Yucca Mountain Review Plan.

Review Responsibilities—High-Level Waste Branch and Environmental and Performance Assessment Branch

2.2.1.3.5.1 Areas of Review

This section reviews climate and net infiltration. Reviewers will also evaluate information, required by 10 CFR 63.21(c)(1), (9)(10), (15), and (19), that is relevant to the abstraction of climate and infiltration.

The staff will evaluate the following parts of the abstraction of climate and infiltration, using the review methods and acceptance criteria in Sections 2.2.1.3.5.2 and 2.2.1.3.5.3:

(1) Description of the climatological, hydrological, geological, and geochemical aspects of net infiltration in the unsaturated zone, and the technical bases the U.S. Department of Energy provides to support model integration across the total system performance assessment abstractions;

(2) Sufficiency of the data and parameters used to justify the model abstraction;

(3) Methods the U.S. Department of Energy uses to characterize data uncertainty, and propagate the effects of this uncertainty through the model abstraction;

(4) Methods the U.S. Department of Energy uses to characterize model uncertainty, and propagate the effects of this uncertainty through the model abstraction;

(5) Approaches the U.S. Department of Energy uses to compare total system performance assessment output to process-level model outputs and empirical studies; and

(6) Use of expert elicitation.

2.2.1.3.5.2 Review Methods

To review the abstraction of climate and infiltration, recognize that models used in the total system performance assessment may range from highly complex process-level models to simplified models, such as response surfaces or look-up tables. Evaluate model adequacy, regardless of the level of complexity.

Review Method 1 Model Integration

Examine the description of physical phenomena and couplings and the descriptions of the geological, hydrological, geochemical, paleohydrological, paleoclimatological, and climatological aspects of the abstraction of the climate and net infiltration that contribute to waste isolation. Assess the adequacy of the technical bases for these descriptions and for incorporating them in this abstraction.

Evaluate whether the description of aspects of geology, hydrology, geochemistry, physical phenomena, and couplings, that may affect climate and net infiltration, is adequate. Verify that conditions and assumptions used in this abstraction are consistent with the body of data presented in the description.

Examine assumptions, technical bases, data, and models used by the U.S. Department of Energy in this abstraction for consistency with other related U.S. Department of Energy abstractions. Evaluate whether the descriptions and technical bases provide transparent and traceable support for this abstraction.

Examine how the features, events, and processes related to climate and net infiltration have been included in the total system performance assessment abstractions.

Confirm that the U.S. Department of Energy abstractions employ adequate spatial and temporal variability of model parameters and boundary conditions to estimate net infiltration flux.

Confirm that averages of parameter estimates used in process-level models over time and space scales are appropriate for the model discretization.

Verify that paleoclimate information is evaluated over the past 500,000 years as the basis for projections of future climate change. For example, confirm that numerical climate models, if used for projection of future climate, are calibrated based on such paleoclimate data.

Verify that the U.S. Department of Energy reviews follow guidance, such as NUREG–1297 and NUREG–1298 (Altman, et al., 1988a,b), or make an acceptable case for using alternative approaches.

Review Method 2 Data and Model Justification

Evaluate the sufficiency of the data used to support conceptual models, process-level models, and alternative conceptual models considered in this abstraction, and the parameters used for each of these models. Evaluate the basis for the data on physical phenomena, couplings, climatology, geology, hydrology, and geochemistry. This basis may include a combination of techniques, such as laboratory experiments, site-specific field measurements, natural analog research, process-level modeling studies, and expert elicitation.

Verify that the mathematical model estimates of net infiltration are at appropriate time and space scales. Confirm that adequate site-specific climatic, surface, and subsurface information is used.

Verify that net infiltration is not underestimated. Confirm that adequate representation of the effects of fracture properties, fracture distributions, matrix properties, heterogeneities, time-varying boundary conditions, evapotranspiration, depth of soil cover, and surface-water runoff and run-on is incorporated in this abstraction.

Confirm the use of adequate sensitivity or uncertainty analyses to assess data sufficiency, and verify the possible need for additional data.

Confirm that adequate accepted and well-documented procedures are applied to develop and calibrate numerical models.

Verify that reasonably complete process-level conceptual and mathematical models are used in the analyses. Verify that the mathematical models are consistent with conceptual models and site characteristics. Confirm that a comparison of the robustness of results from different mathematical models is provided.

Evaluate the methods used by the U.S. Department of Energy in conducting expert elicitation.

Review Method 3 Data Uncertainty

Evaluate the technical bases for parameter values and assumed ranges, probability distributions, and bounding values used in conceptual models, process models, and alternative conceptual models considered in the total system performance assessment for climate and infiltration. The reviewer should verify that the technical bases support the treatment of uncertainty and variability of these parameters in the performance assessment. If conservative values are used as a method for addressing uncertainty and variability, the reviewer should verify that the conservative values result in conservative estimates of risk and do not cause unintended results (i.e., conservative representation of one aspect of the repository behavior that leads to an overall reduction in risk; inappropriate dilution of the risk estimate by assuming an approach is conservative when a parameter range is increased beyond the supporting data).

Confirm if uncertainty in data, because of both temporal and spatial variations in conditions affecting climate and net infiltration, is incorporated into the parameter ranges. For example, evaluate the climatic and hydrostratigraphic parameters used in the abstracted model to verify that they are consistent with site characterization data, and sufficiently detailed to capture heterogeneities that may influence the distribution and rate of liquid-water flux that has moved beyond the zone of evapotranspiration.

Verify that the U.S. Department of Energy appropriately establishes possible statistical correlations between parameters. Verify that an adequate technical basis or bounding argument is provided for neglected correlations.

Verify that the U.S. Department of Energy appropriately establishes possible statistical correlations between parameters. Verify that an adequate technical basis or bounding argument is provided for neglected correlations.

Confirm that performance assessments incorporate the hydrologic effects of future climate change that could alter the rates and patterns of present-day net infiltration into the unsaturated zone.

Review Method 4 Model Uncertainty

Evaluate the U.S. Department of Energy alternative conceptual models used in developing the abstraction for climate and net infiltration. Examine the model parameters, considering available site characterization data, laboratory experiments, field measurements, natural analog research, and process-level modeling studies. Where appropriate, use an alternative total system performance assessment model to evaluate selected parts of the U.S. Department of Energy abstraction of climate and net infiltration.

Verify that the bounds of uncertainty created by the process-level models are adequately reflected in this abstraction. Where appropriate, use an alternative total system performance assessment model to verify that the U.S. Department of Energy total system performance assessment approach reflects or bounds the uncertainties in the process-level models.

Evaluate the treatment of conceptual model uncertainty in light of the available site characterization data, laboratory experiments, field measurements, natural analog information and process-level modeling studies. If adoption of a conservative model is used as an approach for addressing conceptual model uncertainty, the reviewer should verify that the selected conceptual model: (i) is conservative relative to alternative conceptual models that are consistent with the available data and current scientific understanding; and (ii) results in conservative estimates of risk and not cause unintended results (i.e., conservative representation of one aspect of the repository behavior that leads to an overall reduction in the risk estimate).

Review Method 5 Model Support

Evaluate the output from the abstraction of climate and net infiltration. Compare the results with an appropriate combination of site characterization data, process-level modeling, laboratory testing, field measurements, and natural analog data.

Verify adequate justification and technical bases exist to conservatively bound process-level models. In particular, verify that if the U.S. Department of Energy uses an abstracted model to predict water flux into the unsaturated zone, the abstracted model is shown to bound process-level model predictions of the net infiltration flux. Use detailed models of geological, hydrological, geochemical, and climatological processes to evaluate the abstraction of climate and net infiltration.

Evaluate the output of model abstractions against results produced by process-level models. Where practical, use an alternative total system performance assessment model to evaluate selected parts of the U.S. Department of Energy abstraction, and to evaluate the effects of climate and net infiltration on repository performance.

2.2.1.3.5.3 Acceptance Criteria

The following acceptance criteria are based on meeting the requirements of 10 CFR 63.114(a)–(c) and (e)–(g), relating to the climate and net infiltration model abstraction. U.S. Nuclear Regulatory Commission staff should apply the following acceptance criteria, according to the level of importance established in the U.S. Department of Energy risk-informed license application.

Acceptance Criterion 1 System Description and Model Integration Are Adequate.

(1) The total system performance assessment adequately incorporates, or bounds, important design features, physical phenomena, and couplings, and uses consistent and appropriate assumptions throughout the climate and net infiltration abstraction process;

(2) The aspects of geology, hydrology, geochemistry, physical phenomena, and couplings, that may affect climate and net infiltration, are adequately considered. Conditions and assumptions in the abstraction of climate and net infiltration are readily identified and consistent with the body of data presented in the description;

(3) The abstraction of climate and net infiltration uses assumptions, technical bases, data, and models that are appropriate and consistent with other related U.S. Department of Energy abstractions. For example, the assumptions used for climate and net infiltration

are consistent with the abstractions of flow paths in the unsaturated zone and flow paths in the saturated zone (Sections 2.2.1.3.6 and 2.2.1.3.8 of the Yucca Mountain Review Plan, respectively). The descriptions and technical bases provide transparent and traceable support for the abstraction of climate and net infiltration;

(4) Sufficient data and technical bases to assess the degree to which features, events, and processes have been included for this abstraction are provided;

(5) Adequate spatial and temporal variability of model parameters and boundary conditions are employed to model the different parts of the system;

(6) Average parameter estimates are used in process-level models over time and space scales that are appropriate for the model discretization;

(7) Projections of future climate change are based on evaluation of paleoclimate information over the past 500,000 years. For example, numerical climate models, if used for projection of future climate, are calibrated based on such paleoclimate data; and

(8) Guidance in NUREG–1297 and NUREG–1298 (Altman, et al., 1988a,b), or other acceptable approaches for peer reviews and data qualification, is followed.

Acceptance Criterion 2 Data Are Sufficient for Model Justification.

(1) Climatological and hydrological values used in the license application (e.g., time of onset of climate change, mean annual temperature, mean annual precipitation, mean annual net infiltration, etc.) are adequately justified. Adequate descriptions of how the data were used, interpreted, and appropriately synthesized into the parameters are provided;

(2) Estimates of present-day net infiltration using mathematical models at appropriate time and space scales are reasonably verified with site-specific climatic, surface, and subsurface information;

(3) The effects of fracture properties, fracture distributions, matrix properties, heterogeneities, time-varying boundary conditions, evapotranspiration, depth of soil cover, and surface-water runoff and runon are considered, such that net infiltration is not underestimated;

(4) Sensitivity or uncertainty analyses are performed to assess data sufficiency and determine the possible need for additional data;

(5) Accepted and well-documented procedures are used to construct and calibrate numerical models;

(6) Reasonably complete process-level conceptual and mathematical models are used in the analyses. In particular: (i) mathematical models are provided that are consistent with conceptual models and site characteristics; and (ii) the robustness of results from different mathematical models is compared; and

(7) Any expert elicitation conducted is in accordance with NUREG–1563 (Kotra, et al., 1996), or other acceptable approaches.

Acceptance Criterion 3 Data Uncertainty Is Characterized and Propagated Through the Model Abstraction.

(1) Models use parameter values, assumed ranges, probability distributions, and bounding assumptions that are technically defensible, reasonably account for uncertainties and variabilities, and do not result in an under-representation of the risk estimate;

(2) The technical bases for the parameter values used in this abstraction are provided;

(3) Possible statistical correlations are established between parameters in this abstraction. An adequate technical basis or bounding argument is provided for neglected correlations; and

(4) The hydrologic effects of future climate change that may alter the rates and patterns of present-day net infiltration into the unsaturated zone are addressed. Such effects may include changes in soil depths, fracture-fill material, and types of vegetation.

Acceptance Criterion 4 Model Uncertainty Is Characterized and Propagated Through the Model Abstraction.

(1) Alternative modeling approaches of features, events, and processes, consistent with available data and current scientific understanding, are investigated. The results and limitations are appropriately considered in the abstraction;

(2) The bounds of uncertainty created by the process-level models are considered in this abstraction; and

(3) Consideration of conceptual model uncertainty is consistent with available site characterization data, laboratory experiments, field measurements, natural analog information and process-level modeling studies; and the treatment of conceptual model uncertainty does not result in an under-representation of the risk estimate.

Acceptance Criterion 5 Model Abstraction Output Is Supported by Objective Comparisons.

(1) The models implemented in this total system performance assessment abstraction provide results consistent with output from detailed process-level models and/or empirical observations (laboratory and field testing and/or natural analogs);

(2) Abstractions of process-level models may conservatively bound process-level predictions; and

(3) Comparisons are provided of output of abstracted models of climate and net infiltration with output of sensitivity studies, detailed process-level models, natural analogs, and empirical observations, as appropriate.

2.2.1.3.5.4 Evaluation Findings

If the license application provides sufficient information and the regulatory acceptance criteria in Section 2.2.1.3.5.3 are appropriately satisfied, the staff concludes that this portion of the staff evaluation is acceptable. The reviewer writes material suitable for inclusion in the safety evaluation report prepared for the entire application. The report includes a summary statement of what was reviewed and why the reviewer finds the submittal acceptable. The staff can document the review as follows.

If the license application provides sufficient information and the regulatory acceptance criteria in Section 2.1.1.1.3 are appropriately satisfied, the staff concludes that this evaluation is complete. The reviewer writes material suitable for inclusion in the safety evaluation report prepared for the entire application. The report includes a summary statement of what was reviewed and why the reviewer finds the submittal acceptable. The staff can document the review as follows.

U.S. Nuclear Regulatory Commission staff has reviewed the Safety Analysis Report and other information submitted in support of the license application, relevant to climate and infiltration, and has found, with reasonable expectation, that the requirements of 10 CFR 63.114 are satisfied for model abstraction in this section. Technical requirements for conducting a performance assessment in the area of climate and net infiltration have been met. In particular, the U.S. Nuclear Regulatory Commission staff found that:

(1) Appropriate data from the site and surrounding region, uncertainties and variabilities in parameter values, and alternative conceptual models have been used in the analyses, in compliance with 10 CFR 63.114(a)–(c);

(2) Specific features, events, and processes have been included in the analyses, and appropriate technical bases have been provided for inclusion or exclusion, in compliance with 10 CFR 63.114(e);

(3) Specific degradation, deterioration, and alteration processes have been included in the analyses, taking into consideration their effects on annual doses, and appropriate technical bases have been provided for inclusion or exclusion, in compliance with 10 CFR 63.114(f); and

(4) Adequate technical bases have been provided for models used in the performance assessment, as required by 10 CFR 63.114(g).

2.2.1.3.5.5 References

Altman, W.D., J.P. Donnelly, and J.E. Kennedy. NUREG–1297, "Generic Technical Position on Peer-Review for High-Level Nuclear Waste Repositories." Washington, DC: U.S. Nuclear Regulatory Commission. 1988a.

———. NUREG–1298, "Generic Technical Position on Qualification of Existing Data for High-Level Nuclear Waste Repositories." Washington, DC: U.S. Nuclear Regulatory Commission. 1988b.

Kotra, J.P., et al. NUREG–1563, “Branch Technical Position on the Use of Expert Elicitation in the High-Level Radioactive Waste Program.” Washington, DC: U.S. Nuclear Regulatory Commission. 1996.

2.2.1.3.6 Flow Paths in the Unsaturated Zone

To review this model abstraction, consider the degree to which the U.S. Department of Energy relies on flow paths in the unsaturated zone, to demonstrate compliance. Review this model abstraction, considering the risk information evaluated in the “Multiple Barriers” Section (2.2.1.1). For example, if the U.S. Department of Energy relies on flow paths in the unsaturated zone to provide significant delay and/or dilution in the transport of radionuclides to the reasonably maximally exposed individual, then perform a detailed review of this abstraction. If, on the other hand, the U.S. Department of Energy demonstrates this abstraction to have a minor impact on the delay in the transport of radionuclides to the reasonably maximally exposed individual, then conduct a simplified review focusing on the bounding assumptions. The review methods and acceptance criteria provided here are for a detailed review. Some of the review methods and acceptance criteria may not be necessary, in a simplified review, for those abstractions that have a minor impact on performance. The demonstration of compliance with the performance objectives is evaluated using Section 2.2.1.4 of the Yucca Mountain Review Plan.

Review Responsibilities—High-Level Waste Branch and Environmental and Performance Assessment Branch

2.2.1.3.6.1 Areas of Review

This section reviews flow paths in the unsaturated zone. Reviewers will also evaluate information, required by 10 CFR 63.21(c)(1), (9), (10), (15), and (19), that is relevant to the abstraction of flow paths in the unsaturated zone.

The staff will evaluate the following parts of the abstraction of flow paths in the unsaturated zone, using the review methods and acceptance criteria in Sections 2.2.1.3.6.2 and 2.2.1.3.6.3:

(1) Description of the hydrological, geological, and coupled thermal-hydrologic-mechanical-chemical processes of flow paths in the unsaturated zone, and the technical bases the U.S. Department of Energy provides to support model integration across the total system performance assessment abstractions;

(2) Sufficiency of the data and parameters used to justify the total system performance assessment model abstraction;

(3) Methods the U.S. Department of Energy uses to characterize data uncertainty, and propagate the effects of this uncertainty through the total system performance assessment model abstraction;

(4) Methods the U.S. Department of Energy uses to characterize model uncertainty, and propagate the effects of this uncertainty through the total system performance assessment model abstraction;

(5) Approaches the U.S. Department of Energy uses to compare total system performance assessment output to process-level model outputs and empirical studies; and

(6) Use of expert elicitation.

2.2.1.3.6.2 Review Methods

To review the abstraction of flow paths in the unsaturated zone, recognize that models used in the total system performance assessment may range from highly complex process-level models to simplified models, such as response surfaces or look-up tables. Evaluate model adequacy, regardless of the level of complexity.

Review Method 1 Model Integration

Examine the description of physical phenomena and couplings, and the descriptions of the geological, hydrological, geochemical, and thermal-hydrological-mechanical-chemical aspects of the abstraction of flow paths in the unsaturated zone that affect waste isolation. Assess the adequacy of the technical bases for these descriptions and for incorporating them in this abstraction.

Evaluate whether the descriptions of aspects of geology, hydrology, geochemistry, physical phenomena, and couplings that may affect flow paths in the unsaturated zone are adequate. Verify that conditions and assumptions used in this abstraction are consistent with the body of data presented in the description.

Examine assumptions, technical bases, data, and models used by the U.S. Department of Energy in this abstraction for consistency with other related abstractions. Evaluate whether the descriptions and technical bases provide transparent and traceable support for this abstraction.

Confirm that the conditions and assumptions used to describe initial and boundary conditions are consistent with other conditions and assumptions in this abstraction.

Examine how the features, events, and processes related to flow paths in the unsaturated zone have been included in the total system performance assessment abstraction.

Verify that the U.S. Department of Energy abstractions employ adequate spatial and temporal variability of model parameters and boundary conditions to estimate flow paths in the unsaturated zone, percolation flux, and seepage flux.

Verify that appropriate averages of parameter estimates are used in process-level models over time and space scales that are appropriate for the model discretization.

Confirm that potential reduction in unsaturated zone transport distances are accounted for after a climate-induced water table rise.

Verify that the U.S. Department of Energy reviews follow guidance, such as NUREG–1297 and NUREG–1298 (Altman, et al., 1988a,b), or make an acceptable case for using alternative approaches for peer review and data qualification.

Review Method 2 Data and Model Justification

Evaluate the sufficiency of the data used to support conceptual models, process-level models, and alternative conceptual models considered in this abstraction, and the parameters used for each of these models. Evaluate the basis for the data on physical phenomena, couplings, climatology, geology, hydrology, and geochemistry. This basis may include a combination of techniques, such as laboratory experiments, site-specific field measurements, natural analog research, process-level modeling studies, and expert elicitation.

Verify that acceptable techniques, which may include laboratory experiments, site-specific field measurements, natural analog research, and process-level modeling studies, are used in collecting and interpreting the data regarding the geology, hydrology, and geochemistry of the unsaturated zone.

Confirm that estimates of deep-percolation flux rates are conservative or reasonably represent the physical system. Verify that the flow model is calibrated using site-specific hydrologic, geologic, and geochemical data. Confirm that the mathematical model estimates of deep-percolation flux are at appropriate time and space scales.

Verify that appropriate thermal-hydrologic processes are evaluated by testing.

Confirm the use of adequate sensitivity or uncertainty analyses to assess data sufficiency, and verify the possible need for additional data.

Verify that adequate accepted and well-documented procedures are applied to develop and calibrate numerical models.

Verify that reasonably complete process-level conceptual and mathematical models are used in the analyses. Verify that the mathematical models are consistent with conceptual models and site characteristics. Confirm that a comparison of the robustness of results from different mathematical models is provided.

Evaluate the methods used by the U.S. Department of Energy in conducting expert elicitation.

Review Method 3 Data Uncertainty

Evaluate the U.S. Department of Energy assessment of uncertainty and variability in parameters used in the model abstraction. Confirm that uncertainty in data, from both temporal and spatial variations in conditions affecting flow paths in the unsaturated zone, is incorporated into the parameter ranges.

Evaluate the technical bases for parameter values and assumed ranges, probability distributions, and bounding values used in conceptual models, process models, and alternative conceptual models considered in the total system performance assessment for flow paths in the unsaturated zone. The reviewer should verify that the technical bases support the treatment of uncertainty and variability of these parameters in the performance assessment. If conservative values are used as a method for addressing uncertainty and variability, the reviewer should verify that the conservative values result in conservative estimates of risk and do not cause unintended results (i.e., conservative representation of one aspect of the repository behavior that leads to an overall reduction in risk; inappropriate dilution of the risk estimate by assuming an approach is conservative when a parameter range is increased beyond the supporting data).

Confirm that the U.S. Department of Energy appropriately established possible statistical correlations between parameters. Verify that an adequate technical basis or bounding argument is provided for neglected correlations.

Examine the initial conditions, boundary conditions, and computational domain used in sensitivity analyses and/or similar analyses for consistency with available data.

Verify that coupled thermal-hydrologic-mechanical-chemical processes are properly evaluated. Confirm that uncertainties in the characteristics of the natural system and engineered materials are considered.

Verify that the U.S. Department of Energy appropriately establishes possible statistical correlations between parameters. Verify that an adequate technical basis or bounding argument is provided for neglected correlations.

Confirm that parameter values are consistent with the initial and boundary conditions and the assumptions of the conceptual models for the Yucca Mountain site.

Review Method 4 Model Uncertainty

Evaluate the U.S. Department of Energy alternative conceptual models used in developing the abstraction for flow paths in the unsaturated zone. Examine the model parameters, considering available site characterization data, laboratory experiments, field measurements, natural analog research, and process-level modeling studies. Where appropriate, use an alternative total system performance assessment model to evaluate selected parts of the U.S. Department of Energy abstraction of flow paths in the unsaturated zone.

Verify that the bounds of uncertainty created by the process-level models are adequately reflected in this abstraction. Where appropriate, use an alternative total system performance assessment model to verify that the U.S. Department of Energy total system performance assessment approach reflects or bounds the uncertainties in the process-level models.

Evaluate the treatment of conceptual model uncertainty in light of the available site characterization data, laboratory experiments, field measurements, natural analog information and process-level modeling studies. If adoption of a conservative model is used as an approach for addressing conceptual model uncertainty, the reviewer should verify that the selected conceptual model: (i) is conservative relative to alternative conceptual models that are consistent with the available data and current scientific understanding; and (ii) results in conservative estimates of risk and not cause unintended results (i.e., conservative representation of one aspect of the repository behavior that leads to an overall reduction in the risk estimate).

Review Method 5 Model Support

Evaluate the output from the abstraction of flow paths in the unsaturated zone. Compare the results with an appropriate combination of site characterization data, process-level modeling, laboratory testing, field measurements, and natural analog data.

Confirm that adequate justification and technical basis exist to conservatively bound process-level models. Use detailed models of geological, hydrological, geochemical, and thermal-hydrologic-mechanical-chemical processes, to evaluate the total system performance assessment abstractions of flow paths in the unsaturated zone.

Evaluate the output of model abstractions against results produced by process-level models. Where practical, use an alternative total system performance assessment model to evaluate selected parts of the U.S. Department of Energy abstraction, and to evaluate the effects of flow paths in the unsaturated zone on repository performance.

2.2.1.3.6.3 Acceptance Criteria

The following acceptance criteria are based on meeting the requirements of 10 CFR 63.114(a)()–(c) and (e)–(g), relating to the flow paths in the unsaturated zone model abstraction. U.S. Nuclear Regulatory Commission staff should apply the following acceptance criteria, according to the level of importance established in the U.S. Department of Energy risk-informed license application.

Acceptance Criterion 1 System Description and Model Integration Are Adequate.

(1) The total system performance assessment adequately incorporates, or bounds, important design features, physical phenomena, and couplings, and uses consistent and appropriate assumptions throughout the flow paths in the unsaturated zone abstraction process. Couplings include thermal-hydrologic-mechanical-chemical effects, as appropriate;

(2) The aspects of geology, hydrology, geochemistry, physical phenomena, and couplings that may affect flow paths in the unsaturated zone are adequately considered. Conditions and assumptions in the abstraction of flow paths in the unsaturated zone are readily identified and consistent with the body of data presented in the description;

(3) The abstraction of flow paths in the unsaturated zone uses assumptions, technical bases, data, and models that are appropriate and consistent with other related U.S. Department of Energy abstractions. For example, the assumptions used for flow paths in the unsaturated zone are consistent with the abstractions of quantity and chemistry of water contacting waste packages and waste forms, climate and infiltration, and flow paths in the saturated zone (Sections 2.2.1.3.3, 2.2.1.3.5, and 2.2.1.3.8 of the

Yucca Mountain Review Plan, respectively). The descriptions and technical bases are transparent and traceable to site and design data;

(4) The bases and justification for modeling assumptions and approximations of radionuclide transport in the unsaturated zone are consistent with those used in model abstractions for flow paths in the unsaturated zone and thermal-hydrologic-mechanical-chemical effects;

(5) Sufficient data and technical bases to assess the degree to which features, events, and processes have been included in this abstraction are provided;

(6) Adequate spatial and temporal variability of model parameters and boundary conditions are employed in process-level models to estimate flow paths in the unsaturated zone, percolation flux, and seepage flux;

(7) Average parameter estimates used in process-level models are representative of the temporal and spatial discretizations considered in the model;

(8) Reduction in unsaturated zone transport distances, after a climate-induced water table rise, is considered; and

(9) Guidance in NUREG–1297 and NUREG–1298 (Altman, et al., 1988a,b), or other acceptable approaches for peer review and data qualification, is followed.

Acceptance Criterion 2 Data Are Sufficient for Model Justification.

(1) Hydrological and thermal-hydrological-mechanical-chemical values used in the license application are adequately justified. Adequate descriptions of how the data were used, interpreted, and appropriately synthesized into the parameters are provided;

(2) The data on the geology, hydrology, and geochemistry of the unsaturated zone, are collected using acceptable techniques;

(3) Estimates of deep-percolation flux rates constitute an upper bound, or are based on a technically defensible unsaturated zone flow model that reasonably represents the physical system. The flow model is calibrated, using site-specific hydrologic, geologic, and geochemical data. Deep-percolation flux is estimated, using the appropriate spatial and temporal variability of model parameters, and boundary conditions that consider climate-induced change in soil depths and vegetation;

(4) Appropriate thermal-hydrologic tests are designed and conducted, so that critical thermal-hydrologic processes can be observed, and values for relevant parameters estimated;

(5) Sensitivity or uncertainty analyses are performed to assess data sufficiency, and verify the possible need for additional data;

(6) Accepted and well-documented procedures are used to construct and calibrate numerical models;

(7) Reasonably complete process-level conceptual and mathematical models are used in the analyses. In particular: (i) mathematical models are provided that are consistent with conceptual models and site characteristics; and (ii) the robustness of results from different mathematical models is compared; and

(8) Any expert elicitation conducted is in accordance with NUREG–1563 (Kotra, et al., 1996), or other acceptable approaches.

Acceptance Criterion 3 Data Uncertainty Is Characterized and Propagated Through the Model Abstraction.

(1) Models use parameter values, assumed ranges, probability distributions, and bounding assumptions that are technically defensible, reasonably account for uncertainties and variabilities, and do not result in an under-representation of the risk estimate;

(2) The technical bases for the parameter values used in this abstraction are provided;

(3) Possible statistical correlations are established between parameters in this abstraction. An adequate technical basis or bounding argument is provided for neglected correlations;

(4) The initial conditions, boundary conditions, and computational domain used in sensitivity analyses and/or similar analyses are consistent with available data. Parameter values are consistent with the initial and boundary conditions and the assumptions of the conceptual models for the Yucca Mountain site;

(5) Coupled processes are adequately represented; and

(6) Uncertainties in the characteristics of the natural system and engineered materials are considered.

Acceptance Criterion 4 Model Uncertainty Is Characterized and Propagated Through the Model Abstraction.

(1) Alternative modeling approaches of features, events, and processes, consistent with available data and current scientific understanding, are investigated. The results and limitations are appropriately considered in the abstraction;

(2) The bounds of uncertainty created by the process-level models are considered in this abstraction; and

(3) Consideration of conceptual model uncertainty is consistent with available site characterization data, laboratory experiments, field measurements, natural analog information and process-level modeling studies; and the treatment of conceptual model uncertainty does not result in an under-representation of the risk estimate.

Acceptance Criterion 5 Model Abstraction Output Is Supported by Objective Comparisons.

(1) The models implemented in this total system performance assessment abstraction provide results consistent with output from detailed process-level models and/or empirical observations (laboratory and field testing and/or natural analogs);

(2) Abstractions of process-level models conservatively bound process-level predictions; and

(3) Comparisons are provided of output of abstracted model of flow paths in the unsaturated zone with outputs of sensitivity studies, detailed process-level models, natural analogs, and empirical observations, as appropriate.

2.2.1.3.6.4 Evaluation Findings

If the license application provides sufficient information and the regulatory acceptance criteria in Section 2.2.1.3.6.3 are appropriately satisfied, the staff concludes that this portion of the staff evaluation is acceptable. The reviewer writes material suitable for inclusion in the safety evaluation report prepared for the entire application. The report includes a summary statement of what was reviewed and why the reviewer finds the submittal acceptable. The staff can document the review as follows.

U.S. Nuclear Regulatory Commission staff has reviewed the Safety Analysis Report and other information submitted in support of the license application, relevant to flow paths in the unsaturated zone, and has found, with reasonable expectation, that the requirements of 10 CFR 63.114 are satisfied for model abstraction in this section. Technical requirements for conducting a performance assessment in the area of flow paths in the unsaturated zone have been met. In particular, the U.S. Nuclear Regulatory Commission staff found that:

(1) Appropriate data from the site and surrounding region, uncertainties and variabilities in parameter values, and alternative conceptual models have been used in the analyses, in compliance with 10 CFR 63.114(a)–(c);

(2) Specific features, events, and processes have been included in the analyses, and appropriate technical bases have been provided for inclusion or exclusion, in compliance with 10 CFR 63.114(e);

(3) Specific degradation, deterioration, and alteration processes have been included in the analyses, taking into consideration their effects on annual dose, and appropriate technical bases have been provided for inclusion or exclusion, in compliance with 10 CFR 63.114(f); and

(4) Adequate technical bases have been provided for models used in the performance assessment, as required by 10 CFR 63.114(g).

2.2.1.3.6.5 References

Altman, W.D., J.P. Donnelly, and J.E. Kennedy. NUREG–1297, "Generic Technical Position on Peer-Review for High-Level Nuclear Waste Repositories." Washington, DC: U.S. Nuclear Regulatory Commission. 1988a.

———. NUREG–1298, "Generic Technical Position on Qualification of Existing Data for High-Level Nuclear Waste Repositories." Washington, DC: U.S. Nuclear Regulatory Commission. 1988b.

Kotra, J.P., et al. NUREG–1563, "Branch Technical Position on the Use of Expert Elicitation in the High-Level Radioactive Waste Program." Washington, DC: U.S. Nuclear Regulatory Commission. 1996.

2.2.1.3.7 Radionuclide Transport in the Unsaturated Zone

To review this model abstraction, consider the degree to which the U.S. Department of Energy relies on radionuclide transport through the unsaturated zone, to demonstrate compliance. Review this model abstraction, considering the risk information evaluated in the "Multiple Barriers" Section (2.2.1.1). For example, if the U.S. Department of Energy relies on the unsaturated zone to provide significant delay in the transport of radionuclides and/or dilution of concentration to the reasonably maximally exposed individual, then perform a detailed review of this abstraction. If, on the other hand, the U.S. Department of Energy demonstrates this abstraction to have a minor impact on the delay, or a minor impact on the dose to the reasonably maximally exposed individual, then conduct a simplified review, focusing on the bounding assumptions. The review methods and acceptance criteria provided here are for a detailed review. Some of the review methods and acceptance criteria may not be necessary, in a simplified review, for those abstractions that have a minor impact on performance. The demonstration of compliance with the performance objectives is evaluated using Section 2.2.1.4 of the Yucca Mountain Review Plan.

Review Responsibilities—High-Level Waste Branch and Environmental and Performance Assessment Branch

2.2.1.3.7.1 Areas of Review

This section reviews radionuclide transport in the unsaturated zone. Reviewers will also evaluate information, required by 10 CFR 63.21(c)(1), (9), (10), (15), and (19), that is relevant to the abstraction of radionuclide transport in the unsaturated zone.

The staff will evaluate the following parts of the abstraction of radionuclide transport in the unsaturated zone, using the review methods and acceptance criteria in Sections 2.2.1.3.7.2 and 2.2.1.3.7.3

(1) Description of the geological, hydrological, and geochemical aspects of radionuclide transport in the unsaturated zone, and the technical bases the U.S. Department of Energy provides to support model integration across the total system performance assessment abstractions;

(2) Sufficiency of the data and parameters used to justify the total system performance assessment model abstraction;

(3) Methods the U.S. Department of Energy uses to characterize data uncertainty, and propagate the effects of this uncertainty through the total system performance assessment model abstraction;

(4) Methods the U.S. Department of Energy uses to characterize model uncertainty, and propagate the effects of this uncertainty through the total system performance assessment model abstraction;

(5) Approaches the U.S. Department of Energy uses to compare total system performance assessment output to process-level model outputs and empirical studies; and

(6) Use of expert elicitation.

2.2.1.3.7.2 Review Methods

To review the abstraction of radionuclide transport in the unsaturated zone, recognize that models used in the total system performance assessment may range from highly complex process-level models to simplified models, such as response surfaces or look-up tables. Evaluate model adequacy, regardless of the level of complexity.

Review Method 1 Model Integration

Examine the description of design features, physical phenomena, and couplings, and the description of the geological, hydrological, and geochemical aspects of the unsaturated zone included in the abstraction of radionuclide transport in the unsaturated zone, that affect waste isolation. Assess the adequacy of the technical bases for these descriptions, and for incorporating them in the abstraction of radionuclide transport in the unsaturated zone.

Evaluate whether the description of aspects of hydrology, geology, geochemistry, design features, physical phenomena, and couplings, that may affect radionuclide transport in the unsaturated zone, is adequate. Verify that conditions and assumptions used in the total system performance assessment abstraction of radionuclide transport in the unsaturated zone are consistent with the data presented in the description.

Examine assumptions, technical bases, data, and models used by the U.S. Department of Energy in the abstraction of radionuclide transport in the unsaturated zone for consistency with other related U.S. Department of Energy abstractions. Evaluate whether the descriptions and technical bases provide transparent and traceable support for the abstraction of radionuclide transport in the unsaturated zone.

Confirm that the U.S. Department of Energy has propagated boundary and initial conditions, used in the abstraction of radionuclide transport in the unsaturated zone throughout its abstraction approaches.

Examine how the features, events, and processes related to radionuclide transport in the unsaturated zone, have been included in the total system performance assessment abstraction.

Verify that the U.S. Department of Energy follows guidance, such as NUREG–1297 and NUREG–1298 (Altman, et al., 1988a,b), or makes an acceptable case for using alternative approaches for peer review and data qualification.

Review Method 2 Data and Model Justification

Evaluate the sufficiency of the geological, hydrological, and geochemical data used to support parameters, used in conceptual models, process-level models, and alternative conceptual models, considered in the abstraction of radionuclide transport in the unsaturated zone. Assess the sufficiency, transparency, and traceability of the data used to support the technical bases for features, events, and processes that have been included in the abstraction of radionuclide transport in the unsaturated zone.

Verify whether sufficient data have been collected on the characteristics of the geology, hydrology, and geochemistry of the natural system to establish initial and boundary conditions for the abstraction of radionuclide transport in the unsaturated zone.

Evaluate and confirm that data used to support the U.S. Department of Energy abstraction of radionuclide transport in the unsaturated zone are based on appropriate techniques, and are adequate for the accompanying sensitivity/uncertainty analyses. Evaluate the need for additional data based on the sensitivity analyses.

Review Method 3 Data Uncertainty

Evaluate the technical bases for parameter values, assumed ranges, probability distributions, and bounding values used in conceptual models, process models, and alternative conceptual models considered in the abstraction of radionuclide transport in the unsaturated zone. The reviewer should verify that the technical bases support the treatment of uncertainty and variability of these parameters in the performance assessment. If conservative values are used as a method for addressing uncertainty and variability, the reviewer should verify that the conservative values result in conservative estimates of risk and do not cause unintended results (i.e., conservative representation of one aspect of the repository behavior that leads to an overall reduction in risk; inappropriate dilution of the risk estimate by assuming an approach is conservative when a parameter range is increased beyond the supporting data).

Confirm that the U.S. Department of Energy has used flow and transport parameters that are based on techniques that may include laboratory experiments, field measurements, natural analog research, and process-level modeling studies, conducted under conditions relevant to the unsaturated zone at Yucca Mountain. Examine the results of the U.S. Department of Energy field transport tests, and confirm that the U.S. Department of Energy has provided adequate models.

If criticality in the unsaturated zone is included in the total system performance assessment, examine the methods and parameters used by the U.S. Department of Energy to calculate the effective neutron multiplication factor. Evaluate the consequences calculated by the U.S. Department of Energy for criticality in the unsaturated zone.

Verify that the U.S. Department of Energy appropriately establishes possible statistical correlations between parameters. Verify that an adequate technical basis or bounding argument is provided for neglected correlations.

Evaluate the methods used by the U.S. Department of Energy in conducting expert elicitation to define parameter values.

Review Method 4 Model Uncertainty

Evaluate the U.S. Department of Energy alternative conceptual models, used in developing the abstraction for radionuclide transport in the unsaturated zone. Examine the model parameters, considering available site characterization data, laboratory experiments, field measurements, natural analog research, and process-level modeling studies, and evaluate their consistency.

Where appropriate, use an alternative total system performance assessment model to evaluate selected parts of the U.S. Department of Energy abstraction of radionuclide transport in the unsaturated zone. Examine the effects of the alternative conceptual model(s) on repository performance, and evaluate how model uncertainties are defined, documented, and assessed.

Evaluate the treatment of conceptual model uncertainty in light of the available site characterization data, laboratory experiments, field measurements, natural analog information and process-level modeling studies. If adoption of a conservative model is used as an approach for addressing conceptual model uncertainty, the reviewer should verify that the selected conceptual model: (i) is conservative relative to alternative conceptual models that are consistent with the available data and current scientific understanding; and (ii) results in conservative estimates of risk and not cause unintended results (i.e., conservative representation of one aspect of the repository behavior that leads to an overall reduction in the risk estimate).

Examine the mathematical models included in the analyses of radionuclide transport in the unsaturated zone. Examine and evaluate the bases for excluding alternative conceptual models, and the limitations and uncertainties of the chosen model.

Review Method 5 Model Support

Evaluate the output from the abstraction of radionuclide transport in the unsaturated zone, and compare the results with an appropriate combination of site characterization data, process modeling, laboratory testing, field measurements, and natural analog research. Evaluate the sensitivity analyses used to support the abstraction of radionuclide transport in the unsaturated zone in the total system performance assessment.

Use detailed models of geochemical, hydrological, and geological processes to evaluate the total system performance assessment abstractions of radionuclide transport in the unsaturated zone. If practical, use an alternative total system performance assessment model to evaluate selected parts of the U.S. Department of Energy abstraction of radionuclide transport in the unsaturated zone, and evaluate the effects on repository performance. Compare results of the U.S. Department of Energy abstraction to approximations shown to be appropriate for closely analogous natural systems or experimental systems.

Examine the procedures used by the U.S. Department of Energy to develop and test its mathematical and numerical models.

As appropriate, use an alternative total system performance assessment model to evaluate the U.S. Department of Energy sensitivity or bounding analyses, and confirm that the U.S. Department of Energy has used ranges consistent with available site characterization data, field and laboratory tests, and natural analog research.

2.2.1.3.7.3 Acceptance Criteria

The following acceptance criteria are based on meeting the requirements of 10 CFR 63.114(a)()–(c) and (e)–(g), relating to the radionuclide transport in the unsaturated zone model abstraction. U.S. Nuclear Regulatory Commission staff should apply the following acceptance criteria, according to the level of importance established in the U.S. Department of Energy risk-informed license application.

Acceptance Criterion 1 System Description and Model Integration Are Adequate.

(1) Total system performance assessment adequately incorporates important design features, physical phenomena, and couplings, and uses consistent and appropriate assumptions throughout the radionuclide transport in the unsaturated zone abstraction process;

(2) The description of the aspects of hydrology, geology, geochemistry, design features, physical phenomena, and couplings, that may affect radionuclide transport in the unsaturated zone, is adequate. For example, the description includes changes in transport properties in the unsaturated zone, from water-rock interaction. Conditions and assumptions in the total system performance assessment abstraction of radionuclide transport in the unsaturated zone are readily identified, and consistent with the body of data presented in the description;

(3) The abstraction of radionuclide transport in the unsaturated zone uses assumptions, technical bases, data, and models that are appropriate and consistent with other related U.S. Department of Energy abstractions. For example, assumptions used for radionuclide transport in the unsaturated zone are consistent with the abstractions of radionuclide release rates and solubility limits and flow paths in the unsaturated zone (Sections 2.2.1.3.4 and 2.2.1.3.6 of the Yucca Mountain Review Plan, respectively). The descriptions and technical bases provide transparent and traceable support for the abstraction of radionuclide transport in the unsaturated zone;

(4) Boundary and initial conditions used in the abstraction of radionuclide transport in the unsaturated zone are propagated throughout its abstraction approaches. For example, the conditions and assumptions used to generate transport parameter values are

consistent with other geological, hydrological, and geochemical conditions in the total system performance assessment abstraction of the unsaturated zone;

(5) Sufficient data and technical bases for the inclusion of features, events, and processes, related to radionuclide transport in the unsaturated zone in the total system performance assessment abstraction, are provided; and

(6) Guidance in NUREG–1297 and NUREG–1298 (Altman, et al., 1988a,b), or other acceptable approaches, is followed for peer review and data qualification.

Acceptance Criterion 2 Data Are Sufficient for Model Justification.

(1) Geological, hydrological, and geochemical values, used in the license application, are adequately justified (e.g., flow-path length, sorption coefficients, retardation factors, colloid concentrations, etc.). Adequate descriptions of how the data were used, interpreted, and appropriately synthesized into the parameters are provided;

(2) Sufficient data have been collected on the characteristics of the natural system to establish initial and boundary conditions for the total system performance assessment abstraction of radionuclide transport in the unsaturated zone; and

(3) Data on the geology, hydrology, and geochemistry of the unsaturated zone, including the influence of structural features, fracture distributions, fracture properties, and stratigraphy, used in the total system performance assessment abstraction are based on appropriate techniques. These techniques may include laboratory experiments, site-specific field measurements, natural analog research, and process-level modeling studies. As appropriate, sensitivity or uncertainty analyses, used to support the U.S. Department of Energy total system performance assessment abstraction, are adequate to determine the possible need for additional data.

Acceptance Criterion 3 Data Uncertainty Is Characterized and Propagated Through the Model Abstraction.

(1) Models use parameter values, assumed ranges, probability distributions, and bounding assumptions that are technically defensible, reasonably account for uncertainties and variabilities, and do not result in an under-representation of the risk estimate;

(2) For those radionuclides where the total system performance assessment abstraction indicates that transport in fractures and matrix in the unsaturated zone is important to waste isolation: (i) estimated flow and transport parameters are appropriate and valid, based on techniques that may include laboratory experiments, field measurements, natural analog research, and process-level modeling studies, conducted under conditions relevant to the unsaturated zone at Yucca Mountain; and (ii) models are demonstrated to adequately reproduce field transport test results. For example, if a sorption coefficient approach is used, the assumptions implicit in that approach are verified;

(3) If criticality in the unsaturated zone far field is included in the total system performance assessment, an appropriate range of input parameters for calculating the effective neutron multiplication factor is used. The effects on performance of criticality in the unsaturated zone are adequately evaluated;

(4) Uncertainty is adequately represented in parameter development for conceptual models, process-level models, and alternative conceptual models, considered in developing the abstraction of radionuclide transport in the unsaturated zone. This may be done either through sensitivity analyses or use of conservative limits; and

(5) Where sufficient data do not exist, the definition of parameter values and conceptual models is based on appropriate use of expert elicitation, conducted in accordance with NUREG–1563 (Kotra, et al., 1996). If other approaches are used, the U.S. Department of Energy adequately justifies their use.

Acceptance Criterion 4 Model Uncertainty Is Characterized and Propagated Through the Model Abstraction.

(1) Alternative modeling approaches of features, events, and processes are considered and are consistent with available data and current scientific understanding, and the results and limitations are appropriately considered in the abstraction;

(2) Conceptual model uncertainties are adequately defined and documented, and effects on conclusions regarding performance are properly assessed;

(3) Consideration of conceptual model uncertainty is consistent with available site characterization data, laboratory experiments, field measurements, natural analog information and process-level modeling studies; and the treatment of conceptual model uncertainty does not result in an under-representation of the risk estimate; and

(4) Appropriate alternative modeling approaches are consistent with available data and current scientific knowledge, and appropriately consider their results and limitations, using tests and analyses that are sensitive to the processes modeled. For example, for radionuclide transport through fractures, the U.S. Department of Energy adequately considers alternative modeling approaches, to develop its understanding of fracture distributions and ranges of fracture flow and transport properties in the unsaturated zone.

Acceptance Criterion 5 Model Abstraction Output Is Supported by Objective Comparisons.

(1) The models implemented in this total system performance assessment abstraction provide results consistent with output from detailed process-level models and/or empirical observations (laboratory and field testings and/or natural analogs);

(2) Outputs of radionuclide transport in the unsaturated zone abstractions reasonably produce or bound the results of corresponding process-level models, empirical observations, or both. The U.S. Department of Energy abstracted models for

radionuclide transport in the unsaturated zone are based on the same hydrological, geological, and geochemical assumptions and approximations, shown to be appropriate for closely analogous natural systems or experimental systems;

(3) Well-documented procedures that have been accepted by the scientific community to construct and test the mathematical and numerical models are used to simulate radionuclide transport through the unsaturated zone; and

(4) Sensitivity analyses or bounding analyses are provided, to support the total system performance assessment abstraction of radionuclide transport in the unsaturated zone, that cover ranges consistent with site data, field or laboratory experiments and tests, and natural analog research.

2.2.1.3.7.4 Evaluation Findings

If the license application provides sufficient information and the regulatory acceptance criteria in Section 2.2.1.3.7.3 are appropriately satisfied, the staff concludes that this portion of the staff evaluation is acceptable. The reviewer writes material suitable for inclusion in the safety evaluation report prepared for the entire application. The report includes a summary statement of what was reviewed and why the reviewer finds the submittal acceptable. The staff can document the review as follows.

U.S. Nuclear Regulatory Commission staff has reviewed the Safety Analysis Report and other information submitted in support of the license application, relevant to radionuclide transport in the unsaturated zone, and has found, with reasonable expectation, that the requirements of 10 CFR 63.114 are satisfied for model abstraction in this section. Technical requirements for conducting a performance assessment in the area of radionuclide transport in the unsaturated zone have been met. In particular, the U.S. Nuclear Regulatory Commission staff found that:

(1) Appropriate data from the site and surrounding region, uncertainties and variabilities in parameter values, and alternative conceptual models have been used in the analyses, in compliance with 10 CFR 63.114(a)–(c);

(2) Specific features, events, and processes have been included in the analyses, and appropriate technical bases have been provided for inclusion or exclusion, in compliance with 10 CFR 63.114(e);

(3) Specific degradation, deterioration, and alteration processes have been included in the analyses, taking into consideration their effects on annual dose, and appropriate technical bases have been provided for inclusion or exclusion, in compliance with 10 CFR 63.114(f); and

(4) Adequate technical bases have been provided for models used in the performance assessment, as required by 10 CFR 63.114(g).

2.2.1.3.7.5 References

Altman, W.D., J.P. Donnelly, and J.E. Kennedy. NUREG–1297, "Generic Technical Position on Peer-Review for High-Level Nuclear Waste Repositories." Washington, DC: U.S. Nuclear Regulatory Commission. 1988a.

———. NUREG–1298, "Generic Technical Position on Qualification of Existing Data for High-Level Nuclear Waste Repositories." Washington, DC: U.S. Nuclear Regulatory Commission. 1988b.

Kotra, J.P., et al. NUREG–1563, "Branch Technical Position on the Use of Expert Elicitation in the High-Level Radioactive Waste Program." Washington, DC: U.S. Nuclear Regulatory Commission. 1996.

2.2.1.3.8 Flow Paths in the Saturated Zone

To review this model abstraction, consider the degree to which the U.S. Department of Energy relies on flow paths in the saturated zone to demonstrate compliance. Review this model abstraction, considering the risk information evaluated in the "Multiple Barriers" Section (2.2.1.1). For example, if the U.S. Department of Energy relies on saturated zone flow to provide significant delay or dilution in the transport of radionuclides to the reasonably maximally exposed individual, then perform a review of this abstraction. If, on the other hand, the U.S. Department of Energy demonstrates this abstraction to have a minor impact on the dose to the reasonably maximally exposed individual, then conduct a simplified review focusing on the bounding assumptions. The review methods and acceptance criteria provided here are for a detailed review. Some of the review methods and acceptance criteria may not be necessary, in a simplified review, for those abstractions that have a minor impact on performance. The demonstration of compliance with the performance objectives is evaluated, using Section 2.2.1.4 of the Yucca Mountain Review Plan.

Review Responsibilities—High-Level Waste Branch and Environmental and Performance Assessment Branch

2.2.1.3.8.1 Areas of Review

This section reviews flow paths in the saturated zone. Reviewers will also evaluate information, required by 10 CFR 63.21(c)(1), (9), (15), and (19), that is relevant to the abstraction of flow paths in the saturated zone.

The staff will evaluate the following parts of the abstraction of flow paths in the saturated zone, using the review methods and acceptance criteria in Sections 2.2.1.3.8.2 and 2.2.1.3.8.3:

(1) Description of the geological, hydrological, and geochemical aspects of flow paths in the saturated zone, and the technical bases the U.S. Department of Energy provides to support model integration across the total system performance assessment abstractions;

(2) Sufficiency of the data and parameters used to justify the total system performance assessment model abstraction;

(3) Methods the U.S. Department of Energy uses to characterize data uncertainty, and propagate the effects of this uncertainty, through the total system performance assessment model abstraction;

(4) Methods the U.S. Department of Energy uses to characterize model uncertainty, and propagate the effects of this uncertainty, through the total system performance assessment model abstraction;

(5) Approaches the U.S. Department of Energy uses to compare total system performance assessment output to process-level model outputs and empirical studies; and

(6) Use of expert elicitation.

2.2.1.3.8.2 Review Methods

To review the abstraction of flow paths in the saturated zone, recognize that models used in the total system performance assessment may range from highly complex process-level models to simplified models, such as response surfaces or look-up tables. Evaluate model adequacy, regardless of the level of complexity.

Review Method 1 Model Integration

Examine the description of design features, physical phenomena, and couplings, and the description of the geological, hydrological, and geochemical aspects of the saturated zone, included in the abstraction of flow paths, in the saturated zone, that affect waste isolation. Assess the adequacy of the technical bases for these descriptions, and for incorporating them in the abstraction of flow paths in the saturated zone.

Evaluate whether the description of aspects of geology, hydrology, geochemistry, design features, physical phenomena, and couplings, that may affect flow paths in the saturated zone, is adequate. Verify that conditions and assumptions used in the abstraction of flow paths in the saturated zone are consistent with the body of data presented in the description.

Examine assumptions, technical bases, data, and models used by the U.S. Department of Energy in the abstraction of flow paths in the saturated zone for consistency with other related U.S. Department of Energy abstractions. Evaluate whether the descriptions and technical bases provide transparent and traceable support for the abstraction of flow paths in the saturated zone.

Confirm that the U.S. Department of Energy has propagated boundary and initial conditions, used in the abstraction of flow paths in the saturated zone, throughout its abstraction approaches.

Examine how the features, events, and processes, related to flow paths in the saturated zone have been included in the total system performance assessment abstraction.

Verify that the U.S. Department of Energy delineates the flow paths in the saturated zone, considering natural site conditions.

Verify that the U.S. Department of Energy evaluates long-term climate change, based on known patterns of climatic cycles, during the Quaternary period, particularly the last 500,000 years, and other paleoclimate data.

Confirm that the U.S. Department of Energy considers potential geothermal and seismic effects on the ambient saturated zone flow system.

Confirm that the U.S. Department of Energy considers the impact of the expected water table rise on potentiometric heads and flow directions, and consequently on repository performance.

Verify that the U.S. Department of Energy reviews follow guidance, such as NUREG–1297 and NUREG–1298 (Altman, et al., 1988a,b), or make an acceptable case for using alternative approaches.

Review Method 2 Data and Model Justification

Evaluate the sufficiency of the geological, hydrological, geochemical, and climatological data used to support parameters used in conceptual models, process-level models, and alternative conceptual models considered in the total system performance assessment abstraction of flow paths in the saturated zone. Evaluate the basis for the data on physical phenomena, couplings, climatology, geology, hydrology, and geochemistry used in the total system performance assessment abstraction of flow paths in the saturated zone. This basis may include a combination of techniques, such as laboratory experiments, site-specific field measurements, natural analog research, process-level modeling studies, and expert elicitation.

Verify that sufficient data have been collected on the characteristics of the geology, hydrology, and geochemistry of the natural system, to establish initial and boundary conditions for the total system performance assessment abstraction of flow paths in the saturated zone.

Evaluate and confirm that data used to support the U.S. Department of Energy total system performance assessment abstraction of flow paths in the saturated zone are based on appropriate techniques, and are adequate for the accompanying sensitivity/uncertainty analyses. Evaluate the need for additional data, based on sensitivity analyses.

Verify that the U.S. Department of Energy provides sufficient information to substantiate that the proposed mathematical ground-water modeling approach, and proposed model(s) are applicable to site conditions.

Review Method 3 Data Uncertainty

Evaluate the technical bases for parameter values, assumed ranges, probability distributions, and bounding values used in conceptual models, process-level models, and alternative conceptual models, considered in the total system performance assessment abstraction of flow paths in the saturated zone. The reviewer should verify that the technical bases support the treatment of uncertainty and variability of these parameters in the performance assessment. If conservative values are used as a method for addressing uncertainty and variability, the reviewer should verify that the conservative values result in conservative estimates of risk and do not cause unintended results (i.e., conservative representation of one aspect of the repository behavior that leads to an overall reduction in risk; inappropriate dilution of the risk estimate by assuming an approach is conservative when a parameter range is increased beyond the supporting data).

Confirm that model abstractions incorporate uncertainty in hydrologic effects of climate change, based on a reasonably complete search of paleoclimate data.

Verify that the U.S. Department of Energy appropriately establishes possible statistical correlations between parameters. Verify that an adequate technical basis or bounding argument is provided for neglected correlations.

Evaluate the methods used by the U.S. Department of Energy in conducting expert elicitation to define parameter values.

Review Method 4 Model Uncertainty

Evaluate the U.S. Department of Energy alternative conceptual models used in developing the abstraction for flow paths in the saturated zone. Examine the model parameters, considering available site characterization data, laboratory experiments, field measurements, natural analog research, and process-level modeling studies, and evaluate their consistency. As appropriate, confirm that the U.S. Department of Energy has adequately addressed comments from external reviews of the model abstraction.

Where appropriate, use an alternative total system performance assessment model to evaluate selected parts of the U.S. Department of Energy abstraction of flow paths in the saturated zone. Examine the effects of the alternative conceptual model(s) on repository performance, and evaluate how model uncertainties are defined, documented, and assessed.

Evaluate the treatment of conceptual model uncertainty in light of the available site characterization data, laboratory experiments, field measurements, natural analog information and process-level modeling studies. If adoption of a conservative model is used as an approach for addressing conceptual model uncertainty, the reviewer should verify that the selected conceptual model: (i) is conservative relative to alternative conceptual models that are consistent with the available data and current scientific understanding; and (ii) results in conservative estimates of risk and not cause unintended results (i.e., conservative representation of one aspect of the repository behavior that leads to an overall reduction in the risk estimate).

Examine the mathematical models included in the analyses of flow paths in the saturated zone. Also, examine and evaluate the bases for excluding alternative conceptual models, and the limitations and uncertainties of the chosen model.

Review Method 5 Model Support

Evaluate the output from the abstraction of flow paths in the saturated zone, and compare the results with an appropriate combination of site characterization data, process-level modeling, laboratory testing, field measurements, and natural analog research.

Use detailed models of geological, hydrological, and geochemical processes to evaluate the total system performance assessment abstractions of flow paths in the saturated zone. If practical, use an alternative total system performance assessment model to evaluate selected parts of the U.S. Department of Energy abstraction of flow paths in the saturated zone, and evaluate the effects on repository performance. Compare results of the U.S. Department of Energy abstraction to approximations shown to be appropriate for closely analogous natural systems or experimental systems.

Examine the procedures used by the U.S. Department of Energy to develop and test its mathematical and numerical models.

As appropriate, use an alternative total system performance assessment model to evaluate the U.S. Department of Energy sensitivity or bounding analyses, and confirm that the U.S. Department of Energy has used ranges consistent with available site characterization data, field and laboratory tests, and natural analog research.

2.2.1.3.8.3 Acceptance Criteria

The following acceptance criteria are based on meeting the requirements of 10 CFR 63.114(a)–(c) and (e)–(g), relating to the flow paths in the saturated zone model abstraction. U.S. Nuclear Regulatory Commission staff should apply the following acceptance criteria, according to the level of importance established in the U.S. Department of Energy risk-informed license application.

Acceptance Criterion 1 System Description and Model Integration Are Adequate.

(1) Total system performance assessment adequately incorporates important design features, physical phenomena, and couplings, and uses consistent and appropriate assumptions, throughout the flow paths in the saturated zone abstraction process;

(2) The description of the aspects of hydrology, geology, geochemistry, design features, physical phenomena, and couplings, that may affect flow paths in the saturated zone, is adequate. Conditions and assumptions in the abstraction of flow paths in the saturated zone are readily identified, and consistent with the body of data presented in the description;

(3) The abstraction of flow paths in the saturated zone uses assumptions, technical bases, data, and models that are appropriate and consistent with other related U.S. Department of Energy abstractions. For example, the assumptions used for flow paths in the saturated zone are consistent with the total system performance assessment abstraction of representative volume (Section 2.2.1.3.12 of the Yucca Mountain Review Plan). The

descriptions and technical bases provide transparent and traceable support for the abstraction of flow paths in the saturated zone;

(4) Boundary and initial conditions used in the total system performance assessment abstraction of flow paths in the saturated zone are propagated throughout its abstraction approaches. For example, abstractions are based on initial and boundary conditions consistent with site-scale modeling and regional models of the Death Valley ground-water flow system;

(5) Sufficient data and technical bases to assess the degree to which features, events, and processes have been included in this abstraction are provided;

(6) Flow paths in the saturated zone are adequately delineated, considering natural site conditions;

(7) Long-term climate change, based on known patterns of climatic cycles during the Quaternary period, particularly the last 500,000 years, and other paleoclimate data, are adequately evaluated;

(8) Potential geothermal and seismic effects on the ambient saturated zone flow system are adequately described and accounted for;

(9) The impact of the expected water table rise on potentiometric heads and flow directions, and consequently on repository performance, is adequately considered; and

(10) Guidance in NUREG–1297 and NUREG–1298 (Altman, et al., 1988a,b), or other acceptable approaches for peer review and data qualification is followed.

Acceptance Criterion 2 Data Are Sufficient for Model Justification.

(1) Geological, hydrological, and geochemical values used in the license application to evaluate flow paths in the saturated zone are adequately justified. Adequate descriptions of how the data were used, interpreted, and appropriately synthesized into the parameters are provided;

(2) Sufficient data have been collected on the natural system to establish initial and boundary conditions for the abstraction of flow paths in the saturated zone;

(3) Data on the geology, hydrology, and geochemistry of the saturated zone used in the total system performance assessment abstraction are based on appropriate techniques. These techniques may include laboratory experiments, site-specific field measurements, natural analog research, and process-level modeling studies. As appropriate, sensitivity or uncertainty analyses, used to support the U.S. Department of Energy total system performance assessment abstraction, are adequate to determine the possible need for additional data; and

(4) Sufficient information is provided to substantiate that the proposed mathematical ground-water modeling approach and proposed model(s) are calibrated and applicable to site conditions.

Acceptance Criterion 3 Data Uncertainty Is Characterized and Propagated Through the Model Abstraction.

(1) Models use parameter values, assumed ranges, probability distributions, and bounding assumptions that are technically defensible, reasonably account for uncertainties and variabilities, and do not result in an under-representation of the risk estimate;

(2) Uncertainty is appropriately incorporated in model abstractions of hydrologic effects of climate change, based on a reasonably complete search of paleoclimate data;

(3) Uncertainty is adequately represented in parameter development for conceptual models, process-level models, and alternative conceptual models, considered in developing the abstraction of flow paths in the saturated zone. This may be done either through sensitivity analyses or use of conservative limits. For example, sensitivity analyses and/or similar analyses are sufficient to identify saturated zone flow parameters that are expected to significantly affect the abstraction model outcome; and

(4) Where sufficient data do not exist, the definition of parameter values and conceptual models is based on appropriate use of expert elicitation, conducted in accordance with NUREG–1563 (Kotra, et al., 1996). If other approaches are used, the U.S. Department of Energy adequately justifies their uses.

Acceptance Criterion 4 Model Uncertainty Is Characterized and Propagated Through the Model Abstraction.

(1) Alternative modeling approaches of features, events, and processes are considered and are consistent with available data and current scientific understanding, and the results and limitations are appropriately considered in the abstraction;

(2) Conceptual model uncertainties are adequately defined and documented, and effects on conclusions regarding performance are properly assessed. For example, uncertainty in data interpretations is considered by analyzing reasonable conceptual flow models that are supported by site data, or by demonstrating through sensitivity studies that the uncertainties have little impact on repository performance;

(3) Consideration of conceptual model uncertainty is consistent with available site characterization data, laboratory experiments, field measurements, natural analog information and process-level modeling studies; and the treatment of conceptual model uncertainty does not result in an under-representation of the risk estimate; and

(4) Appropriate alternative modeling approaches are consistent with available data and current scientific knowledge, and appropriately consider their results and limitations, using tests and analyses that are sensitive to the processes modeled.

Acceptance Criterion 5 Model Abstraction Output Is Supported by Objective Comparisons.

(1) The models implemented in this total system performance assessment abstraction provide results consistent with output from detailed process-level models and/or empirical observations (laboratory and field testings and/or natural analogs);

(2) Outputs of flow paths in the saturated zone abstractions reasonably produce or bound the results of corresponding process-level models, empirical observations, or both;

(3) Well-documented procedures that have been accepted by the scientific community to construct and test the mathematical and numerical models are used to simulate flow paths in the saturated zone; and

(4) Sensitivity analyses or bounding analyses are provided to support the abstraction of flow paths in the saturated zone, that cover ranges consistent with site data, field or laboratory experiments and tests, and natural analog research.

2.2.1.3.8.4 Evaluation Findings

If the license application provides sufficient information and the regulatory acceptance criteria in Section 2.2.1.3.8.3 are appropriately satisfied, the staff concludes that this portion of the staff evaluation is acceptable. The reviewer writes material suitable for inclusion in the safety evaluation report prepared for the entire application. The report includes a summary statement of what was reviewed and why the reviewer finds the submittal acceptable. The staff can document the review as follows.

U.S. Nuclear Regulatory Commission staff has reviewed the Safety Analysis Report and other information submitted in support of the license application, relevant to flow paths in the saturated zone, and has found, with reasonable expectation, that the requirements of 10 CFR 63.114 are satisfied for model abstraction in this section. Technical requirements for conducting a performance assessment in the area of flow paths in the saturated zone have been met. In particular, the U.S. Nuclear Regulatory Commission staff found that:

(1) Appropriate data from the site and surrounding region, uncertainties and variabilities in parameter values, and alternative conceptual models have been used in the analyses, in compliance with 10 CFR 63.114(a)–(c);

(2) Specific features, events, and processes have been included in the analyses, and appropriate technical bases have been provided for inclusion or exclusion, in compliance with 10 CFR 63.114(e);

(3) Specific degradation, deterioration, and alteration processes have been included in the analyses, taking into consideration their effects on annual dose, and appropriate technical bases have been provided for inclusion or exclusion, in compliance with 10 CFR 63.114(f); and

(4) Adequate technical bases have been provided for models used in the performance assessment, as required by 10 CFR 63.114(g).

2.2.1.3.8.5 References

Altman, W.D., J.P. Donnelly, and J.E. Kennedy. NUREG–1297, "Generic Technical Position on Peer Review for High-Level Nuclear Waste Repositories." Washington, DC: U.S. Nuclear Regulatory Commission. 1988a.

———. NUREG–1298, "Generic Technical Position on Qualification of Existing Data for High-Level Nuclear Waste Repositories." Washington, DC: U.S. Nuclear Regulatory Commission. 1988b.

Kotra, J.P., et al. NUREG–1563, "Branch Technical Position on the Use of Expert Elicitation in the High-Level Radioactive Waste Program." Washington, DC: U.S. Nuclear Regulatory Commission. 1996.

2.2.1.3.9 Radionuclide Transport in the Saturated Zone

To review this model abstraction, consider the degree to which the U.S. Department of Energy relies on radionuclide transport through the saturated zone, to demonstrate compliance. Review this model abstraction considering the risk information evaluated in the "Multiple Barriers" Section (2.2.1.1). For example, if the U.S. Department of Energy relies on the saturated zone to provide significant delay in the transport of radionuclides and/or dilution of concentration to the reasonably maximally exposed individual, then perform a detailed review of this abstraction. If, on the other hand, the U.S. Department of Energy demonstrates this abstraction to have a minor impact on the dose to the reasonably maximally exposed individual, then conduct a simplified review focusing on the bounding assumptions. The review methods and acceptance criteria provided here are for a detailed review. Some of the review methods and acceptance criteria may not be necessary, in a simplified review, for those abstractions that have a minor impact on performance. The demonstration of compliance with the performance objectives is evaluated, using Section 2.2.1.4 of the Yucca Mountain Review Plan.

Review Responsibilities—High-Level Waste Branch and Environmental and Performance Assessment Branch

2.2.1.3.9.1 Areas of Review

This section reviews radionuclide transport in the saturated zone. Reviewers will also evaluate information, required by 10 CFR 63.21(c)(1), (9), (15), and (19), that is relevant to the abstraction of radionuclide transport in the saturated zone.

The staff will evaluate the following parts of the abstraction of radionuclide transport in the saturated zone, using the review methods and acceptance criteria in Sections 2.2.1.3.9.2 and 2.2.1.3.9.3:

(1) Description of the geological, hydrological, and geochemical aspects of radionuclide transport in the saturated zone, and the technical bases the U.S. Department of Energy

provides to support model integration across the total system performance assessment abstractions;

(2) Sufficiency of the data and parameters used to justify the total system performance assessment model abstraction;

(3) Methods the U.S. Department of Energy uses to characterize data uncertainty, and propagate the effects of this uncertainty, through the total system performance assessment model abstraction;

(4) Methods the U.S. Department of Energy uses to characterize model uncertainty, and propagate the effects of this uncertainty, through the total system performance assessment model abstraction;

(5) Approaches the U.S. Department of Energy uses to compare total system performance assessment output with process-level model outputs and empirical studies; and

(6) Use of expert elicitation.

2.2.1.3.9.2 Review Methods

To review the abstraction of radionuclide transport in the saturated zone, recognize that models used in the total system performance assessments may range from highly complex process-level models to simplified models, such as response surfaces or look-up tables. Evaluate model adequacy, regardless of the level of complexity.

Review Method 1 Model Integration

Examine the description of design features, physical phenomena, and couplings, and the description of the geological, hydrological, and geochemical aspects of the saturated zone included in the abstraction of radionuclide transport in the saturated zone that contribute to waste isolation. Assess the adequacy of the technical bases for these descriptions, and for incorporating them in the abstraction of radionuclide transport in the saturated zone.

Evaluate whether the description of aspects of hydrology, geology, geochemistry, design features, physical phenomena, and couplings, that may affect radionuclide transport in the saturated zone, is adequate. Verify that conditions and assumptions used in the abstraction of radionuclide transport in the saturated zone are consistent with the body of data presented in the description.

Examine assumptions, technical bases, data, and models used by the U.S. Department of Energy in the total system performance assessment abstraction of radionuclide transport in the saturated zone for consistency with other related U.S. Department of Energy abstractions. Evaluate whether the descriptions and technical bases provide transparent and traceable support for the abstraction of radionuclide transport in the saturated zone.

Confirm that the U.S. Department of Energy has propagated boundary and initial conditions, used in the abstraction of radionuclide transport in the saturated zone, throughout its abstraction approaches.

Examine how the features, events, and processes, related to radionuclide transport in the saturated zone, have been included in the total system performance assessment abstraction.

Verify that the U.S. Department of Energy follows guidance, such as NUREG–1297 and NUREG–1298 (Altman, et al., 1988a,b), or makes an acceptable case for using alterative approaches to peer review and data qualification.

Review Method 2 Data and Model Justification

Evaluate the sufficiency of the geological, hydrological, and geochemical data used to support parameters used in conceptual models, process-level models, and alternative conceptual models, considered in the abstraction of radionuclide transport in the saturated zone. Assess the sufficiency, transparency, and traceability of the data, used to support the technical bases for features, events, and processes, that have been included in the abstraction of radionuclide transport in the saturated zone.

Verify whether sufficient data have been collected on the characteristics of the geology, hydrology, and geochemistry of the natural system to establish initial and boundary conditions for the abstraction of radionuclide transport in the saturated zone.

Evaluate and confirm that data used to support the U.S. Department of Energy abstraction of radionuclide transport in the saturated zone are based on appropriate techniques, and are adequate for the accompanying sensitivity/uncertainty analyses. Evaluate the need for additional data based on the sensitivity analyses.

Review Method 3 Data Uncertainty

Evaluate the technical bases for parameter values, assumed ranges, probability distributions, and bounding values used in conceptual models, process models, and alternative conceptual models considered in the abstraction of radionuclide transport in the saturated zone. The reviewer should verify that the technical bases support the treatment of uncertainty and variability of these parameters in the performance assessment. If conservative values are used as a method for addressing uncertainty and variability, the reviewer should verify that the conservative values result in conservative estimates of risk and do not cause unintended results (i.e., conservative representation of one aspect of the repository behavior that leads to an overall reduction in risk; inappropriate dilution of the risk estimate by assuming an approach is conservative when a parameter range is increased beyond the supporting data).

Confirm that the U.S. Department of Energy has used flow and transport parameters that are based on techniques that may include laboratory experiments, field measurements, natural analog research, and process-level modeling studies, conducted under conditions relevant to the saturated zone at Yucca Mountain. Examine the results of the U.S. Department of Energy field transport tests, and confirm that the U.S. Department of Energy has provided adequate models.

If criticality in the saturated zone is included in the total system performance assessment, examine the methods and parameters used by the U.S. Department of Energy to calculate the effective neutron multiplication factor. Evaluate the consequences calculated by the U.S. Department of Energy for criticality in the saturated zone.

Verify that the U.S. Department of Energy appropriately establishes possible statistical correlations between parameters. Verify that an adequate technical basis or bounding argument is provided for neglected correlations.

Evaluate the methods used by the U.S. Department of Energy in conducting expert elicitation to define parameter values.

Review Method 4 Model Uncertainty

Evaluate the U.S. Department of Energy alternative conceptual models used in developing the total system performance assessment abstraction for radionuclide transport in the saturated zone. Examine the model parameters, considering available site characterization data, laboratory experiments, field measurements, natural analog research, and process-level modeling studies, and evaluate their consistency.

Where appropriate, use an alternative total system performance assessment model to evaluate selected parts of the U.S. Department of Energy abstraction of radionuclide transport in the saturated zone. Examine the effects of the alternative conceptual model(s) on repository performance, and evaluate how model uncertainties are defined, documented, and assessed.

Evaluate the treatment of conceptual model uncertainty in light of the available site characterization data, laboratory experiments, field measurements, natural analog information and process-level modeling studies. If adoption of a conservative model is used as an approach for addressing conceptual model uncertainty, the reviewer should verify that the selected conceptual model: (i) is conservative relative to alternative conceptual models that are consistent with the available data and current scientific understanding; and (ii) results in conservative estimates of risk and not cause unintended results (i.e., conservative representation of one aspect of the repository behavior that leads to an overall reduction in the risk estimate).

Examine the mathematical models included in the analyses of radionuclide transport in the saturated zone. Examine and evaluate the bases for excluding alternative conceptual models, and the limitations and uncertainties of the chosen model.

Review Method 5 Model Support

Evaluate the output from the abstraction of radionuclide transport in the saturated zone, and compare the results with an appropriate combination of site characterization data, process modeling, laboratory testing, field measurements, and natural analog research. Evaluate the sensitivity analyses used to support the abstraction of radionuclide transport in the saturated zone in the total system performance assessment.

Use detailed models of geochemical, hydrological, and geological processes to evaluate the abstraction of radionuclide transport in the saturated zone. If practical, use an alternative total system performance assessment model to evaluate selected parts of the U.S. Department of Energy abstraction of radionuclide transport in the saturated zone, and evaluate the effects on repository performance. Compare results of the U.S. Department of Energy abstraction with approximations shown to be appropriate for closely analogous natural systems or experimental systems.

Examine the procedures used by the U.S. Department of Energy to develop and test its mathematical and numerical models.

As appropriate, use an alternative total system performance assessment model to evaluate the U.S. Department of Energy sensitivity or bounding analyses, and confirm that the U.S. Department of Energy has used ranges consistent with available site characterization data, field and laboratory tests, and natural analog research.

2.2.1.3.9.3 Acceptance Criteria

The following acceptance criteria are based on meeting the requirements of 10 CFR 63.114(a)–(c) and (e)–(g), relating to the radionuclide transport in the saturated zone model abstraction. U.S. Nuclear Regulatory Commission staff should apply the following acceptance criteria, according to the level of importance established in the U.S. Department of Energy risk-informed license application.

Acceptance Criterion 1 System Description and Model Integration Are Adequate.

(1) Total system performance assessment adequately incorporates important design features, physical phenomena, and couplings, and uses consistent and appropriate assumptions throughout the radionuclide transport in the saturated zone abstraction process;

(2) The description of the aspects of hydrology, geology, geochemistry, design features, physical phenomena, and couplings, that may affect radionuclide transport in the saturated zone, is adequate. For example, the description includes changes in transport properties in the saturated zone, from water-rock interaction. Conditions and assumptions in the abstraction of radionuclide transport in the saturated zone are readily identified, and consistent with the body of data presented in the description;

(3) The abstraction of radionuclide transport in the saturated zone uses assumptions, technical bases, data, and models that are appropriate and consistent with other related U.S. Department of Energy abstractions. For example, assumptions used for radionuclide transport in the saturated zone are consistent with the total system performance assessment abstractions of radionuclide release rates and solubility limits, and flow paths in the saturated zone (Sections 2.2.1.3.4 and 2.2.1.3.8 of the Yucca Mountain Review Plan, respectively). The descriptions and technical bases provide transparent and traceable support for the abstraction of radionuclide transport in the saturated zone;

(4) Boundary and initial conditions used in the abstraction of radionuclide transport in the saturated zone are propagated throughout its abstraction approaches. For example, the conditions and assumptions used to generate transport parameter values are consistent with other geological, hydrological, and geochemical conditions in the total system performance assessment abstraction of the saturated zone;

(5) Sufficient data and technical bases for the inclusion of features, events, and processes related to radionuclide transport in the saturated zone in the total system performance assessment abstraction are provided; and

(6) Guidance in NUREG–1297 and NUREG–1298 (Altman, et al., 1988a,b), or other acceptable approaches for peer review and data qualification is followed.

Acceptance Criterion 2 Data Are Sufficient for Model Justification.

(1) Geological, hydrological, and geochemical values used in the license application are adequately justified (e.g., flow path lengths, sorption coefficients, retardation factors, colloid concentrations, etc.). Adequate descriptions of how the data were used, interpreted, and appropriately synthesized into the parameters are provided;

(2) Sufficient data have been collected on the characteristics of the natural system to establish initial and boundary conditions for the total system performance assessment abstraction of radionuclide transport in the saturated zone; and

(3) Data on the geology, hydrology, and geochemistry of the saturated zone, including the influence of structural features, fracture distributions, fracture properties, and stratigraphy, used in the total system performance assessment abstraction, are based on appropriate techniques. These techniques may include laboratory experiments, site-specific field measurements, natural analog research, and process-level modeling studies. As appropriate, sensitivity or uncertainty analyses used to support the U.S. Department of Energy total system performance assessment abstraction are adequate to determine the possible need for additional data.

Acceptance Criterion 3 Data Uncertainty Is Characterized and Propagated Through the Model Abstraction.

(1) Models use parameter values, assumed ranges, probability distributions, and bounding assumptions that are technically defensible, reasonably account for uncertainties and variabilities, and do not result in an under-representation of the risk estimate;

(2) For those radionuclides where the total system performance assessment abstraction indicates that transport in fractures and matrix in the saturated zone is important to waste isolation: (i) estimated flow and transport parameters are appropriate and valid, based on techniques that may include laboratory experiments, field measurements, natural analog research, and process-level modeling studies conducted under conditions relevant to the saturated zone at Yucca Mountain; and (ii) models are demonstrated to adequately predict field transport test results. For example, if a sorption coefficient approach is used, the assumptions implicit in that approach are validated;

(3) If criticality in the saturated zone is included in the total system performance assessment, an appropriate range of input parameters for calculating the effective neutron multiplication factor is used. The effects on performance of criticality in the saturated zone are adequately evaluated;

(4) Parameter values for processes, such as matrix diffusion, dispersion, and ground-water mixing, are based on reasonable assumptions about climate, aquifer properties, and ground-water volumetric fluxes (Section 2.2.1.3.8 of the Yucca Mountain Review Plan);

(5) Uncertainty is adequately represented in parameter development for conceptual models, process-level models, and alternative conceptual models considered in developing the abstraction of radionuclide transport in the saturated zone. This may be done either through sensitivity analyses or use of conservative limits; and

(6) Where sufficient data do not exist, the definition of parameter values and conceptual models is based on appropriate use of other sources, such as expert elicitation conducted in accordance with NUREG–1563 (Kotra, et al., 1996). If other approaches are used, the U.S. Department of Energy adequately justifies their use.

Acceptance Criterion 4 Model Uncertainty Is Characterized and Propagated Through the Model Abstraction.

(1) Alternative modeling approaches of features, events, and processes are considered and are consistent with available data and current scientific understanding, and the results and limitations are appropriately considered in the abstraction;

(2) Conceptual model uncertainties are adequately defined and documented, and effects on conclusions regarding performance are properly assessed;

(3) Consideration of conceptual model uncertainty is consistent with available site characterization data, laboratory experiments, field measurements, natural analog information and process-level modeling studies; and the treatment of conceptual model uncertainty does not result in an under-representation of the risk estimate; and

(4) Appropriate alternative modeling approaches are consistent with available data and current scientific knowledge, and appropriately consider their results and limitations using tests and analyses that are sensitive to the processes modeled. For example, for radionuclide transport through fractures, the U.S. Department of Energy adequately considers alternative modeling approaches to develop its understanding of fracture distributions and ranges of fracture flow and transport properties in the saturated zone.

Acceptance Criterion 5 Model Abstraction Output Is Supported by Objective Comparisons.

(1) The models implemented in this total system performance assessment abstraction provide results consistent with output from detailed process-level models and/or empirical observations (laboratory and field testings and/or natural analogs);

(2) Outputs of radionuclide transport in the saturated zone abstractions reasonably produce or bound the results of corresponding process-level models, empirical observations, or both. The U.S. Department of Energy-abstracted models for radionuclide transport in the saturated zone are based on the same hydrological, geological, and geochemical assumptions and approximations shown to be appropriate for closely analogous natural systems or laboratory experimental systems;

(3) Well-documented procedures that have been accepted by the scientific community to construct and test the mathematical and numerical models are used to simulate radionuclide transport through the saturated zone; and

(4) Sensitivity analyses or bounding analyses are provided, to support the total system performance assessment abstraction of radionuclide transport in the saturated zone, that cover ranges consistent with site data, field or laboratory experiments and tests, and natural analog research.

2.2.1.3.9.4 Evaluation Findings

If the license application provides sufficient information and the regulatory acceptance criteria in Section 2.2.1.3.9.3 are appropriately satisfied, the staff concludes that this portion of the staff evaluation is acceptable. The reviewer writes material suitable for inclusion in the safety evaluation report prepared for the entire application. The report includes a summary statement of what was reviewed and why the reviewer finds the submittal acceptable. The staff can document the review as follows.

U.S. Nuclear Regulatory Commission staff has reviewed the Safety Analysis Report and other information submitted in support of the license application, relevant to radionuclide transport in the saturated zone ,and has found, with reasonable expectation, that the requirements of 10 CFR 63.114 are satisfied for model abstraction in this section. Technical requirements for conducting a performance assessment in the area of radionuclide transport in the saturated zone have been met. In particular, the U.S. Nuclear Regulatory Commission staff found that:

(1) Appropriate data from the site and surrounding region, uncertainties and variabilities in parameter values, and alternative conceptual models have been used in the analyses, in compliance with 10 CFR 63.114(a)–(c);

(2) Specific features, events, and processes have been included in the analyses, and appropriate technical bases have been provided, for inclusion or exclusion, in compliance with 10 CFR 63.114(e);

(3) Specific degradation, deterioration, and alteration processes have been included in the analyses, taking into consideration their effects on annual dose, and appropriate technical bases have been provided for inclusion or exclusion, in compliance with 10 CFR 63.114(f); and

(4) Adequate technical bases have been provided for models used in the performance assessment, as required by 10 CFR 63.114(g).

2.2.1.3.9.5 References

Altman, W.D., J.P. Donnelly, and J.E. Kennedy. NUREG–1297, "Generic Technical Position on Peer-Review for High-Level Nuclear Waste Repositories." Washington, DC: U.S. Nuclear Regulatory Commission. 1988a.

———. NUREG–1298, "Generic Technical Position on Qualification of Existing Data for High-Level Nuclear Waste Repositories." Washington, DC: U.S. Nuclear Regulatory Commission. 1988b.

Kotra, J.P., et al. NUREG–1563, "Branch Technical Position on the Use of Expert Elicitation in the High-Level Radioactive Waste Program." Washington, DC: U.S. Nuclear Regulatory Commission. 1996.

2.2.1.3.10 Volcanic Disruption of Waste Packages

To review this model abstraction, consider the degree to which the U.S. Department of Energy relies on volcanic disruption of waste packages to demonstrate compliance. Review this model abstraction, considering the risk information evaluated in the "Multiple Barriers" (Section 2.2.1.1). For example, if the U.S. Department of Energy relies on waste package integrity to have a significant effect on dose to the reasonably maximally exposed individual, then perform a detailed review of this abstraction. If, on the other hand, the U.S. Department of Energy demonstrates this abstraction to have a minor impact on dose to the reasonably maximally exposed individual, then conduct a simplified review focusing on the bounding assumptions. The review methods and acceptance criteria provided here are for a detailed review. Some of the review methods and acceptance criteria may not be necessary, in a simplified review, for those abstractions that have a minor impact on performance. The demonstration of compliance with the performance objectives is evaluated, using Section 2.2.1.4 of the Yucca Mountain Review Plan.

Review Responsibilities—High-Level Waste Branch and Environmental and Performance Assessment Branch

2.2.1.3.10.1 Areas of Review

This section reviews volcanic disruption of waste packages. Reviewers will also evaluate information, required by 10 CFR 63.21(c)(1), (9), (15), and (19), that is relevant to the abstraction of volcanic disruption of waste packages.

The staff will evaluate the following parts of the abstraction of volcanic disruption of waste packages, using the review methods and acceptance criteria in Sections 2.2.1.3.10.2 and 2.2.1.3.10.3:

(1) Description of the geological, hydrological, geochemical and design aspects of volcanic disruption of waste packages, and the technical bases the U.S. Department of Energy provides to support model integration across the total system performance assessment abstractions;

(2) Sufficiency of the data and parameters used to justify the total system performance assessment model abstraction;

(3) Methods the U.S. Department of Energy uses to characterize data uncertainty, and propagate the effects of this uncertainty, through the total system performance assessment model abstraction;

(4) Methods the U.S. Department of Energy uses to characterize model uncertainty, and propagate the effects of this uncertainty, through the total system performance assessment model abstraction;

(5) Approaches the U.S. Department of Energy uses to compare total system performance assessment output with process-level model outputs and empirical studies; and

(6) Use of expert elicitation.

2.2.1.3.10.2 Review Methods

To review the abstraction of volcanic disruption of waste packages, recognize that models used in the total system performance assessment may range from highly complex process-level models to simplified models, such as response surfaces or look-up tables. Evaluate model adequacy, regardless of the level of complexity.

Review Method 1 Model Integration

Examine the description of design features, physical phenomena, and couplings, and the description of the geology, geophysics, and geochemistry included in the abstraction of volcanic disruption of waste packages. Assess the adequacy and consistency of the technical bases for these descriptions, and for incorporating them into the total system performance assessment abstraction for volcanic disruption of waste packages. Confirm that models and assumptions used to evaluate volcanic disruption of waste packages are consistent with models and assumptions used elsewhere in the license application.

Verify that models used to assess volcanic disruption of waste packages are consistent with physical processes generally interpreted from igneous features in the Yucca Mountain region. Verify that models of active igneous processes are consistent with processes generally observed at active igneous features.

Evaluate the technical bases used to assess the effects of interactions between engineered repository systems and igneous systems.

Verify that U.S. Department of Energy reviews follow guidance, such as NUREG–1297 and NUREG–1298 (Altman, et al., 1988a,b), or make an acceptable case for using alternative approaches to peer review and data qualification.

Review Method 2 Data and Model Justification

Evaluate the sufficiency of the geological, geophysical, and geochemical data used to support parameters used in conceptual models, process-level models, and alternative conceptual models considered in the total system performance assessment abstraction of volcanic disruption of waste packages.

Confirm that the technical bases for these data are adequately justified, and that data used to model processes affecting volcanic disruption of waste packages are derived, to the extent possible, from adequately documented techniques. Such techniques may include site-specific field measurements, natural analog investigations, and laboratory experiments.

Confirm that sufficient data are available to integrate features, events, and processes relevant to volcanic disruption of waste packages into process-level models. Confirm that appropriate interrelationships and correlations between relevant features, events, and processes are adequately considered in resulting model abstractions.

Evaluate the methods used by the U.S. Department of Energy in conducting expert elicitation to define parameter values.

Review Method 3 Data Uncertainty

Examine the technical bases for parameter values, assumed ranges, probability distributions, and bounding values used in conceptual models, process-level models, and alternative conceptual models, considered in the volcanic disruption of waste packages abstraction. The reviewer should verify that the technical bases support the treatment of uncertainty and variability of these parameters in the performance assessment. If conservative values are used as a method for addressing uncertainty and variability, the reviewer should verify that the conservative values result in conservative estimates of risk and do not cause unintended results (i.e., conservative representation of one aspect of the repository behavior that leads to an overall reduction in risk; inappropriate dilution of the risk estimate by assuming an approach is conservative when a parameter range is increased beyond the supporting data).

Verify that the U.S. Department of Energy appropriately establishes possible statistical correlations between parameters. Verify that an adequate technical basis or bounding argument is provided for neglected correlations.

Evaluate the methods used by the U.S. Department of Energy in conducting expert elicitation to define parameter values.

Review Method 4 Model Uncertainty

Evaluate the alternative conceptual models used in developing the total system performance assessment abstraction for volcanic disruption of waste packages. Examine the model parameters, considering available site characterization data, laboratory experiments, field measurements, natural analog research, and process-level modeling studies, and evaluate their consistency.

Evaluate the treatment of conceptual model uncertainty in light of the available site characterization data, laboratory experiments, field measurements, natural analog information and process-level modeling studies. If adoption of a conservative model is used as an approach for addressing conceptual model uncertainty, the reviewer should verify that the selected conceptual model: (i) is conservative relative to alternative conceptual models that are consistent with the available data and current scientific understanding; and (ii) results in conservative estimates of risk and not cause unintended results (i.e., conservative representation of one aspect of the repository behavior that leads to an overall reduction in the risk estimate).

Review Method 5 Model Support

Evaluate the output from the abstraction of volcanic disruption of waste packages, and compare the results with an appropriate combination of site characterization data, detailed process-level modeling, laboratory testing, field measurements, and natural analog research.

Confirm that inconsistencies between abstracted models and comparative data are explained and quantified. Confirm that the resulting uncertainty is accounted for in the model results.

2.2.1.3.10.3 Acceptance Criteria

The following acceptance criteria are based on meeting the requirements of 10 CFR 63.114(a)–(c) and (e)–(g), relating to the volcanic disruption of waste package model abstraction. U.S. Nuclear Regulatory Commission staff should apply the following acceptance criteria, according to the level of importance established in the U.S. Department of Energy risk-informed license application.

Acceptance Criterion 1 System Description and Model Integration Are Adequate.

(1) Total system performance assessment adequately incorporates important design features, physical phenomena, and couplings, and uses consistent and appropriate assumptions throughout the volcanic disruption of the waste package abstraction process;

(2) Models used to assess volcanic disruption of waste packages are consistent with physical processes generally interpreted from igneous features in the Yucca Mountain region and/or observed at active igneous systems;

(3) Models account for changes in igneous processes, that may occur from interactions with engineered repository systems; and

(4) Guidance in NUREG–1297 and NUREG–1298 (Altman, et al., 1988a,b), or other acceptable approaches is followed.

Acceptance Criterion 2 Data Are Sufficient for Model Justification.

(1) Parameter values used in the license application to evaluate volcanic disruption of waste packages are sufficient and adequately justified. Adequate descriptions of how the data were used, interpreted, and appropriately synthesized into the parameters are provided;

(2) Data used to model processes affecting volcanic disruption of waste packages are derived from appropriate techniques. These techniques may include site-specific field measurements, natural analog investigations, and laboratory experiments;

(3) Sufficient data are available to integrate features, events, and processes, relevant to volcanic disruption of waste packages into process-level models, including determination of appropriate interrelationships and parameter correlations; and

(4) Where sufficient data do not exist, the definition of parameter values and associated conceptual models is based on appropriate use of expert elicitation, conducted in accordance with NUREG–1563 (Kotra, et al., 1996). If other approaches are used, the U.S. Department of Energy adequately justifies their use.

Acceptance Criterion 3 Data Uncertainty Is Characterized and Propagated Through the Model Abstraction.

(1) Models use parameter values, assumed ranges, probability distributions, and bounding assumptions that are technically defensible, and reasonably account for uncertainties and variabilities, and do not result in an under-representation of the risk estimate;

(2) Parameter uncertainty accounts quantitatively for the uncertainty in parameter values observed in site data and the available literature (i.e., data precision), and the uncertainty in abstracting parameter values to process-level models (i.e., data accuracy); and

(3) Where sufficient data do not exist, the definition of parameter values and associated uncertainty is based on appropriate use of expert elicitation, conducted in accordance with NUREG–1563 (Kotra, et al., 1996). If other approaches are used, the U.S. Department of Energy adequately justifies their use.

Acceptance Criterion 4 Model Uncertainty Is Characterized and Propagated Through the Model Abstraction.

(1) Alternative modeling approaches to volcanic disruption of the waste package are considered and are consistent with available data and current scientific understandings, and the results and limitations are appropriately considered in the abstraction;

(2) Uncertainties in abstracted models are adequately defined and documented, and effects of these uncertainties are assessed in the total system performance assessment; and

(3) Consideration of conceptual model uncertainty is consistent with available site characterization data, laboratory experiments, field measurements, natural analog

information and process-level modeling studies; and the treatment of conceptual model uncertainty does not result in an under-representation of the risk estimate.

Acceptance Criterion 5 Model Abstraction Output Is Supported by Objective Comparisons.

(1) Models implemented in the volcanic disruption of waste packages abstraction provide results consistent with output from detailed process-level models and/or empirical observations (laboratory and field testings and/or natural analogs); and

(2) Inconsistencies between abstracted models and comparative data are documented, explained, and quantified. The resulting uncertainty is accounted for in the model results.

2.2.1.3.10.4 Evaluation Findings

If the license application provides sufficient information and the regulatory acceptance criteria in Section 2.2.1.3.10.3 are appropriately satisfied, the staff concludes that this portion of the staff evaluation is acceptable. The reviewer writes material suitable for inclusion in the safety evaluation report prepared for the entire application. The report includes a summary statement of what was reviewed and why the reviewer finds the submittal acceptable. The staff can document the review as follows.

U.S. Nuclear Regulatory Commission staff has reviewed the Safety Analysis Report and other information submitted in support of the license application, relevant to volcanic disruption of waste packages, and has found, with reasonable expectation, that the requirements of 10 CFR 63.114 are satisfied in this section. Technical requirements for conducting a performance assessment in the area of volcanic disruption of waste packages have been met. In particular, the U.S. Nuclear Regulatory Commission staff found that, in regard to volcanic disruption of the waste package:

(1) Appropriate data from the site and surrounding region, uncertainties and variabilities in parameter values, and alternative conceptual models have been used in the analyses, in compliance with 10 CFR 63.114(a)–(c);

(2) Specific features, events, and processes have been included in the analyses, and appropriate technical bases have been provided for inclusion or exclusion, in compliance with 10 CFR 63.114(e);

(3) Specific degradation, deterioration, and alteration processes have been included in the analyses, taking into consideration their effects on annual dose, and appropriate technical bases have been provided for inclusion or exclusion, in compliance with 10 CFR 63.114(f); and

(4) Adequate technical bases have been provided for models used in the performance assessment, as required by 10 CFR 63.114(g).

2.2.1.3.10.5 References

Altman, W.D., J.P. Donnelly, and J.E. Kennedy. NUREG–1297, "Generic Technical Position on Peer-Review for High-Level Nuclear Waste Repositories." Washington, DC: U.S. Nuclear Regulatory Commission. 1988a.

———. NUREG–1298, "Generic Technical Position on Qualification of Existing Data for High-Level Nuclear Waste Repositories." Washington, DC: U.S. Nuclear Regulatory Commission. 1988b.

Kotra, J.P., et al. NUREG–1563, "Branch Technical Position on the Use of Expert Elicitation in the High-Level Radioactive Waste Program." Washington, DC: U.S. Nuclear Regulatory Commission. 1996.

2.2.1.3.11 Airborne Transport of Radionuclides

To review this model abstraction, consider the degree to which the U.S. Department of Energy relies on airborne transport of radionuclides, to demonstrate compliance. Review this model abstraction, considering the risk information evaluated in the "Multiple Barriers" Section(2.2.1.1). For example, if the U.S. Department of Energy relies on waste package integrity to provide significant delay or dilution in the transport of radionuclides to the reasonably maximally exposed individual, then perform a detailed review of this abstraction. If, on the other hand, the U.S. Department of Energy demonstrates this abstraction to have a minor impact on the delay of radionuclides to the reasonably maximally exposed individual, then conduct a simplified review focusing on the bounding assumptions. The review methods and acceptance criteria provided here are for a detailed review. Some of the review methods and acceptance criteria may not be necessary, in a simplified review, for those abstractions that have a minor impact on performance. The demonstration of compliance with the performance objectives is evaluated, using Section 2.2.1.4 of the Yucca Mountain Review Plan.

Review Responsibilities—High-Level Waste Branch and Environmental and Performance Assessment Branch

2.2.1.3.11.1 Areas of Review

This section reviews airborne transport of radionuclides. Reviewers will also evaluate information, required by 10 CFR 63.21(c)(1), (9), (15), and (19), that is relevant to the abstraction of airborne transport of radionuclides.

The staff will evaluate the following parts of the abstraction of airborne transport of radionuclides, using the review methods and acceptance criteria in Sections 2.2.1.3.11.2 and 2.2.1.3.11.3:

(1) Description of the geological, hydrological, geochemical, and meteorological aspects of airborne transport of radionuclides, and the technical bases the U.S. Department of Energy provides to support model integration across the total system performance assessment abstractions;

(2) Sufficiency of the data and parameters used to justify the total system performance assessment model abstraction;

(3) Methods the U.S. Department of Energy uses to characterize data uncertainty, and propagate the effects of this uncertainty, through the total system performance assessment model abstraction;

(4) Methods the U.S. Department of Energy uses to characterize model uncertainty, and propagate the effects of this uncertainty, through the total system performance assessment model abstraction;

(5) Approaches the U.S. Department of Energy uses to compare total system performance assessment output to process-level model outputs and empirical studies; and

(6) Use of expert elicitation.

2.2.1.3.11.2 Review Methods

To review the abstraction of airborne transport of radionuclides, recognize that models used in the total system performance assessment may range from highly complex process-level models to simplified models, such as response surfaces or look-up tables. Evaluate model adequacy, regardless of the level of complexity.

Review Method 1 Model Integration

Examine the description of design features, physical phenomena, and couplings, and the description of the geology, geophysics, geochemistry, and meteorological conditions included in the abstraction of airborne transport of radionuclides. Assess the adequacy and consistency of the technical bases for these descriptions, and for incorporating them into the total system performance assessment abstraction for airborne transport of radionuclides. Confirm that models and assumptions used to evaluate airborne transport of radionuclides are consistent with models and assumptions used elsewhere in the license application.

Verify that models used to assess airborne transport of radionuclides are consistent with physical processes generally interpreted from igneous features in the Yucca Mountain region. Verify that models of active igneous processes are consistent with processes generally observed at active igneous features.

Evaluate the technical bases used to assess the effects of engineered repository systems on the consequences of igneous processes.

Verify that the U.S. Department of Energy reviews follow guidance, such as NUREG–1297 and NUREG–1298 (Altman, et al., 1988a,b), or make an acceptable case for using alternative approaches.

Review Method 2 Data and Model Justification

Evaluate the sufficiency of the geological, geophysical, geochemical, and meteorological data used to support parameters, used in conceptual models, process-level models, and alternative conceptual models, considered in the abstraction of airborne transport of radionuclides.

Verify that the technical bases for these data are adequately justified, and that data used to model processes affecting airborne transport of radionuclides are derived from adequately documented techniques. Such techniques may include site-specific field measurements, natural analog investigations, and laboratory experiments.

Confirm that sufficient data are available to integrate features, events, and processes, relevant to airborne transport of radionuclides into process-level models. Verify that appropriate interrelationships and correlations between relevant features, events, and processes are adequately considered in resulting model abstractions.

Evaluate the methods used by the U.S. Department of Energy in conducting expert elicitation to define parameter values.

Review Method 3 Data Uncertainty

Examine the technical bases for parameter values, assumed ranges, probability distributions, and bounding values used in conceptual models, process models, and alternative conceptual models, considered in the abstraction of airborne transport of radionuclides. The reviewer should verify that the technical bases support the treatment of uncertainty and variability of these parameters in the performance assessment. If conservative values are used as a method for addressing uncertainty and variability, the reviewer should verify that the conservative values result in conservative estimates of risk and do not cause unintended results (i.e., conservative representation of one aspect of the repository behavior that leads to an overall reduction in risk; inappropriate dilution of the risk estimate by assuming an approach is conservative when a parameter range is increased beyond the supporting data).

Verify that the U.S. Department of Energy appropriately establishes possible statistical correlations between parameters. Verify that an adequate technical basis or bounding argument is provided for neglected correlations.

Evaluate the methods used by the U.S. Department of Energy in conducting expert elicitation to define parameter values.

Review Method 4 Model Uncertainty

Evaluate the alternative conceptual models used in developing the abstraction for airborne transport of radionuclides. Examine the model parameters considering available site characterization data, laboratory experiments, field measurements, natural analog research, and process-level modeling studies, and evaluate their consistency.

Evaluate the treatment of conceptual model uncertainty in light of the available site characterization data, laboratory experiments, field measurements, natural analog information and process-level modeling studies. If adoption of a conservative model is used as an approach for addressing conceptual model uncertainty, the reviewer should verify that the selected conceptual model: (i) is conservative relative to alternative conceptual models that are consistent with the available data and current scientific understanding; and (ii) results in conservative estimates of risk and not cause unintended results (i.e., conservative representation of one aspect of the repository behavior that leads to an overall reduction in the risk estimate).

Review Method 5 Model Support

Evaluate the output from the abstraction of airborne transport of radionuclides, and compare the results with an appropriate combination of site characterization data, detailed process-level modeling, laboratory testing, field measurements, and natural analog research.

Confirm that inconsistencies between abstracted models and comparative data are explained and quantified. Confirm that the resulting uncertainty is accounted for in the model results.

2.2.1.3.11.3 Acceptance Criteria

The following acceptance criteria are based on meeting the requirements of 10 CFR 63.114(a)–(c) and (e)–(g), relating to the airborne transport of radionuclide model abstraction. U.S. Nuclear Regulatory Commission staff should apply the following acceptance criteria, according to the level of importance established in the U.S. Department of Energy risk-informed license application.

Acceptance Criterion 1 System Description and Model Integration Are Adequate.

(1) Total system performance assessment adequately incorporates important design features, physical phenomena, and couplings, and uses consistent and appropriate assumptions throughout the airborne transport of radionuclides abstraction process;

(2) Models used to assess airborne transport of radionuclides are consistent with physical processes generally interpreted from igneous features in the Yucca Mountain region and/or observed at active igneous systems;

(3) Models account for changes in igneous processes that may occur from interactions with engineered repository systems; and

(4) Guidance in NUREG–1297 and NUREG–1298 (Altman, et al., 1988a,b), or in other acceptable approaches for peer review and data qualification is followed.

Acceptance Criterion 2 Data Are Sufficient for Model Justification.

(1) Parameter values used in the license application to evaluate airborne transport of radionuclides are sufficient and adequately justified. Adequate descriptions of how the data were used, interpreted, and appropriately synthesized into the parameters are provided;

(2) Data used to model processes affecting airborne transport of radionuclides are derived from appropriate techniques. These techniques may include site-specific field measurements, natural analog investigations, and laboratory experiments;

(3) Sufficient data are available to integrate features, events, and processes, relevant to airborne transport of radionuclides into process-level models, including determination of appropriate interrelationships and parameter correlations; and

(4) Where sufficient data do not exist, the definition of parameter values and associated conceptual models is based on appropriate use of expert elicitation conducted, in accordance with NUREG–1563 (Kotra, et al., 1996). If other approaches are used, the U.S. Department of Energy adequately justifies their use.

Acceptance Criterion 3 Data Uncertainty Is Characterized and Propagated Through the Model Abstraction.

(1) Models use parameter values, assumed ranges, probability distributions, and bounding assumptions that are technically defensible, reasonably account for uncertainties and variabilities, and do not result in an under-representation of the risk estimate;

(2) Parameter uncertainty accounts quantitatively for the uncertainty in parameter values derived from site data and the available literature (i.e., data precision), and the uncertainty introduced by model abstraction (i.e., data accuracy); and

(3) Where sufficient data do not exist, the definition of parameter values and associated uncertainty is based on appropriate use of expert elicitation conducted, in accordance with NUREG–1563 (Kotra, et al., 1996). If other approaches are used, the U.S. Department of Energy adequately justifies their use.

Acceptance Criterion 4 Model Uncertainty Is Characterized and Propagated Through the Model Abstraction.

(1) Alternative modeling approaches to airborne transport of radionuclides are considered and are consistent with available data and current scientific understandings, and the results and limitations are appropriately considered in the abstraction;

(2) Uncertainties in abstracted models are adequately defined and documented, and effects of these uncertainties are assessed in the total system performance assessment; and

(3) Consideration of conceptual model uncertainty is consistent with available site characterization data, laboratory experiments, field measurements, natural analog information and process-level modeling studies; and the treatment of conceptual model uncertainty does not result in an under-representation of the risk estimate.

Acceptance Criterion 5 Model Abstraction Output Is Supported by Objective Comparisons.

(1) Models implemented in the airborne transport of radionuclide abstraction provide results consistent with output from detailed process-level models and/or empirical observations (laboratory and field testings and/or natural analogs); and

(2) Inconsistencies between abstracted models and comparative data are documented, explained, and quantified. The resulting uncertainty is accounted for in the model results.

2.2.1.3.11.4 Evaluation Findings

If the license application provides sufficient information and the regulatory acceptance criteria in Section 2.2.1.3.11.3 are appropriately satisfied, the staff concludes that this portion of the staff evaluation is acceptable. The reviewer writes material suitable for inclusion in the safety evaluation report prepared for the entire application. The report includes a summary statement of what was reviewed and why the reviewer finds the submittal acceptable. The staff can document the review as follows.

U.S. Nuclear Regulatory Commission staff has reviewed the Safety Analysis Report and other information submitted in support of the license application, relevant to the airborne transport of radionuclides and has found, with reasonable expectation, that the requirements of 10 CFR 63.114 are satisfied for model abstraction in this section. Technical requirements for conducting a performance assessment in the area of airborne transport of radionuclides have been met. In particular, the U.S. Nuclear Regulatory Commission staff found that:

(1) Appropriate data from the site and surrounding region, uncertainties and variabilities in parameter values, and alternative conceptual models have been used in the analyses, in compliance with 10 CFR 63.114(a)–(c);

(2) Specific features, events, and processes have been included in the analyses, and appropriate technical bases have been provided for inclusion or exclusion, in compliance with 10 CFR 63.114(e);

(3) Specific degradation, deterioration, and alteration processes have been included in the analyses, taking into consideration their effect on annual dose, and appropriate technical bases have been provided for inclusion or exclusion, in compliance with 10 CFR 63.114(f); and

(4) Adequate technical bases have been provided for models used in the performance assessment, as required by 10 CFR 63.114(g).

2.2.1.3.11.5 References

Altman, W.D., J.P. Donnelly, and J.E. Kennedy. NUREG–1297, "Generic Technical Position on Peer Review for High-Level Nuclear Waste Repositories." Washington, DC: U.S. Nuclear Regulatory Commission. 1988a.

———. NUREG–1298, "Generic Technical Position on Qualification of Existing Data for High-Level Nuclear Waste Repositories." Washington, DC: U.S. Nuclear Regulatory Commission. 1988b.

Kotra, J.P., et al. NUREG–1563, "Branch Technical Position on the Use of Expert Elicitation in the High-Level Radioactive Waste Program." Washington, DC: U.S. Nuclear Regulatory Commission. 1996.

2.2.1.3.12 Concentration of Radionuclides in Ground Water

To review this model abstraction, consider the degree to which estimating the concentration of radionuclides in the annual water demand of ground water affects the U.S. Department of Energy demonstration of compliance. Review this model abstraction, considering the risk information evaluated in the "Multiple Barriers" (Section 2.2.1.1). For example, if the U.S. Department of Energy indicates that a significant portion of the radionuclides that are projected to cross the accessible environment boundary annually will not be included in the determination of the average concentration of radionuclides in 3,000 acre-feet [3.715×10^9 liters], then perform a detailed review of this abstraction. If the U.S. Department of Energy demonstrates this abstraction to have a minor impact on the dose to the reasonably maximally exposed individual, or if the U.S. Department of Energy assumes that all radionuclides that reach the location of the reasonably maximally exposed individual in a given year are included in pumping wells in the annual water demand of 3,000 acre-feet [3.715×10^9 liters], then conduct a simplified review focusing on the bounding assumptions. The review methods and acceptance criteria provided here are for a detailed review. Some of the review methods and acceptance criteria may not be necessary, in a simplified review, for those abstractions that have a minor impact on performance. The demonstration of compliance with the individual protection standard is evaluated using Section 2.2.1.4.1, and compliance with the ground-water protection standard is evaluated using Section 2.2.1.4.3 of the Yucca Mountain Review Plan.

Review Responsibilities—High-Level Waste Branch and Environmental and Performance Assessment Branch

2.2.1.3.12.1 Areas of Review

This section reviews the concentration of radionuclides in ground water abstraction. Reviewers will also evaluate information, required by 10 CFR 63.21(c)(1), (9), (15), and (19), that is relevant to the abstraction.

The staff will evaluate the following parts of the abstraction of the concentration of radionuclides in ground water, using the review methods and acceptance criteria in Sections 2.2.1.3.12.2 and 2.2.1.3.12.3.

(1) Description of the geological and hydrological aspects of the concentration of radionuclides in ground water, and the technical bases the U.S. Department of Energy provides to support model integration across the total system performance assessment abstractions;

(2) Sufficiency of the data and parameters used to justify the total system performance assessment model abstraction;

(3) Methods the U.S. Department of Energy uses to characterize data uncertainty, and propagate the effects of this uncertainty, through the total system performance assessment model abstraction;

(4) Methods the U.S. Department of Energy uses to characterize model uncertainty, and propagate the effects of this uncertainty, through the total system performance assessment model abstraction;

(5) Approaches the U.S. Department of Energy uses to compare total system performance assessment output to process-level model outputs and empirical studies; and

(6) Use of expert elicitation.

2.2.1.3.12.2 Review Methods

To review the abstraction of the concentration of radionuclides in ground water, recognize that models used in the total system performance assessments may range from highly complex process-level models to simplified models, such as response surfaces or look-up tables. Evaluate model adequacy, regardless of the level of complexity.

Review Method 1 Model Integration

Examine the description of design features, physical phenomena, and couplings, and the description of the geological, hydrological, and geochemical aspects of the abstraction of the concentration of radionuclides in ground water that contribute to repository performance.

Assess whether the technical bases for the determination of what portion of the radionuclides that are projected to cross the accessible environment boundary annually is to be included in the calculation of the average concentration of radionuclides in 3,000 acre-feet [3.715×10^9 liters] are adequate.

Evaluate whether the description of the aspects of hydrology and geology that may affect the concentration of radionuclides in ground water is adequate. Evaluate whether the descriptions provide transparent and traceable support for the abstraction.

Examine the assumptions, technical bases, data, and models used by the U.S. Department of Energy in the total system performance assessment abstraction of the concentration of radionuclides in ground water to verify that they are appropriate and consistent with other related U.S. Department of Energy abstractions.

Examine how the features, events, and processes, related to the concentration of radionuclides in ground water have been included in the total system performance assessment abstraction.

Verify that the U.S. Department of Energy has followed the guidance in NUREG–1297 and NUREG–1298 (Altman, et al., 1988a,b), or makes an acceptable case for using alternative approaches.

Review Method 2 Data and Model Justification

Evaluate whether sufficient justification has been provided for climatological and hydrological values used in the license application, and whether the description of how the data are used, interpreted, and appropriately synthesized into the parameters is sufficiently transparent and traceable.

Evaluate whether sufficient data have been used to support the development of conceptual models used in the abstraction of the concentration of radionuclides in ground water as well as the parameters used for each of these models. Verify that sufficient data have been used in characterizing relevant features, events, and processes and incorporating these features, events, and processes into the abstraction of the concentration of radionuclides in ground water.

Confirm that the quality and quantity of data are sufficient for those parameter groups considered important for developing the model abstraction, including groups, such as well classification and design, pumping rates, aquifer parameters, and transport parameters. Where applicable, confirm that reliable statistical estimates can be obtained from the relevant parameter data that can be used to either establish meaningful confidence limits or set meaningful bounding estimates, and confirm that the scales of measured data are appropriately factored into the abstraction.

Review Method 3 Data Uncertainty

Examine the technical bases for parameter values and assumed ranges, probability distributions, and bounding values used in conceptual models, process models, and alternative conceptual models considered in the total system performance assessment abstraction of dilution of radionuclides in ground water due to well pumping. The reviewer should verify that the technical bases support the treatment of uncertainty and variability of these parameters in the performance assessment. If conservative values are used as a method for addressing uncertainty and variability, the reviewer should verify that the conservative values result in conservative estimates of risk and do not cause unintended results (i.e., conservative representation of one aspect of the repository behavior that leads to an overall reduction in risk; inappropriate dilution of the risk estimate by assuming an approach is conservative when a parameter range is increased beyond the supporting data).

Assess whether these parameter values and distributions are consistent with site characterization data, laboratory experiments, field measurements, and natural analog research.

Verify that the U.S. Department of Energy appropriately establishes possible statistical correlations between parameters. Verify that an adequate technical basis or bounding argument is provided for neglected correlations.

Examine the U.S. Department of Energy use of expert elicitation, and confirm that where sufficient data do not exist, the definition of parameter values and conceptual models is based on appropriate use of other sources, such as expert elicitation, conducted in accordance with appropriate guidance, such as NUREG–1563 (Kotra, et al., 1996).

Review Method 4 Model Uncertainty

Evaluate whether appropriate alternative conceptual models are used in developing the abstraction for the concentration of radionuclides in ground water, and examine the model parameters in the context of available data. Compare the results of alternate process models to results from process models used by the U.S. Department of Energy to assess the uncertainty, limitations, and the degree of conservatism present in the U.S. Department of Energy model. Ascertain whether any limitations identified in the U.S. Department of Energy process model, through this comparison, are adequately accounted for in the U.S. Department of Energy abstraction. As appropriate, confirm that the U.S. Department of Energy has adequately addressed comments from external reviews of the model abstraction.

Confirm that the results of plausible alternative conceptual models have been considered appropriately in the abstraction, in the context of site characterization data, laboratory experiments, field measurements, natural analog research, and process modeling studies. In particular, use an alternative total system performance assessment model to evaluate the effect of the alternative conceptual model(s) on repository performance.

Evaluate the treatment of conceptual model uncertainty in light of the available site characterization data, laboratory experiments, field measurements, natural analog information and process-level modeling studies. If adoption of a conservative model is used as an approach for addressing conceptual model uncertainty, the reviewer should verify that the selected conceptual model: (i) is conservative relative to alternative conceptual models that are consistent with the available data and current scientific understanding; and (ii) results in conservative estimates of risk and not cause unintended results (i.e., conservative representation of one aspect of the repository behavior that leads to an overall reduction in the risk estimate).

Review Method 5 Model Support

Evaluate the output from the total system performance assessment model abstraction of the concentration of radionuclides in ground water, and verify that the U.S. Department of Energy compares the results with an appropriate combination of site characterization data, process modeling, laboratory testing, field measurements, and natural analog research. Use detailed models of geochemical, hydrological, and geological processes and an alternative total system performance assessment model to selectively probe the U.S. Department of Energy total system performance assessment analyses, and evaluate selected parts of the U.S. Department of Energy abstraction of the concentration of radionuclides in ground water.

2.2.1.3.12.3 Acceptance Criteria

The following acceptance criteria are based on meeting the requirements of 10 CFR 63.114(a)–(c), (e)–(g), and 63.305 relating to the concentration of radionuclides in ground water abstraction. U.S. Nuclear Regulatory Commission staff should apply the following acceptance criteria, according to the level of importance established in the U.S. Department of Energy risk-informed license application.

Acceptance Criterion 1 System Description and Model Integration Are Adequate.

(1) Total system performance assessment adequately incorporates important design features, physical phenomena, and couplings, and uses consistent and appropriate assumptions throughout the concentration of radionuclides in ground water abstraction process;

(2) The total system performance assessment model abstraction of the concentration of radionuclides in ground water adequately identifies and describes aspects of the determination of what portion of the radionuclides that are projected to cross the accessible environment boundary annually is to be included in the calculation of the average concentration of radionuclides in 3,000 acre-feet [3.715×10^9 liters], and includes the technical bases for these descriptions;

(3) The description of aspects of hydrology and geology that may affect the concentration of radionuclides in ground water is adequate, and identifies those parameters to which the abstraction is sensitive;

(4) The total system performance assessment abstraction of the concentration of radionuclides in ground water uses assumptions, technical bases, data, and models that are appropriate and consistent with other related U.S. Department of Energy abstractions (see Section 2.2.1.4.3 of the Yucca Mountain Review Plan);

(5) Sufficient data and technical bases for the inclusion of features, events, and processes, related to the concentration of radionuclides in ground water in the total system performance assessment abstraction, are provided; and

(6) Guidance in NUREG–1297 and NUREG–1298 (Altman, et al., 1988a,b), or in other acceptable approaches for peer review and data qualification is followed.

Acceptance Criterion 2 Data Are Sufficient for Model Justification.

(1) Climatological and hydrological values used in the license application are adequately justified (e.g., well classification and design, aquifer parameters, transport parameters, etc.). Adequate descriptions of how the data were used, interpreted, and appropriately synthesized into the parameters are provided;

(2) Sufficient data (field, laboratory, and/or natural analog data) are available to adequately define relevant parameters and conceptual models, necessary for developing the concentration of radionuclides in ground water abstraction, in total system performance assessment; and

(3) The quality and quantity of data are sufficient for those parameter groups considered important for developing and calibrating the abstraction model, including groups such as well classification and design, aquifer parameters, and transport parameters.

Acceptance Criterion 3 Data Uncertainty Is Characterized and Propagated Through the Model Abstraction.

(1) Models use parameter values, assumed ranges, probability distributions, and bounding assumptions that are technically defensible, reasonably account for uncertainties and variabilities, and do not result in an under-representation of the risk estimate, and are consistent with the characteristics of the reasonably maximally exposed individual defined in 10 CFR Part 63;

(2) The technical bases for the parameter values and ranges in performance assessment and process models used for estimating the concentration of radionuclides in ground water, characterization data, laboratory experiments, field measurements, and natural analog research, as appropriate;

(3) Uncertainty is adequately represented in parameters of conceptual models, process models, and alternative conceptual models considered in developing the total system performance assessment abstraction of the concentration of radionuclides in ground water either through sensitivity analyses, conservative limits, or bounding values supported by data, as necessary;

(4) Parameters that are important for the abstraction, through total system performance assessment and sensitivity analyses, are identified; and

(5) Where sufficient data do not exist, the definition of parameter values and conceptual models is based on appropriate use of expert elicitation, conducted in accordance with appropriate guidance, such as NUREG–1563 (Kotra, et al., 1996). If other approaches are used, the U.S. Department of Energy adequately justifies their use.

Acceptance Criterion 4 Model Uncertainty Is Characterized and Propagated Through the Model Abstraction.

(1) Alternative modeling approaches of features, events, and processes are considered and are consistent with available data and current scientific understanding, and the results and limitations are appropriately considered in the abstraction;

(2) Sufficient evidence is provided that existing alternative conceptual models of features and processes have been considered, that the models are consistent with available data (e.g., field, laboratory, and natural analog) and current scientific understanding, and that the effects of these alternative conceptual models on total system performance assessment results are adequately evaluated; and

(3) Consideration of conceptual model uncertainty is consistent with available site characterization data, laboratory experiments, field measurements, natural analog information and process-level modeling studies; and the treatment of conceptual model uncertainty does not result in an under-representation of the risk estimate.

Acceptance Criterion 5 Model Abstraction Output Is Supported by Objective Comparisons.

(1) Models implemented in this total system performance assessment abstraction provide results consistent with output from detailed process-level models and/or empirical observations (e.g., laboratory testing, field measurements, and/or natural analogs).

2.2.1.3.12.4 Evaluation Findings

If the license application provides sufficient information and the regulatory acceptance criteria in Section 2.2.1.3.12.3 are appropriately satisfied, the staff concludes that this portion of the staff evaluation is acceptable. The reviewer writes material suitable for inclusion in the safety evaluation report prepared for the entire application. The report includes a summary statement of what was reviewed and why the reviewer finds the submittal acceptable. The staff can document the review as follows.

U.S. Nuclear Regulatory Commission staff has reviewed the Safety Analysis Report and other information submitted in support of the license application, and has found, with reasonable expectation, that the requirements of 10 CFR 63.114 are satisfied. Technical requirements for conducting a performance assessment, with respect to the concentration of radionuclides in ground water, have been met. In particular, the U.S. Nuclear Regulatory Commission staff found that:

(1) Appropriate data from the site and surrounding region, uncertainties and variabilities in parameter values, and alternative conceptual models have been used in the analyses, in compliance with 10 CFR 63.114(a)–(c);

(2) Specific features, events, and processes have been included in the analyses, and appropriate technical bases have been provided for inclusion or exclusion, in compliance with 10 CFR 63.114(e);

(3) Specific degradation, deterioration, and alteration processes have been included in the analyses, taking into consideration their effects on annual dose, and appropriate technical bases have been provided for inclusion or exclusion, in compliance with 10 CFR 63.114(f); and

(4) Adequate technical bases have been provided for models used in the performance assessment, as required by 10 CFR 63.114(g).

U.S. Nuclear Regulatory Commission staff has reviewed the Safety Analysis Report and other information submitted in support of the license application, relevant to the concentration of radionuclides in ground water, and has found, with reasonable expectation, that the requirements of 10 CFR 63.115 are satisfied.

The required characteristics of the reference biosphere have been satisfied. In particular the U.S. Nuclear Regulatory Commission staff found reasonable expectation that:

(1) The features, events, and processes used to describe the reference biosphere, the biosphere pathways, the evolution of climate, and the evolution of the geologic setting

are consistent with present knowledge of the region, conditions, and past processes in the Yucca Mountain region, as required by 10 CFR 63.305(a)–(d).

U.S. Nuclear Regulatory Commission staff has reviewed the Safety Analysis Report and other information submitted in support of the license application, relevant to the concentration of radionuclides in ground water, and has found, with reasonable expectation, that the requirements of 10 CFR 63.332 are satisfied. The specific requirements for the concentration of radionuclides in ground water have been met. In particular, the U.S. Nuclear Regulatory Commission staff found that:

(1) The U.S. Department of Energy uses average hydrologic characteristics to determine the position and dimension of the ground-water aquifers, and projects radionuclide concentrations such that the highest concentration levels in the contaminant plume are included. The annual water demand also contains no more than 3,000 acre-feet [3.715×10^9 liters] and meets any other requirements specified in 10 CFR 63.332(a)(1)–(3).

2.2.1.3.12.5 References

Altman, W.D., J.P. Donnelly, and J.E. Kennedy. NUREG–1297, “Generic Technical Position on Peer Review for High-Level Nuclear Waste Repositories.” Washington, DC: U.S. Nuclear Regulatory Commission. 1988a.

———. NUREG–1298, “Generic Technical Position on Qualification of Existing Data for High-Level Nuclear Waste Repositories.” Washington, DC: U.S. Nuclear Regulatory Commission. 1988b.

Kotra, J.P., et al. NUREG–1563, “Branch Technical Position on the Use of Expert Elicitation in the High-Level Radioactive Waste Program.” Washington, DC: U.S. Nuclear Regulatory Commission. 1996.

2.2.1.3.13 Redistribution of Radionuclides in Soil

To review this model abstraction, consider the degree to which the U.S. Department of Energy relies on redistribution of radionuclides in soil to demonstrate compliance. Review this model abstraction considering the risk information determined in the “Multiple Barriers” Section (2.2.1.1). For example, if the U.S. Department of Energy indicates that redistribution of radionuclides in soil has a strong effect on performance, then perform a detailed review of this abstraction. If, on the other hand, the U.S. Department of Energy demonstrates this abstraction to have a minor impact on the dose to the reasonably maximally exposed individual, then conduct a simplified review focusing on the bounding assumptions. The review methods and acceptance criteria provided here are for a detailed review. Some of the review methods and acceptance criteria may not be necessary, in a simplified review, for those abstractions that have a minor impact on performance. The demonstration of compliance with the individual protection standard is evaluated using Section 2.2.1.4.1, and compliance with the ground-water protection standard is evaluated using Section 2.2.1.4.3 of the Yucca Mountain Review Plan.

Review Responsibilities—High-Level Waste Branch and Environmental and Performance Assessment Branch

2.2.1.3.13.1 Areas of Review

This section reviews redistribution of radionuclides in soil in the biosphere. Reviewers will also evaluate information, required by 10 CFR 63.21(c)(1), (9), (15), and (19), that is relevant to the abstraction of redistribution of radionuclides in soil.

The staff will evaluate the following parts of the abstraction of the redistribution of radionuclides in soil, using the review methods and acceptance criteria in Sections 2.2.1.13.2 and 2.2.1.3.13.3:

(1) Description of the geological, hydrological, pedological, and geochemical aspects of redistribution of radionuclides in soil, and the technical bases the U.S. Department of Energy provides to support model integration across the total system performance assessment abstractions;

(2) Sufficiency of the data and parameters used to justify the total system performance assessment model abstraction;

(3) Methods the U.S. Department of Energy uses to characterize data uncertainty, and propagate the effects of this uncertainty, through the total system performance assessment model abstraction;

(4) Methods the U.S. Department of Energy uses to characterize model uncertainty, and propagate the effects of this uncertainty, through the total system performance assessment model abstraction;

(5) Approaches the U.S. Department of Energy uses to compare total system performance assessment output to process-level model outputs and empirical studies; and

(6) Use of expert elicitation.

2.2.1.3.13.2 Review Methods

To review the abstraction of the redistribution of radionuclides in soil, recognize that models used in the total system performance assessments may range from highly complex process-level models to simplified models, such as response surfaces or look-up tables. Evaluate model adequacy, regardless of the level of complexity.

Review Method 1 Model Integration

Examine the description of features, physical phenomena, and couplings between different models, and confirm that they have been appropriately incorporated in the redistribution of radionuclides in soil abstraction. Confirm that consistent and appropriate assumptions have been made throughout the abstraction.

Examine the aspects of redistribution of radionuclides in soil that have been identified as being important to repository performance, and verify that these aspects are reasonable. Assess the technical bases for these descriptions, and for incorporating them in the total system performance assessment abstraction of redistribution of radionuclides in soil. Evaluate whether the descriptions provide transparent and traceable support for the abstraction.

Examine how the features, events, and processes related to redistribution of radionuclides in soil, have been included in the total system performance assessment abstraction.

Verify that the U.S. Department of Energy reviews follow guidance, such as NUREG–1297 and NUREG–1298 (Altman, et al., 1988a,b), or makes an acceptable case for using alternative approaches for peer review and data qualification.

Review Method 2 Data and Model Justification

Confirm that the data on the pedology, hydrology, and soil chemistry used in the total system performance assessment abstraction are based on a combination of techniques that may include laboratory experiments, site-specific field measurements, natural analog research, and process modeling studies. Examine how data were used, interpreted, and synthesized into parameter values, and verify that it was done appropriately.

Evaluate the sufficiency of the data used to support conceptual models, process-level models, and alternative conceptual models considered in the total system performance assessment abstraction of redistribution of radionuclides in soil. Examine and confirm the sufficiency of the data that support the technical bases, for features, events, and processes related to redistribution of radionuclides in soil, that have been included in the total system performance assessment abstraction.

Review Method 3 Data Uncertainty

Examine the technical bases for parameter values, assumed ranges, probability distributions, and bounding values used in conceptual models, process models, and alternative conceptual models considered in the abstraction of redistribution of radionuclides in soil, and evaluate the assessment of uncertainty and variability in these parameters. The reviewer should verify that the technical bases support the treatment of uncertainty and variability of these parameters in the performance assessment. If conservative values are used as a method for addressing uncertainty and variability, the reviewer should verify that the conservative values result in conservative estimates of risk and do not cause unintended results (i.e., conservative representation of one aspect of the repository behavior that leads to an overall reduction in risk; inappropriate dilution of the risk estimate by assuming an approach is conservative when a parameter range is increased beyond the supporting data).

Evaluate the U.S. Department of Energy input values by comparison with the corresponding input values in the U.S. Nuclear Regulatory Commission data set, to the extent feasible. However, direct comparison of input values may not be possible if the U.S. Nuclear Regulatory Commission and U.S. Department of Energy models are substantially different.

Examine the technical basis used to support parameter values and ranges, and confirm that the selected parameter ranges and distributions adequately represent the conditions in the Yucca Mountain region.

Assess whether uncertainty is adequately represented in parameters of conceptual models, process models, and alternative conceptual models considered in developing the abstraction of dilution of radionuclides in soil, from surface processes, either through sensitivity analyses, conservative limits, or bounding values supported by data. Assess whether correlations between parameters in the abstraction have been appropriately established.

Evaluate the U.S. Department of Energy determination of the sensitivity of the performance of the system to the parameter value or model and verify that the level of adequacy of data required for justification of parameters or models is commensurate with the impact that the parameter or model has on the performance of the system. To the extent feasible, use alternative total system performance assessment code to test the sensitivity of the repository performance to the parameter value or model.

Verify that the U.S. Department of Energy appropriately establishes possible statistical correlations between parameters. Verify that an adequate technical basis or bounding argument is provided for neglected correlations.

Examine the U.S. Department of Energy use of expert elicitation, and confirm that where sufficient data do not exist, the definition of parameter values and conceptual models is based on appropriate use of other sources, such as expert elicitation, conducted in accordance with appropriate guidance, such as NUREG–1563 (Kotra, et al., 1996).

Review Method 4 Model Uncertainty

Verify that the U.S. Department of Energy evaluated all appropriate alternative conceptual models for redistribution of radionuclides in soil. Compare the results of alternate process models to results from process models used by the U.S. Department of Energy to assess the uncertainty, limitations, and the degree of conservatism present in the U.S. Department of Energy model. Ascertain whether any limitations identified in the U.S. Department of Energy process model through this comparison are adequately accounted for in the U.S. Department of Energy abstraction.

Confirm that the results of appropriate alternative conceptual models have been considered in the abstraction in the context of site characterization data, laboratory experiments, field measurements, natural analog research, and process modeling studies. In particular, use an alternative total system performance assessment model to evaluate the effect of the alternative conceptual model(s) on repository performance.

Evaluate the treatment of conceptual model uncertainty in light of the available site characterization data, laboratory experiments, field measurements, natural analog information and process-level modeling studies. If adoption of a conservative model is used as an approach for addressing conceptual model uncertainty, the reviewer should verify that the selected conceptual model: (i) is conservative relative to alternative conceptual models that are consistent with the available data and current scientific understanding; and (ii) results in conservative estimates of risk and not cause unintended results (i.e., conservative representation of one aspect of the repository behavior that leads to an overall reduction in the risk estimate).

Review Method 5 Model Support

Evaluate the output from the abstraction of redistribution of radionuclides in soil and compare the results with an appropriate combination of site characterization data, process modeling, laboratory testing, field measurements, and natural analog research. As appropriate, the reviewer should use an alternative total system performance assessment code to evaluate selected parts of the U.S. Department of Energy abstraction of redistribution of radionuclides in soil.

2.2.1.3.13.3 Acceptance Criteria

The following acceptance criteria are based on meeting the relevant requirements of 10 CFR 63.114(a)–(c), (e)–(g), and 63.305 as they relate to the redistribution of radionuclides in soil abstraction. U.S. Nuclear Regulatory Commission staff should apply the following acceptance criteria, according to the level of importance established in the U.S. Department of Energy risk-informed license application.

Acceptance Criterion 1 System Description and Model Integration Are Adequate.

(1) Total system performance assessment adequately incorporates important features, physical phenomena and couplings between different models, and uses consistent and appropriate assumptions throughout the abstraction of redistribution of radionuclides in the soil abstraction process;

(2) The total system performance assessment model abstraction identifies and describes aspects of redistribution of radionuclides in soil that are important to repository performance, including the technical bases for these descriptions. For example, the abstraction should include modeling of the deposition of contaminated material in the soil and determination of the depth distribution of the deposited radionuclides;

(3) Relevant site features, events, and processes have been appropriately modeled in the abstraction of redistribution of radionuclides, from surface processes, and sufficient technical bases are provided; and

(4) Guidance in NUREG–1297 and NUREG–1298 (Altman, et al., 1988a,b), or other acceptable approaches for peer reviews, is followed.

Acceptance Criterion 2 Data Are Sufficient for Model Justification.

(1) Behavioral, hydrological, and geochemical values used in the license application are adequately justified (e.g., irrigation and precipitation rates, erosion rates, radionuclide solubility values, etc.). Adequate descriptions of how the data were used, interpreted, and appropriately synthesized into the parameters are provided; and

(2) Sufficient data (e.g., field, laboratory, and natural analog data) are available to adequately define relevant parameters and conceptual models necessary for developing the abstraction of redistribution of radionuclides in soil in the total system performance assessment.

Acceptance Criterion 3 Data Uncertainty Is Characterized and Propagated Through the Model Abstraction.

(1) Models use parameter values, assumed ranges, probability distributions, and bounding assumptions that are technically defensible, reasonably account for uncertainties and variabilities, do not result in an under-representation of the risk estimate, and are consistent with the characteristics of the reasonably maximally exposed individual in 10 CFR Part 63;

(2) The technical bases for the parameter values and ranges in the total system performance assessment abstraction are consistent with data from the Yucca Mountain region [e.g., Amargosa Valley survey (Cannon Center for Survey Research, 1997), studies of surface processes in the Fortymile Wash drainage basin: applicable laboratory testings: natural analogs: or other valid sources of data. For example, soil types, crop types, plow depths, and irrigation rates should be consistent with current farming practices, and data on the airborne particulate concentration should be based on the resuspension of appropriate material in a climate and level of disturbance similar to that which is expected to be found at the location of the reasonably maximally exposed individual, during the compliance time period;

(3) Uncertainty is adequately represented in parameters for conceptual models, process models, and alternative conceptual models considered in developing the total system performance assessment abstraction of redistribution of radionuclides in soil, either through sensitivity analyses, conservative limits, or bounding values supported by data, as necessary. Correlations between input values are appropriately established in the total system performance assessment;

(4) Parameters or models that most influence repository performance based on the performance measure and time period of compliance, specified in 10 CFR Part 63, are identified; and

(5) Where sufficient data do not exist, the definition of parameter values and conceptual models on appropriate uses of other sources, such as expert elicitation, are conducted in accordance with appropriate guidance, such as NUREG–1563 (Kotra, et al., 1996).

Acceptance Criterion 4 Model Uncertainty Is Characterized and Propagated Through the Model Abstraction.

(1) Alternative modeling approaches of features, events, and processes are considered and are consistent with available data, and current scientific understanding, and the results and limitations are appropriately considered in the abstraction;

(2) Sufficient evidence is provided that appropriate alternative conceptual models of features, events, and processes have been considered; that the preferred models (if any) are consistent with available data (e.g., field, laboratory, and natural analog) and current scientific understanding; and that the effect on total system performance assessment of uncertainties from these alternative conceptual models has been evaluated; and

(3) Consideration of conceptual model uncertainty is consistent with available site characterization data, laboratory experiments, field measurements, natural analog information and process-level modeling studies; and the treatment of conceptual model uncertainty does not result in an under-representation of the risk estimate.

Acceptance Criterion 5 Model Abstraction Output Is Supported by Objective Comparisons.

(1) Models implemented in the abstraction provide results consistent with output from detailed process-level models and/or empirical observations (e.g., laboratory testing, field measurements, and/or natural analogs).

2.2.1.3.13.4 Evaluation Findings

If the license application provides sufficient information and the regulatory acceptance criteria in Section 2.2.1.3.13.3 are appropriately satisfied, the staff concludes that this portion of the staff evaluation is acceptable. The reviewer writes material suitable for inclusion in the safety evaluation report prepared for the entire application. The report includes a summary statement of what was reviewed and why the reviewer finds the submittal acceptable. The staff can document the review as follows.

These evaluation findings are only with respect to this part of the total system performance assessment model abstraction.

U.S. Nuclear Regulatory Commission staff has reviewed the Safety Analysis Report and other information submitted in support of the license application, relevant to redistribution of radionuclides in soil, and has found, with reasonable expectation, that the requirements of 10 CFR 63.114 are satisfied. Technical requirements for conducting a performance assessment in the area of redistribution of radionuclides in soil have been met. In particular, the U.S. Nuclear Regulatory Commission staff found that:

(1) Appropriate data from the site and surrounding region, uncertainties and variabilities in parameter values, and alternative conceptual models have been used in the analyses, in compliance with 10 CFR 63.114(a)–(c);

(2) Specific features, events, and processes have been included in the analyses, and appropriate technical bases have been provided for inclusion or exclusion, in compliance with 10 CFR 63.114(e);

(3) Specific degradation, deterioration, and alteration processes have been included in the analyses, taking into consideration their effect on annual dose, and appropriate technical bases have been provided for inclusion or exclusion, in compliance with 10 CFR 63.114(f); and

(4) Adequate technical bases have been provided for models used in the performance assessment, as required by 10 CFR 63.114(g).

U.S. Nuclear Regulatory Commission staff has reviewed the Safety Analysis Report and other information submitted in support of the license application, relevant to redistribution of radionuclides in soil, and has found, with reasonable expectation, that the requirements of 10 CFR 63.305 are satisfied. The required characteristics of the reference biosphere have been satisfied. In particular the U.S. Nuclear Regulatory Commission staff found that:

(1) The features, events, and processes used to describe the reference biosphere, the biosphere pathways, the evolution of climate, and the evolution of the geologic setting are consistent with present knowledge of the region, conditions, and past processes in the Yucca Mountain region, as required by 10 CFR 63.305(a)–(d).

2.2.1.3.13.5 References

Altman, W.D., J.P. Donnelly, and J.E. Kennedy. NUREG–1297, "Generic Technical Position on Peer Review for High-Level Nuclear Waste Repositories." Washington, DC: U.S. Nuclear Regulatory Commission. 1988a.

———. NUREG–1298, "Generic Technical Position on Qualification of Existing Data for High-Level Nuclear Waste Repositories." Washington, DC: U.S. Nuclear Regulatory Commission. 1988b.

Cannon Center for Survey Research, University of Nevada. "Identifying and Characterizing the Critical Group Results of a Pilot Study of Amargosa Valley." Las Vegas, Nevada: Cannon Center for Survey Research. 1997.

Kotra, J.P., et al. NUREG–1563, "Branch Technical Position on the Use of Expert Elicitation in the High-Level Radioactive Waste Program." Washington, DC: U.S. Nuclear Regulatory Commission. 1996.

2.2.1.3.14 Biosphere Characteristics

To review this model abstraction, consider the degree to which the U.S. Department of Energy relies on biosphere characteristics to demonstrate compliance. Review this model abstraction considering the risk information evaluated in the "Multiple Barriers" Section (2.2.1.1). For example, if the U.S. Department of Energy indicates that biosphere characteristics have a strong effect on performance, then conduct a detailed review of this abstraction. If, on the other hand, the U.S. Department of Energy demonstrates this abstraction to have a minor impact on the dose to the reasonably maximally exposed individual, then perform a simplified review focusing on the bounding assumptions. The review methods and acceptance criteria provided here are for a detailed review. Some of the review methods and acceptance criteria may not be necessary, in a simplified review, for those abstractions that have a minor impact on performance. The demonstration of compliance with the postclosure individual protection standard is evaluated, using Section 2.2.1.4.1 of the Yucca Mountain Review Plan.

Review Responsibilities—High-Level Waste Branch and Environmental and Performance Assessment Branch

2.2.1.3.14.1 Areas of Review

This section reviews biosphere characteristics that involve application of the characteristics of the reasonably maximally exposed individual and reference biosphere to transforming estimated concentrations of radionuclides in the biosphere to a dose to the reasonably maximally exposed individual. Reviewers will also evaluate information, required by 10 CFR 63.21(c)(1), (9), (15), and (19) that is relevant to the abstraction of the biosphere characteristics modeling.

The staff will evaluate the following parts of the biosphere characteristics, using review methods and acceptance criteria in Sections 2.2.1.3.14.2 and 2.2.1.3.14.3:

(1) Description of the ecological, behavioral, geological, hydrological, geochemical, sociological, and economic aspects of biosphere characteristics, and the technical bases the U.S. Department of Energy provides to support model integration across the total system performance assessment abstractions;

(2) Sufficiency of the data and parameters used to justify the total system performance assessment model abstraction;

(3) Methods the U.S. Department of Energy uses to characterize data uncertainty, and propagate the effects of this uncertainty, through the total system performance assessment model abstraction;

(4) Methods the U.S. Department of Energy uses to characterize model uncertainty, and propagate the effects of this uncertainty, through the total system performance assessment model abstraction;

(5) Approaches the U.S. Department of Energy uses to compare output from the total system performance assessment model abstraction to process-level outputs and empirical studies; and

(6) Use of expert elicitation.

2.2.1.3.14.2 Review Methods

For the abstraction of biosphere characteristics, recognize that models used in the total system performance assessment may range from highly complex process-level models to simplified models, such as response surfaces or look-up tables. Evaluate model adequacy, regardless of the level of complexity.

Review Method 1 System Description and Model Integration

Confirm that the abstraction includes all important site features, physical phenomena, and couplings, and whether consistent and appropriate assumptions have been used through the abstraction.

Verify that the description is adequate, and that the conditions and assumptions in the total system performance assessment abstraction are consistent with the body of data, presented in the description. Confirm that the technical bases for these descriptions, and for incorporating them in the abstraction, are appropriate. Evaluate whether the descriptions provide transparent and traceable support for the abstraction.

Consider important physical phenomena and couplings with other abstractions, and examine them for consistency.

Confirm that the U.S. Department of Energy has used an acceptable approach for peer reviews, such as the guidance in NUREG–1297 and NUREG–1298 (Altman, et al., 1988a,b), or makes an acceptable case for using alternative approaches.

Review Method 2 Data and Model Justification

Confirm that the parameter values used in the license application are adequately justified, and consistent with the definition of the reasonably maximally exposed individual in 10 CFR 63.312. Evaluate how the data were used, interpreted, and appropriately synthesized into the parameters.

Evaluate the sufficiency of the data and parameters used to support the modeling of features, events, and processes in conceptual models, process-level models, and alternative conceptual models considered in the total system performance assessment biosphere characteristics. When evaluating alternate conceptual models of the biosphere or biosphere processes, the reviewer should recognize that 10 CFR 63.305 and 63.312 place a number of constraints on both the biosphere and the characteristics of the reasonably maximally exposed individual. For example, 10 CFR 63.312 limits the diet and living style of the reasonably maximally exposed individual to be representative of the current population of the town of Amargosa Valley, Nevada. Therefore, evaluation of alternate conceptual models should focus on exploring the variability and uncertainty in the features, events, and processes incorporated in the biosphere abstraction , mindful of the regulatory constraints. Evaluation of behavior and characteristics of the reasonably maximally exposed individual should emphasize interpretation of survey studies of the current residents of the Town of Amargosa Valley and how uncertainty and variability in the data are used to derive mean values.

Confirm that the data used in the U.S. Department of Energy total system performance assessment abstraction are based on a combination of techniques, that may include laboratory experiments, site-specific field measurements, natural analog research, and process-level modeling studies. Investigate the effects of any differences in model and implementation approach on dose results by executing an alternative total system performance assessment code with the U.S. Department of Energy input parameters, and by comparing calculated dose results with those reported by the U.S. Department of Energy. Confirm that any differences or identified limitations in model selection and implementation, that significantly decrease dose results, are adequately justified in the U.S. Department of Energy analysis.

Review Method 3 Data Uncertainty

Examine the technical bases for parameter values and ranges used in conceptual models, process-level models, and alternative conceptual models considered in the total system performance assessment biosphere characteristics. When evaluating alternate conceptual models of the biosphere or biosphere processes, the reviewer should recognize that 10 CFR 63.305 and 63.312 put a number of constraints on both the biosphere and the characteristics of the reasonably maximally exposed individual. For example, 10 CFR 63.312 limits the diet and lifestyle of the reasonably maximally exposed individual to be representative of the current population of the Town of Amargosa Valley, Nevada. Therefore, evaluation of alternate conceptual models should focus on exploring the variability and uncertainty in the features, events, and processes incorporated in the biosphere abstraction, mindful of the regulatory constraints. Evaluation of behavior and characteristics of the reasonably maximally exposed individual should emphasize interpretation of local survey studies of the current residents of the Town of Amargosa Valley and how uncertainties and variability in the data are used to derive mean values.

Evaluate the assessment of uncertainty and variability in parameters. Verify that the U.S. Department of Energy has a technically defensible basis to support the determination that the diet and living style of the reasonably maximally exposed individual are based on the mean values of data obtained from surveys of residents of the Town of Amargosa Valley, Nevada, as specified in 10 CFR 63.312.

Examine the technical bases for parameter values, assumed ranges, probability distributions, and bounding values used in conceptual models, process models, and alternative conceptual models considered in the abstraction of biosphere characteristics, and evaluate the assessment of uncertainty and variability in these parameters. The reviewer should verify that the technical bases support the treatment of uncertainty and variability of these parameters in the performance assessment. If conservative values are used as a method for addressing uncertainty and variability, the reviewer should verify that the conservative values result in conservative estimates of risk and do not cause unintended results (i.e., conservative representation of one aspect of the repository behavior that leads to an overall reduction in risk; inappropriate dilution of the risk estimate by assuming an approach is conservative when a parameter range is increased beyond the supporting data).

Evaluate the effects of including uncertainty and variability ranges (for important parameters) in total system performance assessment runs. Tests can provide information on the effects of including these ranges in the total system performance assessment (e.g., sensitivity and uncertainty analyses), and/or demonstrate the effects different ranges may have on dose results. Verify that any differences or identified limitations in the U.S. Department of Energy analysis that significantly decrease dose results are adequately justified.

Evaluate methods used by the U.S. Department of Energy in conducting expert elicitation to define parameter values.

Verify that the U.S. Department of Energy appropriately establishes possible statistical correlations between parameters. Verify that an adequate technical basis or bounding argument is provided for neglected correlations.

Examine the sensitivity of total system performance assessment results to identify parameter differences by comparing total system performance assessment results based on the U.S. Department of Energy and the U.S. Nuclear Regulatory Commission parameter selections. Emphasize those parameters known to be important in biosphere characteristics modeling, such as consumption rates, intake-to-dose conversion factors, plant and animal transfer factors, mass-loading factors, and crop interception fractions.

Review Method 4 Model Uncertainty

Examine the model parameters in the context of available site characterization data, laboratory experiments, field measurements, natural analog research, and process-level modeling studies. To the extent practical and necessary, use an alternative total system performance assessment model to evaluate selected parts of the U.S. Department of Energy biosphere characteristics, and evaluate the effect of the alternative conceptual model(s) on repository performance. When evaluating alternate conceptual models of the biosphere or biosphere processes, the reviewer should recognize that 10 CFR 63.305 and 63.312 put a number of constraints on both the biosphere and selection of the reasonably maximally exposed individual. For example, 10 CFR 63. 312 limits the diet and living style of the reasonably maximally exposed individual to be representative of the current residents of the Town of Amargosa Valley, Nevada. Therefore, evaluation of alternate conceptual models should focus on exploring the variability and uncertainty in the features, events, and processes incorporated in the biosphere abstraction, mindful of the regulatory constraints. Evaluation of behavior and characteristics of the reasonably maximally exposed individual should emphasize interpretation of survey studies of the residents of the Town of Amargosa Valley and how uncertainty and variability in the data are used to derive mean values.

Confirm that sufficient evidence has been presented that existing alternative conceptual models of processes that are important to waste isolation have been considered in the biosphere characteristics.

Evaluate the treatment of conceptual model uncertainty in light of the available site characterization data, laboratory experiments, field measurements, natural analog information and process-level modeling studies. If adoption of a conservative model is used as an approach for addressing conceptual model uncertainty, the reviewer should verify that the selected conceptual model: (i) is conservative relative to alternative conceptual models that are consistent with the available data and current scientific understanding; and (ii) results in conservative estimates of risk and not cause unintended results (i.e., conservative representation of one aspect of the repository behavior that leads to an overall reduction in the risk estimate).

Review Method 5 Model Support

Evaluate the output from the biosphere characteristics modeling and compare the results with an appropriate combination of site characterization data, process-level modeling, laboratory testing, field measurements, and natural analog research. Examine the sensitivity analyses used to support the biosphere characteristics modeling in the total system performance assessment. To the extent practical and necessary, use an alternative total system performance assessment code to evaluate selected parts of the U.S. Department of Energy biosphere characteristics modeling. Compare the U.S. Department of Energy biosphere dose conversion factors with the results of dose modeling using a code, such as GENII-S (Leigh, et al., 1993) and the U.S. Department of Energy input parameter data. The reviewer should conduct confirmatory runs, using alternative dose calculation codes and the U.S. Department of Energy input parameters, as necessary.

2.2.1.3.14.3 Acceptance Criteria

The following acceptance criteria are based on meeting the requirements of 10 CFR 63.114(a)–(c), (e)–(g), 63.305, and 63.312 as they relate to biosphere characteristics modeling.

U.S. Nuclear Regulatory Commission staff should apply the following acceptance criteria, according to the level of importance established in the U.S. Department of Energy risk-informed license application.

Acceptance Criterion 1 System Description and Model Integration Are Adequate.

(1) Total system performance assessment adequately incorporates important site features, physical phenomena, and couplings, and consistent and appropriate assumptions throughout the biosphere characteristics modeling abstraction process;

(2) The total system performance assessment model abstraction identifies and describes aspects of the biosphere characteristics modeling that are important to repository performance, and includes the technical bases for these descriptions. For example, the reference biosphere should be consistent with the arid or semi-arid conditions in the vicinity of Yucca Mountain;

(3) Assumptions are consistent between the biosphere characteristics modeling and other abstractions. For example, the U.S. Department of Energy should ensure that the modeling of features, events, and processes, such as climate change, soil types, sorption coefficients, volcanic ash properties, and the physical and chemical properties of radionuclides are consistent with assumptions in other total system performance assessment abstractions; and

(4) Guidance in NUREG–1297 and NUREG–1298 (Altman, et al., 1988a,b), or in other acceptable approaches for peer reviews, is followed.

Acceptance Criterion 2 Data Are Sufficient for Model Justification.

(1) The parameter values used in the license application are adequately justified (e.g., behaviors and characteristics of the residents of the Town of Amargosa Valley, Nevada, characteristics of the reference biosphere, etc.) and consistent with the definition of the reasonably maximally exposed individual in 10 CFR Part 63. Adequate descriptions of how the data were used, interpreted, and appropriately synthesized into the parameters are provided; and

(2) Data are sufficient to assess the degree to which features, events, and processes related to biosphere characteristics modeling have been characterized and incorporated in the abstraction. As specified in 10 CFR Part 63, the U.S. Department of Energy should demonstrate that features, events, and processes, which describe the biosphere,

are consistent with present knowledge of conditions in the region, surrounding Yucca Mountain. As appropriate, the U.S. Department of Energy sensitivity and uncertainty analyses (including consideration of alternative conceptual models) are adequate for determining additional data needs, and evaluating whether additional data would provide new information that could invalidate prior modeling results and affect the sensitivity of the performance of the system to the parameter value or model.

Acceptance Criterion 3 Data Uncertainty Is Characterized and Propagated Through the Model Abstraction.

(1) Models use parameter values, assumed ranges, probability distributions, and bounding assumptions that are technically defensible, reasonably account for uncertainties and variabilities, do not result in an under-representation of the risk estimate, and are consistent with the definition of the reasonably maximally exposed individual in 10 CFR Part 63;

(2) The technical bases for the parameter values and ranges in the abstraction, such as consumption rates, plant and animal uptake factors, mass-loading factors, and biosphere dose conversion factors, are consistent with site characterization data, and are technically defensible;

(3) Process-level models used to determine parameter values for the biosphere characteristics modeling are consistent with site characterization data, laboratory experiments, field measurements, and natural analog research;

(4) Uncertainty is adequately represented in parameter development for conceptual models and process-level models considered in developing the biosphere characteristics modeling, either through sensitivity analyses, conservative limits, or bounding values supported by data, as necessary. Correlations between input values are appropriately established in the total system performance assessment, and the implementation of the abstraction does not inappropriately bias results to a significant degree;

(5) Where sufficient data do not exist, the definition of parameter values and conceptual models is based on appropriate use of expert elicitation, conducted in accordance with appropriate guidance, such as NUREG–1563 (Kotra, et al., 1996). If other approaches are used, the U.S. Department of Energy adequately justifies their uses; and

(6) Parameters or models that most influence repository performance, based on the performance measure and time period of compliance specified in 10 CFR Part 63, are identified.

Acceptance Criterion 4 Model Uncertainty Is Characterized and Propagated Through the Model Abstraction.

(1) Alternative modeling approaches of features, events, and processes are considered and are consistent with available data and current scientific understanding, and the results and limitations of alternative modeling approaches are appropriately considered in the abstraction. Staff should evaluate alternate conceptual models of the biosphere or

biosphere processes, recognizing that 10 CFR 63.305 and 63.312 place a number of constraints on both the biosphere and the characteristics of the reasonably maximally exposed individual. Alternate conceptual models focus on exploring the variability and uncertainty in the physical features, events, and processes, mindful of the regulatory constraints. Evaluation of behavior and characteristics of the reasonably maximally exposed individual emphasizes understanding the characteristics of the current residents of the Town of Amargosa Valley, and uncertainty and variability in the data used to derive mean values;

(2) Sufficient evidence is provided that existing alternative conceptual models of features and processes that are important to waste isolation, such as plant uptake of radionuclides from soil, soil resuspension, and the inhalation dose model for igneous events, have been considered; and

(3) Consideration of conceptual model uncertainty is consistent with available site characterization data, laboratory experiments, field measurements, natural analog information and process-level modeling studies; and the treatment of conceptual model uncertainty does not result in an under-representation of the risk estimate.

Acceptance Criterion 5 Model Abstraction Output Is Supported by Objective Comparisons.

(1) Dose calculations pertaining to this total system performance assessment abstraction provide results consistent with output from detailed process-level models and/or empirical observations (e.g., laboratory testing, field measurements, and/or natural analogs).

2.2.1.3.14.4 Evaluation Findings

If the license application provides sufficient information and the regulatory acceptance criteria in Section 2.2.1.3.14.3 are appropriately satisfied, the staff concludes that this portion of the staff evaluation is acceptable. The reviewer writes material suitable for inclusion in the safety evaluation report prepared for the entire application. The report includes a summary statement of what was reviewed and why the reviewer finds the submittal acceptable. The staff can document the review as follows.

U.S. Nuclear Regulatory Commission staff has reviewed the Safety Analysis Report and other information submitted in support of the license application, and has found, with reasonable expectation, that the requirements of 10 CFR 63.114 are satisfied, regarding biosphere characteristics modeling in performance assessment. In particular, the U.S. Nuclear Regulatory Commission staff found reasonable expectation that:

(1) Appropriate data from the site and surrounding region, uncertainties and variabilities in parameter values, and alternative conceptual models have been used in the analyses, in compliance with 10 CFR 63.114(a)–(c);

(2) Specific features, events, and processes have been included in the analyses, and appropriate technical bases have been provided for inclusion or exclusion, in compliance with 10 CFR 63.114(e);

(3) Specific degradation, deterioration, and alteration processes have been included in the analyses, taking into consideration their effects on annual dose, and appropriate technical bases have been provided for inclusion or exclusion, in compliance with 10 CFR 63.114(f); and

(4) Adequate technical bases have been provided for models used in the performance assessment, as required by 10 CFR 63.114(h).

U.S. Nuclear Regulatory Commission staff has reviewed the Safety Analysis Report and other information submitted in support of the license application, and has found, with reasonable expectation, that the requirements of 10 CFR 63.305 are satisfied. The required characteristics of the reference biosphere have been justified. In particular, the U.S. Nuclear Regulatory Commission staff found that:

(1) The features, events, and processes used to describe the reference biosphere, the biosphere pathways, the evolution of climate, and the evolution of the geologic setting are consistent with present knowledge of the region, conditions, and past processes in the Yucca Mountain region, as required by 10 CFR 63.305(a);

(2) As required in 10 CFR 63.305(b), the biosphere (other than climate), human biology, and the state of human knowledge and technology are assumed constant from the time of license application, and changes are not projected into the future;

(3) Climate evolution is consistent with the geologic record of natural climate change in the region surrounding the Yucca Mountain site as required by 10 CFR 63.305(c); and

(4) Biosphere pathways are consistent with arid or semi-arid conditions as required by 10 CFR 63.305(d).

U.S. Nuclear Regulatory Commission staff has reviewed the Safety Analysis Report and other information submitted in support of the license application, relevant to biosphere characteristics modeling and the characteristics of the reasonably maximally exposed individual, and has found, with reasonable expectation, that the requirements of 10 CFR 63.312 are satisfied. The required characteristics of the reasonably maximally exposed individual have ben satisfied. In particular, the U.S. Nuclear Regulatory Commission staff found that:

(1) The reasonably maximally exposed individual is a hypothetical person living in the accessible environment above the highest radionuclide concentration in the plume of contamination, with a diet and living style representative of people who now live in the Town of Amargosa Valley, Nevada. The reasonably maximally exposed individual has metabolic and physical characteristics, and well water usage patterns that meet the requirements of 10 CFR 63.312(a)–(e).

2.2.1.3.14.5 References

Altman, W.D., J.P. Donnelly, and J.E. Kennedy. NUREG–1297, "Generic Technical Position on Peer Review for High-Level Nuclear Waste Repositories." Washington, DC: U.S. Nuclear Regulatory Commission. 1988a.

———. NUREG–1298, "Generic Technical Position on Qualification of Existing Data for High-Level Nuclear Waste Repositories." Washington, DC: U.S. Nuclear Regulatory Commission. 1988b.

Kotra, J.P., et al. NUREG–1563, "Branch Technical Position on the Use of Expert Elicitation in the High-Level Radioactive Waste Program." Washington, DC: U.S. Nuclear Regulatory Commission. 1996.

Leigh, C.D., et al. "User's Guide for GENII-S: A Code for Statistical and Deterministic Simulation of Radiation Doses to Humans from Radionuclides in the Environment." SAND 91-0561. Albuquerque, New Mexico: Sandia National Laboratories. 1993.

2.2.1.4 Demonstration of Compliance with the Postclosure Public Health and Environmental Standards

2.2.1.4.1 Demonstration of Compliance with the Postclosure Individual Protection Standard

Review Responsibilities—High-Level Waste Branch and Environmental and Performance Assessment Branch

2.2.1.4.1.1 Areas of Review

This section reviews the analysis of repository performance that demonstrates compliance with the postclosure individual protection standard. Reviewers will also evaluate the information, required by 10 CFR 63.21(c)(11) and (12). The review of compliance with the standards for ground-water protection as required by 10 CFR 63.331 and 63.332 will be conducted using Section 2.2.1.4.3 of the Yucca Mountain Review Plan.

The staff will evaluate the following parts of the analysis of repository performance that demonstrates compliance with the postclosure individual protection standard, using the review methods and acceptance criteria in Sections 2.2.1.4.1.3 and 2.2.1.4.1.4:

(1) Scenario classes that have been included in a set of total system performance assessment calculations;

(2) Calculations of the annual dose curve; and

(3) Credibility of the total system performance assessment results, based on an understanding of assumptions and parameters of the total system performance assessment and consideration of uncertainties of the analysis.

2.2.1.4.1.2 Review Methods

Review Method 1 Scenarios Used in the Calculation of the Annual Dose as a Function of Time

Confirm that the estimates of the annual dose, as a function of time, include all scenario classes that have been determined to be sufficiently probable or to have a sufficient effect on overall performance, that they could not be screened from the total system performance assessment analyses, based on the results of the review conducted using Section 2.2.1.2 of the Yucca Mountain Review Plan.

Confirm that the U.S. Department of Energy calculation of the annual dose curve appropriately sums the contribution of each of the scenario classes consistent with the probability of each scenario class. Verify that the contribution to the annual dose from each scenario class calculation properly accounts for the effects that the time of occurrence of the disruptive events comprising the scenario class has on the consequences. Also, verify that the annual probability of occurrence of the events used to calculate the contribution to the annual dose is consistent with the results of the review conducted, using Section 2.2.1.2.2 of the Yucca Mountain Review Plan. The probabilities of occurrence of all scenario classes included in the annual dose curve should sum to one.

Review Method 2 Demonstration That the Annual Dose to the Reasonably Maximally Exposed Individual in Any Year During the Compliance Period Does Not Exceed the Postclosure Individual Protection Standard

Confirm that the U.S. Department of Energy has conducted a sufficient number of realizations for each scenario class using their total system performance assessment computer code to verify that the results of the total system performance assessment are statistically stable. Use simulations with an alternative total system performance assessment code to help confirm that the appropriate number of realizations were performed to achieve stable results.

Confirm that repository performance and the performance of individual components or subsystems are consistent and reasonable. Verify that results of alternative total system performance assessment code analyses confirm estimates of repository performance. The results should be consistent with the results examined, using Section 2.2.1.1 "System Description and Demonstration of Multiple Barriers" of the Yucca Mountain Review Plan.

Confirm that the total system performance assessment results show that the repository performance results in an annual dose, to the reasonably maximally exposed individual in any year during the compliance period, which does not exceed the postclosure individual protection standard.

Review Method 3 Credibility of the Total System Performance Assessment Code Representation of Repository Performance

In coordination with the reviewers of the model abstractions (using Section 2.2.1.3 of the Yucca Mountain Review Plan), confirm that assumptions and parameter values or distributions used in the total system performance assessment are acceptable.

Verify that the "important" assumptions and parameters identified in each of the abstracted models are adequately captured in the integrated total system performance assessment.

Confirm that the implementation of individual model abstractions are integrated in a manner that does not adversely affect the parameters and assumptions associated with any of the individual model abstractions.

Evaluate how links to other models could affect the acceptability of the assumptions, parameters values, and distributions, or modeling strategies in the model under review.

Confirm that the total system performance assessment code is properly verified, such that there is confidence that the code is modeling the physical processes in the repository system in the manner that was intended (i.e., individual modules of the total system performance assessment code produce results consistent with the results of the reviews of Sections 2.2.1.1, 2.2.1.2, and 2.2.1.3 of the Yucca Mountain Review Plan). Verify that the transfer of data between modules of the code is conducted properly (i.e., units are the same in both modules and the data are assigned to proper variables). Confirm the results from the outputs of individual models using an alternative total system performance assessment code.

Examine the U.S. Department of Energy estimate of the uncertainty in the performance assessment results (i.e., timing and magnitude of the annual dose), and confirm that it is reasonable, considering the uncertainties in modeling assumptions and parameter values reviewed, using Sections 2.2.1.2 and 2.2.1.3 of the Yucca Mountain Review Plan. Use an alternative total system performance assessment code to help confirm the results for the individual modules.

Confirm that the U.S. Department of Energy has used an appropriate approach for sampling parameters in the total system performance assessment code across their ranges of uncertainty.

2.2.1.4.1.3 Acceptance Criteria

The following acceptance criteria are based on meeting the requirements of 10 CFR 63.113(b), 63.114, and 63.312, relating to the analysis of repository performance, that demonstrates compliance with the postclosure individual protection standard.

Acceptance Criterion 1 Scenarios Used in the Calculation of the Annual Dose as a Function of Time Are Adequate.

(1) The annual dose as a function of time includes all scenario classes that have been determined to be sufficiently probable, or to have a sufficient effect on overall performance that they could not be screened from the total system performance assessment analyses; and

(2) The calculation of the annual dose curve appropriately sums the contribution of each of the disruptive event scenario classes. The contribution to the annual dose from each scenario class calculation properly accounts for the effects that the time of occurrence of the disruptive events comprising the scenario class has on the consequences. The annual probability of occurrence of the events used to calculate the contribution to the annual dose is consistent with the results of the scenario analysis. The probabilities of occurrence of all scenario classes, included in calculating the annual dose curve, sum to one.

Acceptance Criterion 2 An Adequate Demonstration Is Provided That the Annual Dose to the Reasonably Maximally Exposed Individual in Any Year During the Compliance Period Does Not Exceed the Exposure Standard.

(1) A sufficient number of realizations has been obtained, for each scenario class, using the total system performance assessment code, to ensure that the results of the calculations are statistically stable;

(2) The annual dose curve includes confidence intervals (e.g., 95th and 5th percentile) to represent the uncertainty in the dose calculations;

(3) Repository performance and the performance of individual components or subsystems are consistent and reasonable; and

(4) The total system performance assessment results confirm that the repository performance results in annual dose, to the reasonably maximally exposed individual, in any year, during the compliance period, that does not exceed the postclosure individual protection standard.

Acceptance Criterion 3 The Total System Performance Assessment Code Provides a Credible Representation of Repository Performance.

(1) Assumptions made within the total system performance assessment code are consistent among different modules of the code. The use of assumptions and parameter values that differ among modules of the code is adequately documented;

(2) The total system performance assessment code is properly verified, such that there is confidence that the code is modeling the physical processes in the repository system in the manner that was intended. The transfer of data between modules of the code is conducted properly;

(3) The estimate of the uncertainty in the performance assessment results is consistent with the model and parameter uncertainty; and

(4) The total system performance assessment sampling method ensures that sampled parameters have been sampled across their ranges of uncertainty.

2.2.1.4.1.4 Evaluation Findings

If the license application provides sufficient information and the regulatory acceptance criteria in Section 2.2.1.4.1.3 are appropriately satisfied, the staff concludes that this portion of the staff evaluation is acceptable. The reviewer writes material suitable for inclusion in the safety evaluation report prepared for the entire application. The report includes a summary statement of what was reviewed and why the reviewer finds the submittal acceptable. The staff can document the review as follows.

U.S. Nuclear Regulatory Commission staff has reviewed the Safety Analysis Report and other information submitted in support of the license application, and has found, with reasonable expectation, that the requirements of 10 CFR 63.113(b) are satisfied. The performance objectives for the geologic repository after permanent closure have been met. In particular:

(1) The engineered barrier system is designed so that, working in combination with the natural barriers, the annual dose to the reasonably maximally exposed individual meets the postclosure individual protection standard during the first 10,000 years after permanent closure, as required by 10 CFR 63.113(b); and

(2) The ability of the geologic repository to limit radiological exposures has been demonstrated, through a performance assessment, meeting the requirements of 10 CFR 63.114, and uses the reference biosphere defined in 10 CFR 63.305(a)–(e), the reasonably maximally exposed individual as defined in 10 CFR 63.312(a)–(e), and excludes the effects of human intrusion.

U.S. Nuclear Regulatory Commission staff has reviewed the Safety Analysis Report and other information submitted in support of the license application, and has found that the requirements of 10 CFR 63.114(a) are satisfied. Technical requirements for conducting a performance assessment have been met. In particular:

(1) Appropriate data from the site and surrounding region, uncertainties and variabilities in parameter values, and alternate conceptual models have been used in the analyses, in compliance with 10 CFR 63.114(a)–(c);

(2) The U.S. Department of Energy has considered those events that have at least one chance in 10,000 of occurring over 10,000 years, in compliance with 10 CFR 63.114(d);

(3) Specific features, events, and processes have been included in the analyses, and appropriate technical bases have been provided for inclusion or exclusion, in compliance with 10 CFR 63.114(e);

(4) Specific degradation, deterioration, and alteration processes have been included in the analyses, taking into consideration their effects on annual dose, and appropriate technical bases have been provided for inclusion or exclusion, in compliance with 10 CFR 63.114(f); and

(5) Adequate technical bases are provided for models used in the performance assessment, as required by 10 CFR 63.114(g).

2.2.1.4.1.5 References

None.

2.2.1.4.2 Demonstration of Compliance with the Human Intrusion Standard

Review Responsibilities—High-Level Waste Branch and Environmental and Performance Assessment Branch

2.2.1.4.2.1 Areas of Review

This section reviews the analysis of performance in the event of limited human intrusion. Reviewers will also evaluate the information, required by 10 CFR 63.21(c)(13).

The staff will evaluate the following parts of the analysis of performance, in the event of limited human intrusion, using the review methods and acceptance criteria in Sections 2.2.1.4.2.2 and 2.2.1.4.2.3:

(1) Results of the separate total system performance assessment performed for human intrusion;

(2) Technical bases and associated analyses used to determine the time of occurrence of human intrusion without recognition by the drillers; and

(3) Credibility of the evaluation of human intrusion based on an understanding of assumptions and parameters of the total system performance assessment, characteristics of the intrusion event, and consideration of uncertainties in the analysis.

2.2.1.4.2.2 Review Methods

Review Method 1 Evaluation of the Time of Occurrence of an Intrusion Event

Verify that the technical bases and associated analyses used to determine the time of occurrence of human intrusion (by drilling through a degraded engineered barrier without recognition by the drillers) are adequate and appropriate. For example, the technical bases include analyses of the time to which the engineered barrier system has degraded to the point at which a driller can intercept a waste package, but not recognize it has occurred.

Review Method 2 Evaluation of An Intrusion Event That Demonstrates That the Annual Dose to the Reasonably Maximally Exposed Individual in Any Year During the Compliance Period Is Acceptable

Confirm that the total system performance assessment for human intrusion is performed separately from the overall total system performance assessment, and meets the requirements for performance assessments, specified in 10 CFR 63.114. If exposures to the reasonably maximally exposed individual occur more than 10,000 years after permanent closure, as a result of human intrusion occurring at any time after disposal, confirm that results of the total system performance assessment for human intrusion and its bases are provided in the environmental impact statement for Yucca Mountain.

Verify that the total system performance assessment for human intrusion is identical to the total system performance assessment for individual protection, except that it assumes the occurrence of a postulated human intrusion event with characteristics, as defined in 10 CFR 63.322 and excludes the consideration of unlikely natural features, events, and processes, as specified at 10 CFR 63.342.

Confirm that a sufficient number of realizations has been run, for each scenario class, using the total system performance assessment code to ensure that the results of the calculations are statistically stable.

Verify that the estimated repository performance is reasonable and consistent with the results evaluated during the review, using Section 2.2.1.4 of the Yucca Mountain Review Plan, and with the characteristics of the postulated intrusion event. Use results of an alternative total system performance assessment code to confirm repository performance with the postulated intrusion event.

Verify that the annual dose curve for limited human intrusion confirms that the repository system meets performance objectives, specified in 10 CFR 63.321, for limited human intrusion events.

Review Method 3 The Total System Performance Assessment Code Representation of the Intrusion Event

In coordination with the reviewers of the model abstractions (using Section 2.2.1.3 of the Yucca Mountain Review Plan), confirm that assumptions made within the total system performance assessment for evaluating the postulated intrusion event are consistent among different modules of the code. Verify that any use of assumptions and parameter values that differ among modules of the code is adequately documented.

Confirm that the total system performance assessment code is properly verified, such that there is confidence that the code is modeling the physical processes in the repository system in the manner that is consistent with the characteristics of the postulated intrusion event. Verify that the transfer of data between modules of the code is conducted properly (i.e., units are the same in both modules and the data are assigned to proper variables). Use an alternative total system performance assessment code to confirm the U.S. Department of Energy results for the outputs of individual modules.

Verify that the estimate of the uncertainty in the performance assessment results (i.e., timing and magnitude of annual dose) is consistent with the uncertainties considered in the characteristics of the postulated intrusion event and the uncertainties (i.e., model and parameter uncertainty) evaluated, using Sections 2.2.1.2 and 2.2.1.3 of the Yucca Mountain Review Plan.

Confirm that the total system performance assessment sampling method ensures that sampled parameters of the postulated intrusion event have been sampled across their ranges of uncertainty.

2.2.1.4.2.3 Acceptance Criteria

The following acceptance criteria are based on meeting the requirements of 10 CFR 63.113(d), relating to analysis of performance in the event of limited human intrusion.

Acceptance Criterion 1 Evaluation of the Time of an Intrusion Event.

(1) The technical basis and associated analyses adequately support the selection of time of occurrence of human intrusion, as specified in 10 CFR 63.321.

Acceptance Criterion 2 Evaluation of an Intrusion Event Demonstrates That the Annual Dose to the Reasonably Maximally Exposed Individual in Any Year During the Compliance Period Is Acceptable.

(1) The total system performance assessment for human intrusion is performed separately from the overall total system performance assessment, and meets the requirements for performance assessments, specified in 10 CFR 63.114;

(2) The total system performance assessment for human intrusion is identical to the total system performance assessment for individual protection, except that it assumes the occurrence of a postulated human intrusion event with characteristics, as defined in 10 CFR 63.322 and excludes the consideration of unlikely natural features, events, and processes;

(3) A sufficient number of realizations has been run using the total system performance assessment code, to ensure that the results of the calculations are statistically stable;

(4) The estimated repository performance is reasonable and consistent with the analysis of overall repository performance, and with the characteristics of the postulated intrusion event; and

(5) The annual dose curve for limited human intrusion confirms that the repository system meets performance objectives, specified in 10 CFR 63.321, for limited human intrusion events.

Acceptance Criterion 3 The Total System Performance Assessment Code Provides a Credible Representation of the Intrusion Event.

(1) Assumptions made on the method of transport from a breached waste package within the total system performance assessment for evaluating the postulated intrusion event are consistent among different modules of the code. The use of assumptions and parameter values that differ among modules of the code is adequately documented;

(2) The total system performance assessment code for evaluating human intrusion is properly verified, such that there is confidence that the code is modeling the physical processes in the repository system in the manner that is consistent with the characteristics of the postulated intrusion event. The transfer of data between modules of the code is conducted properly;

(3) The estimate of the uncertainty in the performance assessment results is consistent with the uncertainties considered in the characteristics of the postulated intrusion event, and with model and parameter uncertainty; and

(4) The sampling method used in the total system performance assessment ensures that sampled parameters of the postulated intrusion event have been sampled across their ranges of uncertainty.

2.2.1.4.2.4 Evaluation Findings

If the license application provides sufficient information and the regulatory acceptance criteria in Section 2.2.1.4.2.3 are appropriately satisfied, the staff concludes that this portion of the staff evaluation is acceptable. The reviewer writes material suitable for inclusion in the safety evaluation report prepared for the entire application. The report includes a summary statement of what was reviewed and why the reviewer finds the submittal acceptable. The staff can document the review as follows.

U.S. Nuclear Regulatory Commission staff has reviewed the Safety Analysis Report and other information submitted in support of the license application, and has found, with reasonable expectation, that the requirements of 10 CFR 63.113(d) are satisfied. The requirements for demonstrating repository performance, in the event of limited human intrusion, have been met.

2.2.1.4.2.5 References

None.

2.2.1.4.3 Analysis of Repository Performance that Demonstrates Compliance with the Separate Ground-Water Protection Standards

Review Responsibilities—High-Level Waste Branch and Environmental and Performance Assessment Branch

2.2.1.4.3.1 Areas of Review

This section reviews analysis of repository performance that demonstrates compliance with the separate ground-water protection standards. Reviewers will also evaluate information, required by 10 CFR 63.21(c)(1), (9), (14), and (15).

The staff will evaluate the following parts of the analysis of repository performance that demonstrate compliance with the separate ground-water protection standards, using the review methods and acceptance criteria in Sections 2.2.1.4.3.2 and 2.2.1.4.3.3:

(1) Calculations of the concentrations of specified radionuclides and doses as functions of time; and

(2) Credibility and consistency of the methods and assumptions used to identify the location of highest concentration of radionuclides in the accessible environment and to estimate the physical dimensions of the 3,000 acre-feet [3.715×10^9 liters] representative volume of ground water.

To review this analysis, evaluate the adequacy of the U.S. Department of Energy license application, relative to the degree to which estimating the concentration of radionuclides in ground water affects the U.S. Department of Energy license application. Review this analysis considering the risk information evaluated in the Multiple Barriers section of the Yucca Mountain Review Plan. For example, if the U.S. Department of Energy indicates that a significant portion of the radionuclides that are projected to cross the accessible environment boundary annually will not be included in the determination of the average concentration of radionuclides in 3,000 acre-feet [3.715×10^9 liters], then perform a detailed review of this analysis. If the U.S. Department of Energy assumes that all radionuclides that reach the compliance location in a given year are included in the annual water demand of 3,000 acre-feet [3.715×10^9 liters], then conduct a simplified review focusing on the bounding assumptions. The review methods and acceptance criteria provided here are for a detailed review. Some of the review methods and acceptance criteria may not be necessary.

2.2.1.4.3.2 Review Methods

Review Method 1 Demonstration that the Ground-Water Radioactivity and Doses at Any Year During the Compliance Period Do Not Exceed the Separate Ground-Water Protection Standard

Confirm that the U.S. Department of Energy has provided an estimate of ground-water radioactivity for the representative volume of ground water that includes combined radium-226 and radium-228, gross alpha activity (including radium-226 but excluding radon and uranium), and combined beta and photon emitting radionuclides.

Verify that the average level of radioactivity in the representative volume of ground water is calculated using methods, assumptions, models, and data that are consistent with the performance assessment calculations for the undisturbed case over the period of 10,000 years after disposal (the separate ground-water protection standard does not consider unlikely events), per 10 CFR 63.342. Confirm that the calculated ground-water radioactivity is supported by adequate technical bases that are consistent with those evaluated in the performance assessment abstractions for “Flow Paths in the Saturated Zone” (Section 2.2.1.3.8), “Radionuclide Transport in the Saturated Zone” (Section 2.2.1.3.9), and the “Concentration of Radionuclides in Ground Water” (Section 2.2.1.3.12).

Compare the calculated level of ground-water radioactivity at any year during the 10,000-year compliance period to the limits established in 10 CFR 63.331.

Review Method 2 Methods and Assumptions used to Determine the Location and Shape of the Representative Volume of Ground Water

Verify that the representative volume of ground water is located along the radionuclide migration path from the proposed repository at Yucca Mountain to the accessible environment.

Compare the hydrologic and transport parameters used to determine the location of the representative volume of ground water. Confirm that assumptions, methods, models, and data are consistent with those used in repository performance assessment calculations for the undisturbed case over the period of 10,000 years after disposal. Verify that the calculations are supported by an adequate technical basis that is consistent with performance assessment abstractions for “Flow Paths in the Saturated Zone” (Section 2.2.1.3.8) and “Radionuclide Transport in the Saturated Zone” (Section 2.2.1.3.9).

Confirm that the representative volume of ground water is located in such a way that it includes the highest concentration level in the plume of contamination. Verify that the location of the highest concentration of radionuclides in the plume of contamination used for the location of the representative volume of ground water is consistent with the requirements used to define characteristics of the reasonably maximally exposed individual in 10 CFR 63.312(a). Confirm that the locations of the representative volume of ground water and the highest concentration level in the plume of contamination are consistent with the performance assessment model abstraction "Concentration of Radionuclides in Ground Water" (Section 2.2.1.3.12).

Review Method 3 Methods and Assumptions Used in Calculating the Physical Dimensions of the Representative Volume of Ground Water

Confirm that the representative volume of ground water is within an aquifer containing less than 10,000 milligrams of total dissolved solids per liter of water, and contains 3,000 acre-feet (3.714×10^9 liters).

Verify that the physical dimensions of the representative volume are determined using one of the methods defined in 10 CFR 63.332. Depending on the method selected, confirm that information such as well characteristics, ground-water flow direction, and screening intervals are consistent with those used in repository performance assessment calculations for the undisturbed case over the period of 10,000 years after disposal, and the calculations are supported by an adequate technical basis. For example, evaluate whether the levels of ground-water radioactivity in the representative volume of ground water are determined using modeling approaches and parameters that are consistent with those used in the performance assessment abstractions for "Flow Paths in the Saturated Zone" (Section 2.2.1.3.8) and "Radionuclide Transport in the Saturated Zone" (Section 2.2.1.3.9).

2.2.1.4.3.3 Acceptance Criteria

The following acceptance criteria are based on meeting the requirements of 10 CFR 63.331, relating to compliance with the separate standards for protection of ground water, and 10 CFR 63.332, relating to the representative volume of ground water.

Acceptance Criterion 1 An Adequate Demonstration is Provided That the Expected Concentration of Combined Radium-226 and Radium-228, Expected Concentration of Specified Alpha-Emitting Radionuclides, and Expected Whole Body or Organ-Specific Doses from any Photon- or Beta-Emitting Radionuclides at Any Year During the Compliance Period Do Not Exceed the Separate Ground-Water Protection Standards.

(1) The U.S. Department of Energy has provided an estimate of ground-water radioactivity for the representative volume of ground water that includes combined radium-226 and radium-228, gross alpha activity (including radium-226 but excluding radon and uranium), and combined beta- and photon-emitting radionuclides;

(2) The level of ground-water radioactivity in the representative volume of ground water is calculated using methods, assumptions, models, and data that are consistent with the repository performance assessment calculations for the undisturbed case over the period of 10,000 years after disposal, and the calculations are supported by an adequate

technical basis. For example, the level of ground-water radioactivity in the representative volume of ground water is determined using modeling approaches and parameters that are consistent with those evaluated in the performance assessment abstractions for "Flow Paths in the Saturated Zone" (Section 2.2.1.3.8) and "Radionuclide Transport in the Saturated Zone" (Section 2.2.1.3.9), and using analyses consistent with those in the "Concentration of Radionuclides in Ground Water" (Section 2.2.1.3.12) model abstraction, and excludes the consideration of unlikely natural features, events, and processes; and

(3) The average level of ground-water radioactivity at any year during the 10,000-year compliance period meets the limits specified in 10 CFR 63.331.

Acceptance Criterion 2 The Methods and Assumptions Used to Determine the Position of the Representative Volume of Ground Water are Credible and Consistent, and the Representative Volume of Ground Water Includes the Highest Concentration Level in the Plume of Contamination in the Accessible Environment.

(1) The representative volume of ground water is located along the radionuclide migration path from the proposed repository at Yucca Mountain to the accessible environment as defined in 10 CFR 63.302;

(2) The location of the representative volume of ground water is determined using average hydrologic parameters that are consistent with those used in repository performance assessment calculations for the undisturbed case over the period of 10,000 years after disposal, and the calculations are supported by an adequate technical basis. For example, the levels of ground-water radioactivity in the representative volume of ground water are determined using modeling approaches and parameters that are consistent with those evaluated in the performance assessment abstractions for "Flow Paths in the Saturated Zone" (Section 2.2.1.3.8) and "Radionuclide Transport in the Saturated Zone" (Section 2.2.1.3.9), and excludes the consideration of unlikely natural features, events, and processes; and

(3) The representative volume of ground water is located in such a way that it includes the highest concentration level in the plume of contamination. In this respect, the location of the highest concentration of radionuclides in the plume of contamination used for the location of the representative volume of ground water is consistent with the requirements used to define characteristics of the reasonably maximally exposed individual in 10 CFR 63.312(a). For example, the locations of the representative volume of ground water and the highest concentration level in the plume of contamination are consistent with those used in analyses in "Concentration of Radionuclides in Ground Water" (Section 2.2.1.3.12).

Acceptance Criterion 3 The Methods and Assumptions Used to Calculate the Physical Dimensions of the Representative Volume of Ground Water are Credible and Consistent.

(1) The representative volume of ground water is within an aquifer containing less than 10,000 milligrams of total dissolved solids per liter of water, and contains no more than 3,000 acre-feet [3.714×10^9 liters];

(2) The physical dimensions of the representative volume are determined using one of the methods defined in 10 CFR 63.332. Depending on the method selected, information including, but not limited to, ground-water flow direction, and screening intervals are consistent with those used in repository performance assessment calculations for the undisturbed case over the period of 10,000 years after disposal, and the calculations are supported by an adequate technical basis. For example, the levels of ground-water radioactivity in the representative volume of ground water are determined using modeling approaches and parameters that are consistent with those evaluated in the performance assessment abstractions for "Flow Paths in the Saturated Zone" (Section 2.2.1.3.8) and "Radionuclide Transport in the Saturated Zone" (Section 2.2.1.3.9), and excludes the consideration of unlikely natural features, events, and processes; and

(3) The representative volume of ground water is consistent with the water usage characteristics of the reasonably maximally exposed individual defined in 10 CFR 63.312(c) and 63.312(d). For example, the representative volume of ground water is consistent with the annual water demand used in analyses in "Concentration of Radionuclides in Ground Water" (Section 2.2.1.3.12).

2.2.1.4.3.4 Evaluation Findings

If the license application provides sufficient information and the regulatory acceptance criteria in Section 2.2.1.4.3.3 are appropriately satisfied, the staff concludes that this portion of the staff evaluation is acceptable. The reviewer writes material suitable for inclusion in the safety evaluation report prepared for the entire application. The report includes a summary statement of what was reviewed and why the reviewer finds the submittal acceptable. The staff can document the review as follows.

U.S. Nuclear Regulatory Commission staff has reviewed the Safety Analysis Report and other information submitted in support of the license application, and has found that the requirements of 10 CFR 63.331 and 10 CFR 63.332 are satisfied. The requirements for demonstrating compliance with the ground-water protection standards have been met. In particular, the U.S. Nuclear Regulatory Commission staff found that:

(1) The average concentrations of combined radium-226 and radium-228, gross alpha activity (including radium-226 but excluding radon and uranium) and combined beta and photon emitting radionuclides meet the limits required by 10 CFR 63.331.

U.S. Nuclear Regulatory Commission staff has reviewed the Safety Analysis Report and other information submitted in support of the license application, relevant to the concentration of radionuclides in ground water, and has found that the requirements of 10 CFR 63.332 are satisfied. The specific requirements for the representative volume of ground water have been met. In particular, the U.S. Nuclear Regulatory Commission staff found that:

(1) The representative volume of ground water is within an aquifer containing less than 10,000 milligrams of total dissolved solids per liter of water to meet a given water demand. Average hydrologic characteristics that are consistent with the repository performance assessment calculations are used to determine the position and dimension of the ground-water aquifers, and projects average radionuclide concentrations for the representative volume such that the highest concentration levels in the contaminant plume are included. The representative volume should also contain no more than 3,000 acre-feet [3.714×10^9 liters] and meet any other requirements specified in 10 CFR 63.332(a)(1)–(3).

(2) The dimensions of the representative volume of ground water are calculated using one of the alternative methods specified in 10 CFR 63.332(b)(1)–(2).

2.2.1.4.3.5 References

None.

2.3 Research And Development Program to Resolve Safety Questions

Review Responsibilities—High-Level Waste Branch and Environmental and Performance Assessment Branch

2.3.1 Areas of Review

This section reviews the research and development program for resolving safety questions related to structures, systems, and components important to safety and engineered or natural barriers important to waste isolation. Reviewers will evaluate the information, required by 10 CFR 63.21(c)(16). The program is required to identify, describe, and discuss those safety features or components for which further technical information is required, to confirm the adequacy of design, and engineered or natural barriers.

The staff will evaluate the following parts of the research and development program to resolve safety questions, using the review methods and acceptance criteria in Sections 2.3.2 and 2.3.3:

(1) Identification and description of safety questions;

(2) Identification and description of the research and development programs that will be conducted to resolve any safety questions for structures, systems, and components important to safety and the engineered and natural barriers important to waste isolation;

(3) A schedule for completion of the program, as related to the projected startup date of repository operation; and

(4) The design alternatives or operational restrictions available, if the results of the program do not demonstrate acceptable resolution of the safety question problem(s).

2.3.2 Review Methods

Review Method 1 Identification and Description of Safety Questions

Verify that the license application identifies safety questions. If there are deficiencies, examine the rationale for them to verify that it is adequate.

Review Method 2 Identification and Detailed Description of the Research and Development Programs to Resolve Any Safety Questions for Structures, Systems, and Components Important to Safety and the Engineered and Natural Barriers Important to Waste Isolation

Verify that for each safety question identified, a detailed research and development program has been established. Verify there is a description of the specific technical information that must be obtained to demonstrate acceptable resolution of the safety question. The description of the program should be of sufficient detail to show how the information will be obtained. Verify that criteria described in the research and development program to resolve safety questions incorporate appropriate scientific or engineering techniques to address the scope of the issues. Examine the specific programs to verify that appropriate analyses, experiments, data collection, field tests, or other techniques have been identified, and that the timing and sequence of these activities have been specified.

Review Method 3 Schedule for Completion of the Program as Related to the Projected Startup Date of Repository Operation, and Commitment to Include Resolved Questions in Amendments to the License Application

Verify schedules for resolution of safety questions specify a date by which the issues should be resolved. Schedules should include intermediate dates or events at which decisions relating to the issue resolution program implementation will be made, if appropriate. The program and schedule should be detailed enough to show the interface with the repository design, construction activities, schedule proposed for receipt and emplacement of wastes, and any other related activities. In conducting this verification, consider the accessibility of underground locations, conditions that are likely to exist at the geologic repository operations area, and other interferences that might exist during construction. Evaluate the research and development program for compatibility with other site activities and any schedule proposed for receipt and emplacement of wastes. The schedule must be compatible with: (i) other site activities and schedules, including the performance confirmation program (10 CFR Part 63, Subpart F); (ii) repository design; and (iii) site characteristics. It should also satisfy the requirements of any license conditions, established under 10 CFR 63.32 and 63.42.

Verify a commitment in the license application to include resolved questions in amendments to the license application.

Review Method 4 Design Alternatives or Operational Restrictions Available in the Event That the Results of the Program Do Not Demonstrate Acceptable Resolution of the Problem

Verify there is an alternative plan to demonstrate acceptable resolution of the safety questions. Design alternatives or operational restrictions should be discussed in the alternative plan. Confirm there is a discussion of any programs that will be conducted during operation to demonstrate the acceptability of contemplated future changes in design or operation.

2.3.3 Acceptance Criteria

The following acceptance criteria meet the requirements of 10 CFR 63.21(c)(16).

Acceptance Criterion 1 The Identification and Descriptions of Safety Questions Are Adequate.

Acceptance Criterion 2 The U.S. Department of Energy Adequately Identifies, and Describes in Detail, a Research and Development Program That Will Be Conducted to Resolve Any Safety Questions, in a Reasonable Time Period, for Structures, Systems, and Components Important to Safety, and the Engineered and Natural Barriers Important to Waste isolation.

Acceptance Criterion 3 The U.S. Department of Energy Provides a Reasonable Schedule for the Completion of the Program, as Related to the Projected Startup Date of Repository Operation, and the Date When Items Are Expected to Be Resolved. The U.S. Department of Energy Makes a Commitment to Include Resolved Questions in Requested Amendments to the License Application, as Appropriate.

Acceptance Criterion 4 The U.S. Department of Energy Provides the Design Alternatives or Operational Restrictions Available, If the Results of the Program Do Not Demonstrate Acceptable Resolution of the Problem.

2.3.4 Evaluation Findings

If the license application provides sufficient information and the regulatory acceptance criteria in Section 2.3.3 are appropriately satisfied, the staff concludes that this portion of the staff evaluation is acceptable. The reviewer writes material suitable for inclusion in the safety evaluation report prepared for the entire application. The report includes a summary statement of what was reviewed and why the reviewer finds the submittal acceptable. The staff can document the review as follows.

U.S. Nuclear Regulatory Commission staff has reviewed the Safety Analysis Report and other information submitted in support of the license application, and has found, with reasonable assurance, that the requirements of 10 CFR 63.21(c)(16) are satisfied. Requirements for identification and description of safety questions related to structures, systems, and components and the engineered and natural barriers have been met. The U.S. Department of Energy has provided a detailed description of the programs designed to resolve safety questions, including a schedule indicating when these questions would be resolved. The design alternatives or operational restrictions available, if the results of the program do not demonstrate acceptable resolution of the problem, have been provided. Repository construction can proceed, considering the scope of the safety questions and the programs and schedules for their resolution.

2.3.5 References

None.

2.4 Performance Confirmation Program

Review Responsibilities—High-Level Waste Branch and Environmental and Performance Assessment Branch

2.4.1 Areas of Review

Subpart F of 10 CFR Part 63 provides the requirements for the performance confirmation program. The staff defines performance confirmation as the program of tests, experiments, and analyses that is conducted to evaluate the adequacy of the information used to demonstrate compliance with the performance objectives in Subpart E (refer to 10 CFR 63.2). The need for a performance confirmation program is unique to high-level radioactive waste disposal. This reflects the uncertainties in estimating geologic repository performance over thousands of years. At permanent closure, 10 CFR 63.51(a)(1) requires the U.S. Department of Energy to present an update of the postclosure performance assessment. The updated assessment includes any performance confirmation data collected and relevant to postclosure performance. The U.S. Nuclear Regulatory Commission will then decide whether the U.S. Department of Energy comprehensive program of testing, monitoring, and confirmation suggests the repository will work as planned. Unless the U.S. Department of Energy designs the repository to preserve the option to retrieve the waste before permanent closure, an action reserved to the U.S. Nuclear Regulatory Commission could be foreclosed, and an unsafe condition could be transmitted to future generations. Therefore, the broad reference to the performance objectives under Subpart E in the performance confirmation definition reflects the need to consider retrievability when monitoring subsurface conditions, and that preserving the retrieval option is a preclosure performance requirement. The general requirements for the performance confirmation program do not require testing and monitoring to confirm preclosure performance in other contexts (that is, testing and monitoring structures, systems, and components important to safety). The general requirements at 10 CFR 63.131 focus on subsurface conditions, as well as the natural and engineered systems and components required for repository operation and that are designed or assumed to operate as barriers after permanent closure. The bases for the acceptance criteria are the requirements for performance confirmation, in 10 CFR Part 63, that are performance-based. Where suitable, the acceptance criteria are also risk informed, because performance confirmation focuses on those parameters and natural and engineered barriers important to waste isolation.

The staff will confirm that the submittal complies with the requirements for tests, specified by 10 CFR 63.74(b) and 10 CFR Part 63, Subpart F, "Performance Confirmation Program." The staff will evaluate the information that is relevant to the performance confirmation program and is in the Safety Analysis Report, as required by 10 CFR 63.21(c)(17).

The staff will evaluate the following parts of the performance confirmation program, using the review methods and acceptance criteria in Sections 2.4.2 and 2.4.3:

(1) General requirements for the performance confirmation program, including:

(a) Objectives of the performance confirmation program to acquire data by identified *in situ* monitoring, laboratory, and field testing, and *in situ* experiments, to indicate whether: (i) actual subsurface conditions (i.e., specific geotechnical and design parameters, including natural processes, pertaining to the geologic setting) encountered and changes in those conditions (including any interactions between natural and engineered systems) during construction and waste emplacement operations are within the limits assumed in the licensing review; and (ii) natural and engineered systems and components that are designed or assumed to operate as barriers after permanent closure are functioning as intended and anticipated;

(b) Overall schedule for performance confirmation; and

(c) Plans to implement the performance confirmation program, so the program: (i) does not adversely affect the ability of the geologic and engineered elements of the geologic repository to meet the performance objectives; (ii) provides baseline information and analysis of that information on those parameters and natural processes of the geologic setting that may change because of site characterization, construction, and operations; and (iii) monitors and analyzes changes from the baseline condition of parameters that could affect the performance of the geologic repository.

(2) Confirmation of geotechnical and design parameters, including:

(a) Technical measuring, testing, and geologic mapping program during repository construction and operation to confirm geotechnical and design parameters pertaining to natural systems and components that are designed or assumed to operate as barriers after permanent closure, to verify they are functioning, as intended and expected;

(b) Technical program to monitor, *in situ*, the thermomechanical response of the underground facility until permanent closure to ensure the performance of the geologic and engineering features is within design limits; and

(c) Surveillance program to evaluate subsurface conditions against design assumptions, including procedures to: (i) compare measurements and observations with original design bases and assumptions; (ii) determine the need for changes to the design or construction methods, if significant differences exist between the measurements and observations and the original design bases and assumptions; and (iii) report significant differences between measurements and observations and the original design bases and assumptions, their significance to health and safety, and recommended changes, to the U.S. Nuclear Regulatory Commission.

(3) Design testing including

(a) Technical program to test engineered systems and components, other than waste packages, used in the design during the early or developmental stages of construction. This includes, for example, borehole and shaft seals, backfill, and drip shields;

(b) Technical program to evaluate the thermal interaction effects of waste packages, backfill, drip shields, rock, and unsaturated zone and saturated zone water;

(c) Schedule for starting tests of engineered systems and components used in the design;

(d) Plan to conduct a test, before permanent backfill placement begins, to evaluate the effectiveness of backfill placement and compaction procedures against design requirements, if the U.S. Department of Energy includes backfill in the repository design; and

(e) Plan for conducting tests to evaluate the effectiveness of borehole, shaft, and ramp seals before full-scale sealing.

(4) Monitoring and testing waste packages, including:

(a) Plan for monitoring the condition of waste packages at the geologic repository operations area, including an evaluation of the: (i) representativeness of those waste packages chosen for monitoring, and (ii) representativeness of the waste package environment of the waste packages chosen for monitoring;

(b) Plan for laboratory experiments that focus on the internal condition of the waste packages, including an evaluation of the degree the environment experienced by the emplaced waste packages within the underground facility is duplicated in the laboratory experiments; and

(c) Duration of the waste package monitoring and testing program.

2.4.2 Review Methods

Review Method 1 Compliance with General Requirements for the Performance Confirmation Program

(1) Verify that the U.S. Department of Energy performance confirmation plan provides the program objectives. Confirm that those objectives are sufficient to meet the general requirements for the performance confirmation program. This includes verifying that enough technical information exists, and plans for specific *in situ* monitoring, laboratory, and field testing, and *in situ* experiments are identified to carry out stated objectives. Specifically, verify that the U.S. Department of Energy performance confirmation plan:

(a) Identifies the natural and engineered systems and components that are designed or assumed to operate as barriers after permanent closure, including their specific functions, the U.S. Department of Energy selected to monitor and test, to ensure they are functioning as intended and expected;

(b) Includes the method used to select the natural and engineered systems and components, that are designed or assumed to operate as barriers after permanent closure; the U.S. Department of Energy will monitor and test to ensure they are functioning as intended and expected;

(c) Identifies specific geotechnical and design parameters, pertaining to natural systems and components that are assumed to operate as barriers after permanent closure, including natural processes and considering any interactions between natural and engineered systems and components, the U.S. Department of Energy has selected to measure or observe;

(d) Includes the method used to select specific geotechnical and design parameters to be measured or observed, including natural processes and considering any interactions between natural and engineered systems and components;

(e) Includes specific *in situ* monitoring, laboratory and field testing, and *in situ* experiments to acquire needed data;

(f) Specifies which *in situ* monitoring, laboratory and field testing, or *in situ* experimental methods the U.S. Department of Energy will apply to the selected: (i) geotechnical and design parameters, including natural processes, pertaining to natural systems and components that are assumed to operate as barriers after permanent closure, including natural processes and considering; (ii) engineered systems and components that are designed or assumed to operate as barriers after permanent closure; and (iii) interactions between natural and engineered systems and components;

(g) Includes the expected changes (that is, design bases and assumptions) from baseline for the selected geotechnical and design parameters, including natural processes, pertaining to natural systems and components that are assumed to operate as barriers after permanent closure, including natural processes and considering that will result from construction and waste emplacement operations or interactions between natural and engineered systems; and

(h) Includes the intended and expected design bases for the selected natural and engineered systems and components, which are designed or assumed to operate as barriers after permanent closure.

(2) Verify the U.S. Department of Energy performance confirmation plan includes a schedule for planned activities, and assess whether the schedule is sufficient to meet the general requirements for the performance confirmation program;

(3) Assess the U.S. Department of Energy approach to implement the performance confirmation program. This includes verifying that the U.S. Department of Energy performance confirmation plan includes the information necessary to determine whether the U.S. Department of Energy will implement the program, as required, and to complete

the detailed technical reviews, using Review Methods 2, 3, and 4 of this section. Specifically, verify the U.S. Department of Energy performance confirmation plan includes:

(a) Provisions to ensure that performance confirmation activities do not adversely affect the ability of the natural and engineered elements of the geologic repository to meet the performance objectives;

(b) Baseline information for selected geotechnical and design parameters, including natural processes, pertaining to natural systems and components that are assumed to operate as barriers after permanent closure;

(c) Methods used to establish the baseline information for selected geotechnical and design parameters, including natural processes, pertaining to natural systems and components that are assumed to operate as barriers after permanent closure;

(d) A commitment to monitor and analyze changes from the baseline condition of selected geotechnical and design parameters, including natural processes, pertaining to natural systems and components that are assumed to operate as barriers after permanent closure that could affect the waste isolation of a geologic repository;

(e) A commitment to monitor engineered systems and components that are designed or assumed to operate as barriers after permanent closure to indicate whether they are functioning as intended and expected; and

(f) Terms for periodic assessment and update of the performance confirmation plan.

(4) Verify the U.S. Department of Energy performance confirmation plan includes or cites administrative procedures related to records and reports, construction records, reports of deficiencies, and inspections. Confirm that the U.S. Department of Energy administrative procedures to implement the performance confirmation program are adequate.

Review Method 2 Compliance with Requirements to Confirm Geotechnical and Design Parameters

(1) Confirm that the U.S. Department of Energy performance confirmation plan provides an acceptable program of measuring, testing, and geologic mapping, during repository construction and operation, to confirm geotechnical and design parameters, including natural processes, pertaining to natural systems and components that are assumed to operate as barriers after permanent closure. Specifically:

(a) Evaluate the adequacy of the method the U.S. Department of Energy used to select the geotechnical and design parameters to monitor and analyze;

(b) Verify that the U.S. Department of Energy list of selected geotechnical and design parameters is reasonable and complete;

(c) Evaluate the adequacy of the method the U.S. Department of Energy used to establish the baseline of the selected geotechnical and design parameters;

(d) Verify that the baseline for each of the selected geotechnical and design parameters, established by the U.S. Department of Energy, is reasonable;

(e) Confirm that the U.S. Department of Energy estimates of the expected changes (that is, original design bases and assumptions) from baseline for the selected geotechnical and design parameters are reasonable; and

(f) Verify that the monitoring, testing, or experimental methods are suitable for each geotechnical or design parameter the U.S. Department of Energy will monitor and analyze.

(2) Verify the U.S. Department of Energy performance confirmation program includes plans to monitor, *in situ*, the thermomechanical response of the underground facility until permanent closure, and evaluate the adequacy of those plans. Specifically:

(a) Evaluate the adequacy of the method the U.S. Department of Energy used to select the *in situ* thermomechanical response parameters to monitor and analyze;

(b) Verify that the U.S. Department of Energy list of selected *in situ* thermomechanical response parameters is reasonable and complete;

(c) Evaluate the adequacy of the method the U.S. Department of Energy used to establish the baseline of the selected *in situ* thermomechanical response parameters;

(d) Confirm that the baseline of *in situ* thermomechanical response parameters, established by the U.S. Department of Energy, are reasonable;

(e) Verify that the U.S. Department of Energy estimates of the anticipated changes (i.e., original design bases and assumptions) from baseline for the selected *in situ* thermomechanical response parameters are reasonable; and

(f) Confirm that the monitoring, testing, or experimental methods are suitable for each *in situ* thermomechanical response parameter the U.S. Department of Energy will monitor and analyze.

(3) Verify that the U.S. Department of Energy performance confirmation plan provides an adequate surveillance program to monitor and evaluate subsurface conditions against design assumptions. Specifically:

  (a) Verify the U.S. Department of Energy performance confirmation plan includes provisions to compare measurements and observations with original design bases and assumptions. Evaluate the adequacy of those procedures;

  (b) Verify the U.S. Department of Energy performance confirmation plan includes provisions to determine the need for modifications to the design or construction methods, if significant differences exist between the measurements and observations and the original design bases and assumptions. Evaluate the adequacy of those procedures; and

  (c) Verify the U.S. Department of Energy performance confirmation plan includes provisions to report significant differences between measurements and observations and the original design bases and assumptions, their significance to health and safety, and recommended changes to the U.S. Nuclear Regulatory Commission. Evaluate the adequacy of those procedures.

Review Method 3 Compliance with Requirements for Design Testing

(1) Confirm that the U.S. Department of Energy performance confirmation plan provides an adequate program of testing engineered systems and components, other than waste packages, used in the design. Specifically:

  (a) Evaluate the adequacy of the method the U.S. Department of Energy used to select the engineered systems and components, that are designed or assumed to operate as barriers after permanent closure, that the U.S. Department of Energy will monitor and test;

  (b) Verify that the U.S. Department of Energy list of selected engineered systems and components is reasonable and complete;

  (c) Confirm that the monitoring, testing, or experimental methods are suitable for each engineered system or component the U.S. Department of Energy will monitor or test; and

  (d) Verify that the intended and expected design bases for the selected engineered systems and components are reasonable.

(2) Verify whether the U.S. Department of Energy included thermal interaction effects of waste packages, rock, unsaturated zone and saturated zone water, and other engineered systems and components in the design testing program. confirm that the testing program for thermal interaction effects is adequate. Specifically:

  (a) Evaluate the adequacy of the method the U.S. Department of Energy used to select the thermal interaction effects of waste packages, rock, unsaturated zone and saturated zone water, and other engineered systems and components in the design testing program;

(b) Verify that the U.S. Department of Energy list of selected thermal interaction effects of waste packages, rock, unsaturated zone and saturated zone water, and other engineered systems and components is reasonable and complete;

(c) Confirm that the monitoring, testing, or experimental methods are suitable for each thermal interaction effect of waste packages, rock, unsaturated zone and saturated zone water, and other engineered systems and components the U.S. Department of Energy will monitor or test; and

(d) Confirm that the intended and expected design bases for the selected thermal interaction effects of waste packages, rock, unsaturated zone and saturated zone water, and other engineered systems and components are reasonable.

(3) Verify that the schedule for testing engineered systems and components used in the design is sufficient to meet the requirements for the design testing program;

(4) Confirm that the U.S. Department of Energy performance confirmation plan provides an adequate program of tests to evaluate the effectiveness of backfill placement and compaction procedures against design requirements (only if the U.S. Department of Energy included backfill in the repository design). Specifically:

(a) Evaluate the adequacy of the method the U.S. Department of Energy used to select the backfill placement and compaction procedures in the design testing program;

(b) Confirm that the U.S. Department of Energy list of selected backfill placement and compaction procedures is reasonable and complete;

(c) Verify that the monitoring, testing, or experimental methods are suitable for the backfill placement and compaction procedures the U.S. Department of Energy will monitor or test; and

(d) Confirm that the intended and expected design bases for the selected backfill placement and compaction procedures are reasonable.

(5) Verify that the U.S. Department of Energy performance confirmation plan provides an adequate program of tests to evaluate the effectiveness of borehole, shaft, and ramp seals before full-scale sealing (only if the U.S. Department of Energy included seals for borehole, shaft, and ramp in the repository design). Specifically:

(a) Evaluate the adequacy of the method the U.S. Department of Energy used to select the program of tests to evaluate the effectiveness of borehole, shaft, and ramp seals before full-scale sealing in the design testing program;

(b) Confirm that the U.S. Department of Energy program of tests to evaluate the effectiveness of borehole, shaft, and ramp seals before full-scale sealing is reasonable and complete;

(c) Verify that the monitoring, testing, or experimental methods are suitable for the program of tests to evaluate the effectiveness of borehole, shaft, and ramp seals before full-scale sealing; and

(d) Confirm that the intended and expected design bases for the selected program of tests to evaluate the effectiveness of borehole, shaft, and ramp seals before full-scale sealing are reasonable.

Review Method 4 Compliance with Requirements for Monitoring and Testing Waste Packages

(1) Confirm that the U.S. Department of Energy performance confirmation plan provides an adequate program for monitoring the condition of waste packages at the geologic repository operations area. Verify the plan requires an evaluation of the: (i) representativeness of those waste packages chosen for monitoring; and (ii) representativeness of the waste package environment of the waste packages chosen for monitoring. Specifically:

(a) Evaluate the waste packages the U.S. Department of Energy will monitor and test to verify that they are representative of those to be emplaced in terms of materials, design, structure, fabrication, and inspection methods;

(b) Verify that the environment of the waste packages the U.S. Department of Energy will monitor and test is representative of the emplacement environment and is consistent with safe operations;

(c) Confirm the environmental conditions the U.S. Department of Energy will monitor and evaluate include, but are not limited to, those describing water chemistry;

(d) Verify that monitoring and testing includes evaluation of closure welds, fabrication defects, and post-fabrication damage, in particular damage that may occur during handling operations; and

(e) Verify the program is technically feasible, taking into consideration whether the methods proposed are suitable and practicable, and the sensors and devices to be used are either able to sustain the prevailing environmental conditions (e.g., temperature, humidity, radiation) during the required period of repository operation or will be replaceable.

(2) Confirm that the U.S. Department of Energy performance confirmation plan provides an adequate program of laboratory experiments that focus on the internal condition of the waste packages. Verify the plan includes an evaluation of the degree the environment experienced by the emplaced waste packages within the underground facility is

duplicated in the laboratory, as well as confirm that this evaluation is adequate. Specifically:

(a) Verify that the program and plan provide data needed to confirm waste package performance assessment models and assumptions; and

(b) Confirm that corrosion monitoring and testing includes, but is not limited to, the use of corrosion coupons.

(3) Confirm that the schedule for the waste package monitoring and testing program is sufficient to meet the requirements for such a program.

2.4.3 Acceptance Criteria

The following acceptance criteria are based on meeting the requirements of 10 CFR 63.131, 63.132, 63.133, and 63.134 for the performance confirmation program.

Acceptance Criterion 1 The Performance Confirmation Program Meets the General Requirements Established for Such a Program.

(1) The objectives of the performance confirmation program are consistent with the general requirements in that the program will provide data to indicate whether: (i) actual subsurface conditions encountered and changes in those conditions during construction and waste emplacement operations are within the limits assumed in the licensing review; and (ii) natural and engineered systems and components that are designed or assumed to operate as barriers after permanent closure are functioning as intended and expected. The performance confirmation plan provides sufficient technical information and plans for *in situ* monitoring, laboratory and field testing, and *in situ* experiments to carry out the objectives in that:

(a) It identifies the natural and engineered systems and components that are designed or assumed to operate as barriers after permanent closure, including their specific functions, the U.S. Department of Energy selected to monitor and test, to ensure they are functioning as intended and expected;

(b) It includes the method used to select the natural and engineered systems and components, which are designed or assumed to operate as barriers after permanent closure, the U.S. Department of Energy will monitor and test, to ensure they are functioning as intended and expected;

(c) It identifies specific geotechnical and design parameters, pertaining to natural systems and components that are assumed to operate as barriers after permanent closure including natural processes and any interactions between natural and engineered systems and components, the U.S. Department of Energy selected to be measured or observed;

(d) It includes the method used to select the geotechnical and design parameters including any interactions between natural and engineered systems and components, the U.S. Department of Energy will measure or observe;

(e) It includes specific *in situ* monitoring, laboratory and field testing, and *in situ* experiments to acquire needed data;

(f) It specifies which *in situ* monitoring, laboratory and field testing, or *in situ* experimental methods the U.S. Department of Energy will apply to the selected: (i) geotechnical and design parameters, including natural processes, pertaining to natural systems and components that are assumed to operate as barriers after permanent closure; (ii) engineered systems and components that are designed or assumed to operate as barriers after permanent closure; and (iii) interactions between natural and engineered systems and components;

(g) It includes the expected changes (i.e., design bases and assumptions) from baseline for the selected geotechnical and design parameters, including natural processes, pertaining to natural systems and components that are assumed to operate as barriers after permanent closure that will result from construction and waste emplacement operations; and

(h) It includes the intended and expected design bases for the selected natural and engineered systems and components, which are designed or assumed to operate as barriers after permanent closure.

(2) The schedule for the performance confirmation program is consistent with the general requirements. The program started during site characterization and will continue until permanent closure.

(3) The U.S. Department of Energy will implement the performance confirmation program in a manner consistent with the general requirements in that:

(a) Procedures require the U.S. Department of Energy to consider adverse effects on the ability of the natural and engineered elements of the geologic repository to meet the performance objectives before initiating any *in situ* monitoring, tests, or experiments to acquire data;

(b) It provides baseline information and analysis of that information on those parameters and natural processes pertaining to pertaining to natural systems and components that are assumed to operate as barriers after permanent closure that may be changed by site characterization, construction, and operations;

(c) It commits to monitoring and analyzing changes from the baseline condition for those parameters and natural processes that could affect health and safety. Exceptions from this commitment for any particular parameter are identified and technically justified (refer to Acceptance Criterion 2 of this section);

(d) It commits to monitoring engineered systems and components that are designed or assumed to operate as barriers after permanent closure to indicate whether they are functioning as intended and expected. Exceptions from this commitment for any particular system or component are identified and technically justified (refer to Acceptance Criterion 2 of this section); and

(e) It provides terms for periodic assessment and update of the performance confirmation plan.

(4) The performance confirmation plan includes or cites procedures to meet the requirements for records and reports, specified at 10 CFR 63.71.

Acceptance Criterion 2 The Performance Confirmation Program to Confirm Geotechnical and Design Parameters Meets the Requirements Established for Such a Program.

(1) The performance confirmation plan establishes a program for measuring, testing, and geologic mapping to confirm geotechnical and design parameters, including natural processes, pertaining to natural systems and components that are assumed to operate as barriers after permanent closure. The U.S. Department of Energy will implement the program during repository construction and operation. The program is consistent with the requirements in that:

(a) Geotechnical and design parameters the U.S. Department of Energy will monitor and analyze are selected using a performance-based method that focuses on those parameters that could affect health and safety. The U.S. Department of Energy also considered the need to preserve the retrieval option;

(b) Results of performance assessments confirm the list of selected geotechnical and design parameters is reasonable and complete. The U.S. Department of Energy has justified excluding any geotechnical and design parameter that is important to waste isolation. Acceptable justification factors include the certainty provided by existing baseline information and the low likelihood of changes in that parameter as a result of construction, waste emplacement operations, or interactions between natural and engineered systems;

(c) The baseline of selected geotechnical and design parameters was determined using analytical or statistical methods appropriate for the particular parameter;

(d) The baseline of selected geotechnical and design parameters considered all data available at the time of the submittal;

(e) The effects of construction, waste emplacement operations, and interactions between natural and engineered systems are considered in the original design bases and assumptions for the geotechnical and design parameters; and

(f) Monitoring, testing, and experimental methods are suitable for the nature of individual parameters in terms of time, space, resolution, and technique. Instrumentation reliability and replacement requirements are considered;

(2) The program includes adequate plans to monitor, *in situ*, the thermomechanical response of the underground facility until permanent closure. The program is consistent with the requirements in that:

(a) *In situ* thermomechanical response parameters that the U.S. Department of Energy will monitor and analyze are selected using a performance-based method that focuses on those parameters that could affect health and safety. The U.S. Department of Energy also considered the need to preserve the retrieval option;

(b) Results of performance assessments confirm that the list of selected *in situ* thermomechanical response parameters is reasonable and complete. The U.S. Department of Energy has justified excluding any *in situ* thermomechanical response parameter that is important to waste isolation. Acceptable justification factors include the certainty provided by existing baseline information and the low likelihood of changes in that parameter as a result of construction, waste emplacement operations, or interactions between natural and engineered systems;

(c) The baseline of selected *in situ* thermomechanical response parameters was determined using analytical or statistical methods appropriate for the particular parameter;

(d) The baseline of selected *in situ* thermomechanical response parameters considered all data available at the time of the submittal;

(e) The effects of construction, waste emplacement operations, and interactions between natural and engineered systems are considered in the original design bases and assumptions for the *in situ* thermomechanical response parameters; and

(f) Monitoring, testing, and experimental methods are suitable for the nature of individual parameters in terms of time, space, resolution, and technique. Instrumentation reliability and replacement requirements are considered.

(3) The performance confirmation plan sets up a surveillance program to evaluate subsurface conditions against design assumptions. The program is consistent with the requirements in that:

(a) It includes provisions for comparing measurements and observations with original design bases and assumptions. Comparisons are done routinely and in a timely manner to ensure that if any significant differences exist between the measurements and observations and the original design bases and assumptions, their significance to health and safety, and the need for design changes can be determined quickly and efficiently;

(b) It includes provisions for determining the need for modifications to the design or construction methods if significant differences exist between measurements and observations and original design bases and assumptions. Acceptable variations in the design bases and assumptions the design would accommodate without an adverse impact on health and safety have been provided. If construction methods or design needs to be modified to address changed conditions, the U.S. Department of Energy design control process used in the design phase may be used; and

(c) It includes provisions for reporting significant differences between measurements and observations and the original design bases and assumptions, their significance to health and safety and recommended changes to the Commission. These provisions meet the requirements for reports of deficiencies specified at 10 CFR 63.73.

Acceptance Criterion 3 The Performance Confirmation Program for Design Testing Meets the Requirements Established for Such a Program.

(1) The performance confirmation plan establishes a program for design testing. The program is consistent with the requirements in that:

(a) Engineered systems and components the U.S. Department of Energy will test are selected using a performance-based method that focuses on those systems and components important to waste isolation;

(b) Results of performance assessments confirm that the list of selected engineered systems and components is reasonable and complete. The U.S. Department of Energy has justified excluding any engineered system or component that is important to waste isolation from this program. An acceptable justification factor is the certainty that the system or component can perform its intended function;

(c) Testing methods are suitable for the particular engineered system or component being tested in terms of time, space, resolution, and technique. Testing methods are selected, in part, by considering the data needed to design the engineered systems and components. Test locations are selected considering compatibility with the environment in which the components or systems are to function. Instrumentation reliability and replacement requirements have been considered; and

(d) The effects of waste emplacement operations and interactions between natural and engineered systems are considered in estimates of the intended and expected design bases.

(2) Thermal interaction effects of waste packages, rock, unsaturated zone and saturated zone water, and other engineered systems and components used in the design are

included in the design testing program. The program is consistent with the requirements in that:

(a) Thermal interaction effects of waste packages, rock, unsaturated zone and saturated zone water, and other engineered systems and components the U.S. Department of Energy will test are selected using a performance-based method that focuses on those systems and components important to health and safety;

(b) Results of performance assessments confirm that the list of selected thermal interaction effects of waste packages, rock, unsaturated zone and saturated zone water, and other engineered systems and components is reasonable and complete. The U.S. Department of Energy has justified excluding any thermal interaction effects of waste packages, rock, unsaturated zone and saturated zone water, and other engineered systems and components that are important to waste isolation from this program. An acceptable justification factor is the certainty that the system or component can perform its intended function;

(c) Testing methods are suitable for the particular thermal interaction effects of waste packages, rock, unsaturated zone and saturated zone water, and other engineered systems and components being tested in terms of time, space, resolution, and technique. Testing methods are selected, in part, by considering the data needed to design the thermal interaction effects of waste packages, rock, unsaturated zone and saturated zone water, and other engineered systems and components. Test locations are selected considering compatibility with the environment in which the components or systems are to function. Instrumentation reliability and replacement requirements have been considered; and

(d) The effects of waste emplacement operations and interactions between natural and engineered systems are considered in estimates of the intended and anticipated performance limits (that is, design assumptions).

(3) The design testing program requires that the effectiveness of backfill placement and compaction procedures against design requirements be demonstrated in an *in situ* test if backfill is included in the design. The importance of the contribution of the backfill to the long-term health and safety is considered in specifying testing requirements such as backfill material, gradation, and placement density, which are an indication of the water tightness or permeability of the backfill. Specifically:

(a) Backfill placement and compaction procedures the U.S. Department of Energy will test are selected using a performance-based method that focuses on those systems and components important to waste isolation;

(b) Results of performance assessments confirm that the list of selected backfill placement and compaction procedures is reasonable and complete. The U.S. Department of Energy has justified excluding any backfill placement and compaction procedures that are important to waste isolation from this program.

An acceptable justification factor is the experience base in implementing placement and compaction procedures and the certainty of achieving the design bases for placement and compaction;

(c) Testing methods are suitable for the particular backfill placement and compaction procedures being tested in terms of time, space, resolution and technique. Testing methods are selected, in part, by considering the data needed to design the backfill placement and compaction procedures. Test locations are selected considering compatibility with the environment in which the components or systems are to function. Instrumentation reliability and replacement requirements have been considered; and

(d) The effects of waste emplacement operations and backfill placement and compaction procedure interactions between natural and engineered systems are considered in estimates of the intended and anticipated design bases.

(4) The design testing program requires that the effectiveness of borehole, shaft, and ramp seals be demonstrated in a test before full-scale sealing. The importance of seals to the long-term performance of the repository is considered in planning the seal test program. Specifically:

(a) The program of tests to evaluate the effectiveness of borehole, shaft, and ramp seals before full-scale sealing was selected, using a performance-based method that focuses on those systems and components important to waste isolation;

(b) Results of performance assessments confirm that the program of tests to evaluate the effectiveness of borehole, shaft, and ramps seals, before full-scale sealing, is reasonable and complete. The U.S. Department of Energy has justified excluding any tests to evaluate the effectiveness of borehole, shaft, and ramp seals, before full-scale sealing, that are important to waste isolation, from this program. An acceptable justification factor is the certainty that the seals can perform their intended function considering the available experience base related to seals and the likelihood of achieving the design bases for seals;

(c) Testing methods are suitable for the particular program of tests to evaluate the effectiveness of borehole, shaft, and ramps seals before full-scale sealing, in terms of time, space, resolution and technique. Testing methods are selected, in part, by considering the data needed to design the program of tests to evaluate the effectiveness of borehole, shaft, and ramp seals before full-scale sealing. Test locations are selected considering compatibility with the environment in which the components or systems are to function. Instrumentation reliability and replacement requirements have been considered; and

(d) The effects of waste emplacement operations, and the program of tests to evaluate the effectiveness of borehole, shaft, and ramp seals, before full-scale sealing, on interactions between natural and engineered systems, are considered in estimates of the intended and anticipated design bases.

Acceptance Criterion 4 The Performance Confirmation Program for Monitoring and Testing Waste Packages Meets the Requirements Established for Such a Program.

(1) The performance confirmation plan establishes a program for monitoring and testing the condition of waste packages at the geologic repository operations area. Further, the program is adequate because:

(a) The waste packages the U.S. Department of Energy will monitor and test are representative of those to be emplaced in terms of materials, design, structure, fabrication, and inspection methods;

(b) The environment of the waste packages the U.S. Department of Energy will monitor and test is representative of the emplacement environment, and is consistent with safe operations;

(c) The environmental conditions the U.S. Department of Energy will monitor and evaluate include, but are not limited to, those describing water chemistry;

(d) Monitoring and testing include evaluation of closure welds, fabrication defects, and post-fabrication damage, in particular damage that may occur during handling operations; and

(e) The program is technically feasible, taking into consideration that the methods proposed are suitable and practicable and the sensors and devices to be used are either able to sustain the prevailing environmental conditions (e.g., temperature, humidity, radiation) during the required period of repository operation, or are replaceable.

(2) The performance confirmation plan establishes a program of laboratory experiments that focuses on the internal condition of the waste packages. The environment experienced by the emplaced waste packages is duplicated in the laboratory experiments to the extent practicable. The laboratory experiments are adequate because:

(a) They provide data needed to design the waste package and confirm performance assessment models and assumptions; and

(b) Corrosion monitoring and testing include, but are not limited to, the use of corrosion coupons.

(3) The schedule for the waste package program requires monitoring and testing to begin as soon as practicable. Monitoring and testing will continue as long as practical up to the time of permanent closure.

2.4.4 Evaluation Findings

If the license application provides sufficient information and the regulatory acceptance criteria in Section 2.4.3 are appropriately satisfied, the staff concludes that this portion of the staff evaluation is acceptable. The reviewer writes material suitable for inclusion in the safety evaluation report prepared for the entire application. The report includes a summary statement of what was reviewed and why the reviewer finds the submittal acceptable. The staff can document the review as follows.

U.S. Nuclear Regulatory Commission staff has reviewed the Safety Analysis Report and other information submitted in support of the license application and has found, with reasonable assurance, that the requirements of 10 CFR 63.74(b) and 10 CFR Part 63, Subpart F—"Performance Confirmation Program" are satisfied. The performance objectives of Subpart E are met. In particular, the staff found reasonable assurance that an acceptable performance confirmation program will be conducted to evaluate the adequacy of information supporting the granting of the license.

U.S. Nuclear Regulatory Commission staff has reviewed the Safety Analysis Report and other information submitted in support of the license application and has found, with reasonable assurance, that the requirements of 10 CFR 63.131 are satisfied. The general requirements for a performance confirmation program will be met. In particular, the staff found that:

(1) The performance confirmation program will provide data to indicate whether: (i) actual subsurface conditions encountered and changes in those conditions during construction and waste emplacement are within limits assumed in the licensing review; and (ii) natural and engineered systems and components that are designed or assumed to operate as barriers after permanent closure are functioning as intended and expected;

(2) The performance confirmation program will include *in situ* monitoring, laboratory and field testing, and *in situ* experiments, as appropriate;

(3) The performance confirmation program was started during site characterization and will continue until permanent closure; and

(4) The performance confirmation program will be implemented such that it: (i) does not adversely affect the performance of the geologic and engineered elements of the repository; (ii) provides adequate baseline information on parameters and natural processes pertaining to the geologic setting that may be changed by site characterization, construction, and operational activities; (iii) monitors and analyzes changes from the baseline condition of parameters that could affect the performance of a geologic repository; and (iv) monitors natural and engineered systems and components that are designed or assumed to operate as barriers after permanent closure.

U.S. Nuclear Regulatory Commission staff has reviewed the Safety Analysis Report and other information submitted in support of the license application and has found, with reasonable assurance, that the requirements of 10 CFR 63.132 are satisfied. The requirements to confirm geotechnical and design parameters will be met. In particular, the staff found that:

(1) An adequate continuing program of measuring, testing, and geologic mapping, during repository construction and operation, will be conducted to confirm geotechnical and design parameters (including natural processes) pertaining to the geologic setting;

(2) An adequate program to monitor or test natural systems and components that are designed or assumed to operate as barriers after permanent closure will be conducted, to ensure they are functioning as intended and expected;

(3) An adequate program to monitor, *in situ*, the thermomechanical response of the underground facility will be conducted until permanent closure; and

(4) An adequate surveillance program will be conducted to monitor and evaluate subsurface conditions against design assumptions. The surveillance program will: (i) compare measurements and observations with original design bases and assumptions; (ii) determine the need for modifications to the design or construction methods if significant differences exist between measurements and observations and the original design bases and assumptions; and (iii) report significant differences between measurements and observations and the original design bases and assumptions, their significance to health and safety, and recommended changes to the Commission.

U.S. Nuclear Regulatory Commission staff has reviewed the Safety Analysis Report and other information submitted in support of the license application and has found, with reasonable assurance, that the requirements of 10 CFR 63.133 are satisfied. The requirements for design testing will be met. In particular, the staff found that:

(1) An adequate program for testing engineered systems and components will be conducted;

(2) An adequate program for evaluating the thermal interaction effects of waste packages, rock, unsaturated zone and saturated zone water, and other engineered systems and components used in the design will be conducted;

(3) Testing will begin during the early or developmental stages of construction;

(4) Backfill placement and compaction procedures will be tested against design requirements before permanent backfill placement begins; and

(5) The effectiveness of borehole, shaft, and ramp seals will be tested before full-scale sealing proceeds.

U.S. Nuclear Regulatory Commission staff has reviewed the Safety Analysis Report and other information submitted in support of the license application and has found, with reasonable assurance, that the requirements of 10 CFR 63.134 are satisfied. The requirements for monitoring and testing waste packages will be met. In particular, the staff found that:

(1) An adequate program for monitoring and testing the condition of waste packages at the geologic repository operations area will be conducted. Waste packages will be representative of those to be emplaced and the environment will be representative of the emplacement environment;

(2) The waste package monitoring and testing program will include appropriate laboratory experiments that focus on the internal condition of the waste packages. The laboratory experiments will duplicate the environment of the emplaced waste packages to the extent practicable; and

(3) The waste package monitoring program will continue as long as practical up to the time of permanent closure.

2.4.5 References

None.

2.5 Administrative and Programmatic Requirements

2.5.1 Quality Assurance Program

Review Responsibilities—High-Level Waste Branch and Environmental and Performance Assessment Branch

Quality assurance comprises all those planned and systematic actions necessary to provide adequate confidence that the geologic repository and its structures, systems, and components important to safety, the design and characterization of engineered and natural barriers important to waste isolation, and activities related thereto will perform satisfactorily in service. Quality assurance includes quality control, which comprises those quality assurance actions related to the physical characteristics of a material, structure, system, or component that provide a means to control the quality of the material, structure, system, or component to predetermined requirements.

The purpose of this review is to verify that the U.S. Department of Energy has a quality assurance program that complies with the requirements of 10 CFR Part 63. Additionally, this Section (2.5.1) of the Yucca Mountain Review Plan will be used to confirm that changes to the U.S. Nuclear Regulatory Commission-approved quality assurance program meet the specific quality assurance program change control requirements of 10 CFR 63.144. The basis for these determinations is a review and evaluation of the U.S. Department of Energy quality assurance program and changes to it submitted in accordance with 10 CFR Part 63. The results of the review and evaluation will be documented in the safety evaluation report.

This review plan is written to accommodate the use of graded quality assurance controls for structures, systems, and components and barriers, important to safety or waste isolation, that have been categorized as low-safety-risk-significant. If a graded quality assurance process is selected by the U.S. Department of Energy, the review provisions contained in this Yucca Mountain Review Plan section must be applied to structures, systems, and components and barriers categorized as high-safety-risk-significant. As provided for in Acceptance Criterion 2 of this Section (2.5.1.2), the U.S. Department of Energy may propose reduced quality assurance controls for selected elements of the quality assurance program, for structures, systems, and components and barriers categorized as low-safety-risk-significant. This categorization process must be risk-informed. If graded quality assurance is not used, the review provisions contained in this Section (2.5.1) of the Yucca Mountain Review Plan would apply to all structures, systems, and components and barriers subject to the quality assurance requirements contained in 10 CFR Part 63. As provided for in this Section (2.5.1) of the Yucca Mountain Review Plan, the U.S. Department of Energy may propose alternatives to these review provisions.

2.5.1.1 Areas of Review

This section addresses review of the U.S. Department of Energy quality assurance program. In determining compliance with the requirements specified in 10 CFR 63.21(c)(20) and 10 CFR Part 63, Subpart G (10 CFR 63.141–144), the reviewers will evaluate information specified in 10 CFR 63.21(c)(20).

The following elements of the quality assurance program will be evaluated using the review methods and acceptance criteria in Sections 2.5.1.2 and 2.5.1.3:

(1) Quality Assurance Organization;
(2) Quality Assurance Program;
(3) Design Control;
(4) Procurement Document Control;
(5) Instructions, Procedures, and Drawings;
(6) Document Control;
(7) Control of Purchased Material, Equipment, and Services;
(8) Identification and Control of Materials, Parts, and Components;
(9) Control of Special Processes;
(10) Inspection;
(11) Test Control;
(12) Control of Measuring and Test Equipment;
(13) Handling, Storage, and Shipping;
(14) Inspection, Test, and Operating Status;
(15) Nonconforming Materials, Parts, or Components;
(16) Corrective Action;
(17) Quality Assurance Records; and
(18) Audits.

2.5.1.2 Review Methods

The review should be conducted as follows.

Each element of the quality assurance program description will be reviewed against the acceptance criteria specified in Section 2.5.1.3 of the Yucca Mountain Review Plan and the documents and positions contained in Section 2.5.1.5 of the Yucca Mountain Review Plan. The assigned High-Level Waste Branch quality assurance program reviewer will interface with the other High-Level Waste Branch reviewers to verify that they have documented the acceptability of the identification of structures, systems, and components and barriers covered by the quality assurance program (e.g., the identification of these structures, systems, and components and barriers is typically compiled in a list referred to as the Q-List). Further, if the graded quality assurance process is used, the assigned reviewer will interface with other High-Level Waste Branch reviewers to verify that they have documented the acceptability of any safety-risk-significance categorization process used to support the graded quality assurance process.

If required, the High-Level Waste Branch will process the necessary request(s) for additional information to the U.S. Department of Energy and coordinate the response with the appropriate branches for acceptance. Changes to the quality assurance program will be evaluated to assure at a minimum that such changes have not degraded the previously approved program. Consideration should be given to the current regulatory position(s) in the area of the change in determining acceptability of the change. The reviewer's judgment during the evaluation process is to be based on an assessment of the material presented. Any exceptions or proposed alternatives to the Yucca Mountain Review Plan section, including the documents and positions cited in Section 2.5.1.5 of the Yucca Mountain Review Plan, will be carefully reviewed to verify that they are clearly defined and that an adequate basis exists for acceptance.

The acceptability of the quality assurance program is evaluated by the following review procedures:

(1) The quality assurance program description should be reviewed in detail to confirm that each of the criteria of 10 CFR 63.142 has been acceptably addressed (by the quality assurance program describing how the applicable criteria are satisfied) and if there is an adequate commitment to comply with the documents and positions contained in Section 2.5.1.5 of the Yucca Mountain Review Plan. The quality assurance program description should also be reviewed to verify that the U.S. Department of Energy approach to meeting the quality assurance criteria and commitments is acceptable;

(2) The measures described to implement 10 CFR 63.142 should be evaluated to confirm that management support exists (e.g., does it appear that the quality assurance program controls have adequate review, approval, and endorsement of management?);

(3) The duties, responsibilities, and authority of personnel performing quality assurance functions should be reviewed to verify that they provide sufficient independence to effectively perform these functions;

(4) Based on: (i) review of information provided in the license application and any subsequent quality assurance program changes; (ii) meetings with the U.S. Department of Energy; (iii) assessment of the ongoing quality assurance program activities; and (iv) the results of inspections, a judgment is made and documented in the safety evaluation report that the U.S. Department of Energy is capable of implementing quality assurance responsibilities in accordance with an effective quality assurance program; and

(5) The review of program commitments and descriptions of how the commitments will be met, organizational arrangements, and capabilities to fulfill quality assurance requirements should lead to a conclusion regarding acceptability of the program, as described in Section 2.5.1.4.

    The review will confirm that the commitments and the description of how the commitments are implemented, to the extent necessary, are objective and stated in inspectable terms.

2.5.1.3 Acceptance Criteria

General Acceptance Criteria

The criteria in the following introductory paragraphs and the 18 numbered acceptance criteria are based on meeting the relevant requirements of 10 CFR Parts 21, 63.21(c)(20), 63.44, 63.73, and 63.141–144, as they relate to the quality assurance program.

The U.S. Department of Energy quality assurance program description document must describe how the applicable requirements of 10 CFR 63.142 will be satisfied.

The U.S. Department of Energy quality assurance program and associated quality assurance program controls and implementing procedures regarding activities performed must be in place before activities begin.

It is not sufficient for the U.S. Department of Energy documents to assert that particular requirements are met or provided for. The description of the quality assurance program submitted in the license application and any subsequent quality assurance program changes must identify individual position titles and organizations that are responsible for meeting particular requirements, in order to allow the reviewer to understand the process by which the U.S. Department of Energy expects to meet specific requirements and to evaluate whether or not following that process would lead to compliance with requirements. Defining a process involves establishing authorities, assigning responsibilities, and issuing instructions and procedures.

The U.S. Department of Energy shall establish a quality assurance program for site characterization; acquisition, control, and analysis of samples and data; tests and experiments; scientific studies; facility and equipment design and construction; facility operation; performance confirmation; permanent closure; and decontamination and dismantling of surface facilities in accordance with 10 CFR 63.21(c)(20) and 63.142. Applicable provisions contained in the U.S. Department of Energy quality assurance program must be incorporated into the quality assurance programs of the principal contractors as related to their applicable scope of work. The U.S. Department of Energy quality assurance program must describe how each criterion of 10 CFR 63.142 will be met. Further, if the U.S. Department of Energy chooses to implement a graded quality assurance program, the specific graded quality assurance controls for each quality assurance program element would need to be identified. The acceptance criteria used by the High-Level Waste Branch to evaluate this quality assurance program are specified in this Section (2.5.1.3) of the Yucca Mountain Review Plan. Acceptance Criteria 1–18 are organized to reflect the 18 criteria contained in 10 CFR 63.142.

The U.S. Department of Energy shall establish a quality assurance program to include all activities up to the time of receipt of high-level radioactive waste for disposal in the geologic repository. These activities include site characterization; acquisition, control, and analysis of samples and data; tests and experiments; scientific studies; facility and equipment design and construction; and performance confirmation. The Yucca Mountain Review Plan will be modified, at the appropriate time, to include facility operation, permanent closure, and decontamination and dismantling of surface facilities. The U.S. Nuclear Regulatory Commission staff should confirm that the scope of the Yucca Mountain Review Plan includes those activities described in the U.S. Department of Energy quality assurance program under review. Appropriate conditions should be imposed on quality assurance program and Yucca Mountain project approvals that reflect the scope of activities described in the quality assurance programs and applications submitted for U.S. Nuclear Regulatory Commission staff and approval by the U.S. Department of Energy.

The acceptance criteria include a commitment to comply with the documents and positions contained in Section 2.5.1.5 of the Yucca Mountain Review Plan. Where appropriate, the quality assurance program description may reference a commitment to comply with certain provisions of a document identified in Section 2.5.1.5 of the Yucca Mountain Review Plan and not repeat the text of the document in the quality assurance program. For example, it may be appropriate for the U.S. Department of Energy to indicate compliance with NQA–1–1983 and the exceptions noted in Acceptance Criterion 17 of this Section (2.5.1) of the Yucca Mountain Review Plan for the section of its quality assurance program that addresses records. In certain instances, when the quality assurance program description section references other documents (e.g., NQA–1–1983) as commitments, additional text may be needed because there may be provisions of the Yucca Mountain Review Plan section that are not addressed in the referenced documents. Thus, the commitment constitutes an integral part of the quality assurance program description and requirements.

Exceptions and alternatives to these acceptance criteria and the documents and positions contained in Section 2.5.1.5 of the Yucca Mountain Review Plan may be adopted by the U.S. Department of Energy, provided the applicant can otherwise demonstrate that it satisfies the quality assurance program requirements in 10 CFR Part 63. The High-Level Waste Branch review allows for flexibility in defining methods and controls while still satisfying pertinent regulations. If the quality assurance program description meets the applicable acceptance criteria of this Section (2.5.1.3) of the Yucca Mountain Review Plan and the commitments contained in Section 2.5.1.5 of the Yucca Mountain Review Plan or provides acceptable exceptions or alternatives, that satisfy regulatory requirements, then the program will be considered to be in compliance with pertinent U.S. Nuclear Regulatory Commission regulations.

<u>Specific Acceptance Criteria</u>

Acceptance Criterion 1 The organizational elements responsible for the quality assurance program are acceptable provided that:

(1) Responsibility for the overall quality assurance program is retained and exercised by the U.S. Department of Energy;

(2) The U.S. Department of Energy identifies and describes major delegation of work involved in establishing and implementing the quality assurance program or any part thereof to other organizations;

(3) When major portions of the U.S. Department of Energy quality assurance program are delegated:

(a) The U.S. Department of Energy describes how responsibility is exercised for the overall program. The extent of management oversight is addressed, including the location, qualifications, and number of personnel performing these functions, and the bases for them;

(b) The U.S. Department of Energy evaluates the performance (frequency and method stated once per year, although a longer cycle may be acceptable with other evaluations of individual elements) of work by the delegated organization; and

(c) Qualified individual position titles or organizational element(s) are identified within the U.S. Department of Energy organization as responsible for the quality of the delegated work before initiation of activities.

(4) Clear management controls and effective lines of communication exist for quality assurance activities among the U.S. Department of Energy and the principal contractors to assure proper management, direction, and implementation of the quality assurance program;

(5) Organization charts clearly identify all on-site and off-site organizational elements that function under the cognizance of the quality assurance program (e.g., design, engineering, procurement, shipping, receiving, storage, manufacturing, construction, inspection, auditing, testing, instrumentation and control), engineering, maintenance and preclosure (operations), modifications, dismantling, etc.; the lines of responsibility; and a description of the bases for determining the size of the quality assurance organization, including the inspection staff;

(6) The U.S. Department of Energy (and principal contractors) describe the quality assurance responsibilities of each of the organizational elements noted on the organization charts. The authorities and duties of individual positions and organizations performing activities important to safety or waste isolation are clearly established and delineated in writing;

(7) The U.S. Department of Energy (and principal contractors) identify a management position that retains overall authority and direct responsibility for the definition, direction, and effectiveness of the overall quality assurance program. (Normally, this position is the quality assurance manager.) This position has the following characteristics:

    (a) Is at the same or higher organization level as the highest line manager directly responsible for performing activities affecting quality (such as engineering, procurement, construction, and operation) and is sufficiently independent from cost and schedule;

    (b) Has effective communication channels with other senior management positions;

    (c) Has responsibility for approval of quality assurance manual(s);

    (d) Has no other duties or responsibilities, unrelated to quality assurance, that would prevent his/her full attention to quality assurance matters;

    (e) Has sufficient authority to effectively implement responsibilities; and

    (f) Is sufficiently free from cost and schedule responsibilities.

Qualification requirements for this position are established in a position description that includes the following prerequisites: management experience through assignments to responsible positions; in-depth knowledge of quality assurance regulations, policies, practices, and standards; and appropriate experience working in quality assurance or related activity in nuclear-related design, construction, or operation or in a similar technically based industry. The qualifications for this position should be at least equivalent to those described in American National Standards Institute/American Nuclear Society, American Nuclear Society–3.1–1993, "Selection and Training of

Nuclear Power Plant Personnel" [American National Standards Institute/American Nuclear Society, as endorsed by the regulatory positions in Regulatory Guide 1.8, Revision 3 (U.S. Nuclear Regulatory Commission, 2000)].

(8) Verification of conformance to established requirements is accomplished by individuals or groups, within the quality assurance organization, that do not have direct responsibility for performing the work being verified, or by individuals or groups trained and qualified in quality assurance concepts and practices and independent of the organization responsible for performing the task;

(9) Individuals and organizations performing quality assurance functions have direct access to management levels that will assure the ability to identify quality problems; initiate, recommend, or provide solutions through designated channels; and verify implementation of solutions;

(10) The organizations and the position titles of those within the organization having the above authority are identified, and clear lines of authority are provided;

(11) Designated quality assurance personnel, sufficiently free from direct pressures for cost/schedule, have the responsibility delineated in writing to stop unsatisfactory work and control further processing, delivery, installation, or use of nonconforming material until proper disposition of a nonconformance, deficiency, or unsatisfactory condition has occurred;

(12) The organizational positions with stop-work authority are identified;

(13) Provisions are established for the resolution of disputes involving quality, arising from a difference of opinion between quality assurance personnel and other department (e.g., engineering, procurement, construction, etc.) representatives;

(14) Designated quality assurance individuals are involved in day-to-day facility activities important to safety or important to waste isolation. For example, the quality assurance organization routinely attends and participates in daily work schedule and status meetings to assure that it is kept abreast of day-to-day work assignments. There is adequate quality assurance coverage relative to procedural and inspection controls, acceptance criteria, and quality assurance staffing and qualification of personnel to carry out quality assurance assignments;

(15) Policies regarding the implementation of the quality assurance program are documented and made mandatory. These policies are established at the U.S. Department of Energy Office of Civilian Radioactive Waste Management level; and

(16) If the quality assurance organizational structure of the U.S. Department of Energy or its principal contractors identifies a position for an individual, at the construction site, or the geologic repository operations area, that is responsible for directing and managing the site quality assurance program, there must be controls identified for this position in the quality assurance program. These controls must assure that the individual assigned to this position has: (i) an appropriate level within the organizational structure, (ii) identified

responsibilities, and (iii) authority to exercise proper control over the quality assurance program. These controls must also assure that this individual is free from nonquality assurance duties and can thus give full attention to ensuring that the quality assurance program at the repository site is being effectively implemented.

Acceptance Criterion 2 The activities related to the quality assurance program are acceptable provided that:

(1) The scope of the quality assurance program includes:

(a) A commitment that structures, systems, and components important to safety, design and characterization of engineered and natural barriers important to waste isolation, and activities related thereto, will be subject to the applicable controls of the quality assurance program. Such activities include, but are not limited to: site characterization; acquisition and analyses of samples and data; tests and experiments; scientific studies; facility and equipment design and construction; facility operation; performance confirmation; permanent closure; and decontamination and dismantlement of surface facilities. The structures, systems, and components, barriers, and related consumables covered by the quality assurance program are identified in the Q-list as addressed in Section 2.1.1.6, "Identification of Structures, Systems, and Components Important to Safety, Safety Controls, and Measures to Ensure Availability of the Safety Systems"; and Section 2.2.1, "Performance Assessment," of the Yucca Mountain Review Plan;

(b) A commitment that the preoperational test program (before the start of preclosure operations) will be conducted in accordance with the quality assurance program and a description of how the quality assurance program will be applied;

(c) A commitment that the development, control, and use of computer software supporting a safety or waste isolation function will be conducted in accordance with the quality assurance program and a description of how the quality assurance program will be applied; and

(d) A commitment that special equipment, environmental conditions, skills, or processes will be provided as necessary.

(2) A brief summary of the U.S. Department of Energy Office of Civilian Radioactive Waste Management quality assurance policies is given. The organizational group or individual position having responsibility for each policy statement is identified;

(3) Provisions are established to assure that quality-affecting procedures required to implement the quality assurance program are: (i) consistent with quality assurance program commitments and corporate policies, (ii) are properly documented and controlled, and (iii) made mandatory through a policy statement or equivalent document signed by the responsible official;

(4) The quality assurance organization reviews and documents concurrence with these quality-related procedures;

(5) The quality-affecting procedural controls and changes to procedural controls, of the principal contractors should be provided for the applicant's review with documented agreement of acceptance before initiation of activities affected by the quality assurance program;

(6) Provisions are included for notifying the U.S. Nuclear Regulatory Commission of changes for review and acceptance of the accepted description of the quality assurance program, in accordance with 10 CFR 63.144. Changes to the U.S. Nuclear Regulatory Commission-approved quality assurance program must be processed in accordance with the applicable requirements of 10 CFR 63.144, and revisions to the U.S. Department of Energy quality assurance program documentation should be forwarded to the U.S. Nuclear Regulatory Commission;

The U.S. Department of Energy should inform the High-Level Waste Branch of changes in the quality assurance program organizational elements, when possible, within 30 days after announcement.

(7) The U.S. Department of Energy (and its principal contractors) commit to comply with: (i) the requirements in 10 CFR 63.44, 63.73, and 63.141–144; and (ii) the documents and regulatory positions and documents contained in Section 2.5.1.5 of the Yucca Mountain Review Plan and any exceptions contained in the acceptance criteria. Further, the U.S. Department of Energy (and its principal contractors) commit to conduct activities under 10 CFR 63.73 and 10 CFR Part 21 commercial-grade-item dedication activities, in accordance with the quality assurance program;

The quality assurance organization and the necessary technical organizations should participate early in the quality assurance program definition stage to assess and identify the extent that quality assurance controls are to be applied to specific structures, systems, and components and barriers important to waste isolation. This effort may involve applying a defined graded approach to certain structures, systems, and components in accordance with their safety/risk significance and affects such disciplines as design, procurement, document control, inspection tests, special processes, records, and audits.

(8) The Graded Quality Assurance Process: A graded application of quality assurance, if used, requires U.S. Department of Energy justification and U.S. Nuclear Regulatory Commission reviewer acceptance. A graded quality assurance program is structured to apply quality assurance measures and controls to all items and activities in proportion to their importance to safety or importance to waste isolation. The graded approach for the application of quality assurance controls must be adequately described. The quality assurance program should identify items and activities that are important to safety or important to waste isolation and their degree of importance based on the safety/risk significance of the items and activities. High-safety-risk-significant items and activities should have a high level of control (e.g., the full application of the quality assurance controls), and less-safety-risk-significant items and activities may have reduced quality

assurance controls applied. However, the U.S. Department of Energy may chose to apply the highest level of quality assurance controls to all items and activities.

If the U.S. Department of Energy decides to apply quality assurance controls in a graded manner, its quality assurance program must address the various elements of the graded quality assurance process. The activities related to the graded quality assurance process include:

(a) The safety-risk-significance categorization process is adequately described and is subject to review in accordance with Section 2.1.1.6 (for preclosure) and Section 2.2.1 (for postclosure) of the Yucca Mountain Review Plan (U.S. Nuclear Regulatory Commission, 2001). Although this review is performed using other sections of the Yucca Mountain Review Plan, the quality assurance program should describe, at a high level, the safety-risk-significance categorization process;

Provisions for reassessing the safety-risk-significance categorization when new information becomes available should be appropriately described.

(b) The U.S. Department of Energy may select two or more safety-risk-significance categories (e.g., high, low, or medium). The quality assurance program describes each safety-risk-significance category selected;

(c) The selection of graded quality assurance controls to be applied to each safety-risk-significant category must be described in adequate detail. Section 3.2, "Potential Areas for Implementing Graded Quality Assurance Program Controls," of Regulatory Guide 1.176, "An Approach for Plant-Specific, Risk-Informed Decision-Making: Graded Quality Assurance" (U.S. Nuclear Regulatory Commission, 1998), provides guidance on acceptable application of graded quality assurance controls. In proposing reduced quality assurance controls, the following two basic objectives should be kept in mind: (i) the graded quality assurance program should be sufficient to reasonably ensure the design integrity and ability of the structures, systems, and components or barrier to successfully perform its intended important to safety or waste-isolation function, and (ii) the graded quality assurance program should include processes and documentation that support an effective corrective action program. The selection of graded quality assurance controls may be applied to any element of the quality assurance program;

(d) Provisions for a feedback process to adjust graded quality assurance controls should be described. Provisions for reassessing the quality assurance controls when new information becomes available through adverse trends or nonconformance reporting should be described;

The U.S. Department of Energy quality assurance program description should discuss elements specifically related to effective corrective actions and causal analysis. Because it is not completely understood at the onset of the graded quality assurance program how changes will ultimately affect structures, systems, and components fabrication, construction, installation, testing, and performance, and given that the categorization process cannot address these changes in a quantitative manner, it is important that the U.S. Department of Energy have an effective process in place so that adjustments can be made in the graded quality assurance program on the basis of repository and industry experiences. Within this area, the U.S. Department of Energy process controls should have the capability to determine whether structures, systems, and components have been treated properly in the graded quality assurance program. Failures, or adverse performance degradations, of low-safety-risk-significant structures, systems, and components should be identified in accordance with the U.S. Department of Energy corrective action programs, so that the U.S. Department of Energy can ascertain whether the reduction of the quality assurance controls has resulted in excessive nonconformances and an unacceptable decrease in performance of structures, systems, and components and barriers.

The U.S. Department of Energy should employ techniques such as monitoring, surveillance, and trend analysis to identify when a structure, system, and component is found to be unacceptable or the reliability and availability of low-safety-risk-significant structures, systems, and components are trending toward unacceptable levels. Structure, system, and component monitoring approaches should be used to accomplish this goal.

(e) Provisions for an effective root-cause analysis and corrective action as a result of the feedback process should be described. Provisions should also be described for evaluating common cause/mode failures. The U.S. Department of Energy corrective action efforts should determine, as a minimum, the apparent cause of repetitive failures of structures, systems, and components under the graded quality assurance controls so that it can be decided whether graded quality assurance controls should be adjusted. In some instances, a failure may result in an unanticipated event and may cause the categorization of the structures, systems, and components to be changed;

(f) Provisions should also be in place for the U.S. Department of Energy to obtain documented U.S. Nuclear Regulatory Commission approval before implementing any quality assurance program changes that reduce previous commitments; and

(g) The use of reduced sampling plans for low-safety-risk-significant structures, systems, and components and related activities is required to be documented in accordance with Acceptance Criterion 3 of this section.

(9) Existing or proposed quality assurance procedures are identified that reflect the documents and regulatory positions contained in Section 2.5.1.5 of the Yucca Mountain Review Plan. The requirements in 10 CFR Parts 21 and 63.73, and each criterion of 10 CFR 63.142 will be met by documented procedures. In addition, activities conducted under 10 CFR 63.73 and commercial-grade-item-dedication activities conducted under 10 CFR Part 21 must conform to the applicable provisions of the quality assurance program;

(10) A description is provided that emphasizes how the docketed quality assurance program description controls, particularly the requirements in 10 CFR 63.21(c)(20), 63.44, 63.73, and 63.141–144 and the regulatory positions and documents contained in Section 2.5.1.5 of the Yucca Mountain Review Plan, will be implemented properly;

(11) A description is provided of how management (either above or outside the quality assurance organization) regularly assesses the scope, status, and adequacy of the quality assurance program and its compliance with 10 CFR Part 63, Subpart G. These assessments should include: (i) frequent review of program status through reports, meetings, audits, surveillance, and observations; and (ii) performance of an annual assessment that is preplanned and documented, with corrective action identified and tracked;

(12) Quality-related activities (such as design and procurement) initiated before the U.S. Nuclear Regulatory Commission issuance of the license are controlled under a U.S. Nuclear Regulatory Commission-approved quality assurance program in accordance with the requirements of 10 CFR Part 63, Subpart G. Approved procedures and a sufficient number of trained personnel should be available to implement the applicable portion of the quality assurance program before the initiation of the activity;

(13) A summary description is provided on how responsibilities and control of quality-related activities are transferred from principal contractors to the U.S. Department of Energy during any phase out of principal contractor activities;

(14) A provision is included to establish any additional quality assurance program provisions for preclosure operations and to establish that such provisions should be implemented before commencement of startup activities and startup testing;

(15) Confirmation is provided to: (i) commit to continued implementation of the quality assurance program for any design or site modification or construction activities that occur during preclosure; and (ii) commit that the preoperational test program or an acceptable alternative will continue to be applied during preclosure after site modification or construction activities;

(16) Indoctrination, training, and qualification programs are established such that:

(a) Personnel responsible for performing quality-affecting activities are instructed as to the purpose, scope, and implementation of quality-related manuals, instructions, and procedures;

(b) Personnel verifying activities affecting quality are trained and qualified in the principles, techniques, and requirements of the activity being performed;

(c) For formal training and qualification programs, documentation includes the objective, content of the program, attendees, and date of attendance;

(d) Proficiency tests are given to personnel performing and verifying activities affecting quality, and acceptance criteria are developed to determine if individuals are properly trained and qualified;

(e) A certificate of qualifications clearly delineates: (i) the specific functions personnel are qualified to perform; and (ii) the criteria used to qualify personnel in each function;

(f) Proficiency of personnel performing and verifying activities affecting quality is maintained by retraining, reexamining, and/or recertifying as determined by management or program commitment;

(g) Appropriate management personnel monitor the performance of individuals involved in activities affecting quality and determine the need for retraining. A system of annual appraisal and evaluation can satisfactorily accomplish this;

(h) Qualified personnel, when required, are certified in accordance with applicable codes and standards; and

(i) For the qualification of inspection and test personnel, Appendix 2A–1 of NQA–1–1983, "Nonmandatory Guidance on the Qualification of Inspection and Test Personnel" (American Society of Mechanical Engineers, 1983), provides guidance. The provisions of Appendix 2A–1 (or acceptable alternatives) must be met as part of Supplement 2-1, "Supplementary Requirements for the Qualification of Inspection and Test Personnel."

(17) A readiness review program has been established and procedures are in place to assure that the program is executed at appropriate major milestones to complement the inspection program; and

(18) Provisions are established that effectively demonstrate through a matrix system or alternative means that each criterion of 10 CFR 63.142 is properly documented, described, and addressed by implementing procedures and/or instructions.

Acceptance Criterion 3 The activities related to design control are acceptable provided that:

(1) The scope of the design control program includes design activities associated with the preparation and review of design documents, including the correct translation of applicable regulatory requirements and design bases into design, procurement, and procedural documents. Included in the scope are such activities as field design engineering; physics (including criticality physics), seismic, stress, thermal, and hydraulic analyses; radiation shielding; compatibility of materials; delineation of acceptance criteria for inspections and tests; the Safety Analysis Report accident analyses; associated computer software; features to facilitate decontamination; suitability; accessibility for in-service inspection, maintenance, and repair; and quality standards;

(2) The term "design" includes specifications; drawings; design criteria; design bases; structures, systems, and components performance requirements for preclosure; and natural and engineered barriers of the repository system. It also includes inputs and outputs at each stage of design development (e.g., from conceptual design to final design). Design information and design activities also refer to data collection and analyses and computer software that are used in supporting design development and verification. Design information and activities include general plans and detailed procedures for data collection and analyses and related information such as test and analyses results. Data analyses include the initial step, data reduction, as well as broad system analyses (such as performance assessments), that integrate other data and analyses for individual parameters;

(3) The design control program provides for the correct translation of applicable regulatory requirements and design bases into design, procurement, and procedural documents;

(4) Measures are established to assure that applicable regulatory requirements, design bases, and design features developed through the site characterization phase activities for structures, systems, and components and software supporting a safety or waste isolation function are correctly translated into specifications, drawings, instructions, and plans;

(5) Design control measures are established and are applied to: (i) the design of structures, systems, and components that are important to safety; (ii) engineered and natural barriers that are important to waste isolation; (iii) the description of the geologic setting and the plans for data collection and analysis activities that will generate information pertinent to the repository design and that will be relied on in licensing and performance confirmation; and (iv) computer software used in such activities. These design measures must apply to the design inputs, outputs, and site characterization activities and performance-confirmation activities;

(6) Organizational responsibilities are described for preparing, reviewing, approving, and verifying design documents such as system descriptions, design input and criteria, design drawings, design analyses, related computer software supporting a safety or waste isolation function, specifications, and procedures;

(7) Errors and deficiencies in approved design documents, including design methods (such as computer software supporting a safety or waste isolation function), that could adversely affect structures, systems, and components important to safety or waste isolation are documented, and action is taken to assure that all errors and deficiencies are corrected;

(8) Deviations from specified quality standards are identified and formally documented, and procedures are established to assure their control;

(9) Internal and external design interface controls, procedures, and lines of communication among participating design organizations and across technical disciplines are established and described for the review, approval, release, distribution, and revision of documents involving design interfaces to assure that structures, systems, and

components are compatible geometrically, functionally, and with processes and environment;

(10) Procedures are established and described, requiring a documented check to verify the dimensional accuracy and completeness of design drawings and specifications;

(11) Procedures are established and described, requiring that design drawings and specifications be reviewed by the quality assurance organization to assure that the documents: (i) are prepared, reviewed, and approved in accordance with the U.S. Department of Energy procedures; and (ii) contain the necessary quality assurance requirements such as inspection and test requirements, acceptance requirements, and the extent to which inspection and test results are required to be documented;

(12) Guidelines or criteria are established and described for determining the method of design verification (e.g., design review, alternate calculations, or tests);

(13) Procedures are established and described, for design verification activities, that assure the following:

(a) The verifier is qualified and is not directly responsible for the design (i.e., neither the performer nor his/her immediate supervisor). In exceptional circumstances, the designer's immediate supervisor can perform the verification provided: the supervisor is the only technically qualified individual; the need is individually documented and approved in advance by the supervisor's management; and quality assurance audits cover frequency and effectiveness of the use of supervisors as design verifiers, to guard against abuse;

(b) Design verification, if other than by qualification testing of a prototype, is completed before release: (i) for procurement, manufacturing, or construction; or (ii) to another organization for use in other design activities. In cases where this timing cannot be satisfied, the design verification may be deferred, providing that the justification for this action is documented and the unverified portion of the design output document and all design output documents, based on the unverified data, are appropriately identified and controlled. Construction site activities associated with a design or design change should not proceed without verification past the point where the installation would become irreversible (i.e., require extensive demolition and rework). In all cases, the design verification must be complete before waste package placement in the repository, or before reliance on the structure, system, or component to perform its function;

(c) Procedural control is established for design documents that reflect the commitments of the Safety Analysis Report; this control differentiates between documents that receive formal design verification by interdisciplinary or multiorganizational teams and those that can be reviewed by a single individual (a signature and date are acceptable documentation for personnel certification). Design documents subject to procedural control include, but are not limited to, specifications, calculations, associated computer software supporting a safety or waste isolation function, system descriptions, parts of the Safety Analysis Report when used as a design document, and drawings, including flow diagrams, piping and instrument diagrams, control logic diagrams, electrical single line diagrams, structural systems for major facilities, site arrangements, and equipment locations. Specialized reviews should be used when uniqueness or special design considerations warrant; and

(d) The responsibilities of the verifier, the areas and features to be verified, the pertinent considerations to be verified, and the extent of documentation are identified in procedures.

(14) The following provisions are included if the design verification method is by test only:

(a) Procedures provide criteria that specify when verification should be by test;

(b) Prototype, component, or feature testing is performed as early as possible before installation of facility equipment, before the installation would become irreversible; and

(c) Verification by test is performed under conditions that simulate the full range, including the most adverse anticipated, design conditions, as determined by analysis.

(15) Requirements relating to scientific investigation include the following:

(a) Independent review is performed of scientific notebooks, which are required to include:

(i) Statement of objective and description of work performed;

(ii) Identification of method(s) and computer software used;

(iii) Identification of samples and measuring and test equipment used;

(iv) Description of work as it was performed, results obtained, names of individuals performing the work, and dated initials or signatures, as appropriate, of individuals making entries; and

(v) Description of changes made to methods used, as appropriate.

(b) Data are identified in a manner that facilitates traceability to associated documentation and qualification status of the data. Identification and traceability are maintained throughout the lifetime of the data. Requirements for data reduction are described in sufficient detail, to permit independent reproducibility by another qualified individual. Data that are directly relied on to address safety or waste-isolation issues must be qualified from origin or classified as accepted data. Unqualified data directly relied on to address safety or waste-isolation issues must be qualified or it cannot be used in the license application;

(c) Documentation is transparent, identifies principal lines of investigation considered, and is legible and in a form suitable for reproduction, filing, and retrieval; and

(d) As applicable, other requirements of the U.S. Department of Energy quality assurance program apply to the control of scientific investigations.

(16) Model development and approaches to validation are planned, controlled, and documented. Procedures are established for model validation [NUREG–1636 (U.S. Nuclear Regulatory Commission, 1999)];

(17) Procedures are established to assure that certified computer software supporting a safety or waste isolation function is qualified for use in design and that such use is specified, in accordance with the following requirements:

(a) Software is defined as computer programs, procedures, rules, and associated documentation developed to support a safety or waste isolation function;

(b) Software supporting a safety or waste isolation function should perform all intended functions, provide correct solutions, and not perform or cause any adverse unintended functions;

(c) Controls should be established to permit authorized access and prevent unauthorized access to computer systems;

(d) Software verification and validation activities supporting a safety or waste isolation function are planned, documented, and performed for each item of software, software changes, and system configurations that are determined to affect software. Specifically:

(i) Software verification of the various software life cycle phases (e.g., the requirement, design, implementation, and testing life cycle phases, as discussed below) is performed to assure that the products of a given life cycle phase are traceable and fulfill the requirements of the previous phase and/or previous phases;

(ii) Verification reviews identify reviewers and their specific review responsibilities; and

(iii) Individuals not directly involved with the development of the software perform software verification and validation activities. In cases where this level of independence may not be achieved, an individual associated with the development of the software may perform these activities with a higher level of management approval and documented justification.

(e) A plan or similar document addressing software quality assurance supporting a safety or waste isolation function is in existence for each new software project at the start of the software life cycle. The plan for software identifies:

(i) A description of the overall nature and purpose of the software;

(ii) The software products to which it applies;

(iii) The organization responsible for performing the work and achieving software quality and the tasks and responsibilities of that organization;

(iv) Required documentation;

(v) Standards, conventions, techniques, or methodologies that should guide the software activity;

(vi) Required software reviews; and

(vii) Methods for error reporting and corrective action.

(f) The software development and maintenance process supporting a safety or waste isolation function should proceed in a planned, traceable, and orderly manner, using a defined software life-cycle methodology, which should address the following phases:

(i) Requirement Phase

(A) Software requirements such as functionality, performance, design constraints, attributes, and external interfaces are specified, documented, and reviewed.

(ii) Design Phase

(A) Software design is developed, documented, and reviewed based on the requirements depicted in the requirements document.

(iii) Implementation Phase

(A) The design is translated into source code and resulting executables necessary to perform the functions required;

(B) The source code and resulting executables should adhere to the design specifications; and

(C) User information is developed, documented, and reviewed in accordance with the design to delineate how the software is to be used.

(iv) Testing Phase

(A) Software activities are performed, documented, and verified at the end of the implementation phase to assure that the software installs properly and satisfies the requirements for its intended use;

(B) Testing to an approved plan or process is the primary method of software validation to assure adherence to requirements and to assure that the software produces correct results for test cases;

(C) Software validation documentation describes the task and specifies criteria for accomplishing the validation of the software at the end of the development cycle; and

(D) Modifications to released software are subjected to regression testing to detect errors introduced during modification of the software, to verify that modifications have not caused unintended adverse affects, and to verify that modified software still meets specified requirements.

(v) Operations and Maintenance Phase

(A) On acceptable validation of the software, the software is designated as baselined and placed under configuration management controls.

(vi) Installation and Checkout Phase

(A) Software installation and checkout activities are performed and documented when the software is installed on a computer, or when there are changes in the operating system, to assure that the software installs properly and satisfies requirements for its intended use.

(vii) Retirement Phase

(A) The support for a software product is terminated and use of the software is prevented.

(g) A software configuration management system supporting a safety or waste isolation function should be established that consists of the following:

(i) A configuration identification that includes:

(A) A definition of the baseline elements of each software baseline;

(B) A unique identification of each software item, including version or revision, to be placed under software configuration management; and

(C) Assignment of unique identifiers that relate baseline documents to their associated software items. Cross-references between baseline documents and associated software should be maintained.

(ii) A configuration change control that includes:

(A) A release and control process for baseline elements;

(B) A formal process to control and document changes to baseline elements;

(C) A formal evaluation of the baseline element or change to the baseline element, and approval by the organization responsible for approving the baseline element;

(D) A process of transmitting information concerning approved changes to all organizations affected by the changes; and

(E) A software verification and validation process to assure that software changes are appropriately reflected in software documentation and to assure that document traceability is maintained.

(iii) A configuration status accounting that includes:

(A) A listing of approved baseline elements and unique identifiers;

(B) The status of proposed, in-process, or approved changes to baseline elements; and

(C) A history of changes to software items, including descriptions of changes between versions of software items.

(h) Requirements controlling software procurement and services supporting a safety or waste isolation function are established to assure proper verification and validation support, software maintenance, configuration control, and performance of software audits, assessments, or surveys. Requirements for the supplier's reporting of software errors to the purchaser and, as appropriate, the purchaser's reporting of software errors to the supplier are identified;

(i) Software engineering elements supporting a safety or waste isolation function must define the baseline documents that are to be maintained as records;

(j) Provisions for defect reporting and resolution specify that:

(i) A software defect reporting and resolution system is implemented for software errors and failures supporting a safety or waste isolation function, to assure that problems are promptly reported to affected organizations and to assure formal processing of problem resolutions; and

(ii) If a defect is identified in software supporting a safety or waste isolation function that adversely affects previous applications, the condition adverse to quality is documented and controlled in accordance with Acceptance Criterion 16 of this section.

(k) Provisions for control of the use of software supporting a safety or waste isolation function specify that:

(i) Affected organizations control and document the use of released software items such that comparable results can be obtained, with any differences explained, through independent replication of the process;

(ii) Use of software is independently reviewed and approved to assure that the software selected is suitable to the problem being solved; and

(iii) Documentation for the receipt of software is obtained from software configuration management and maintained for all software in operation or use.

(l) Procedures are established describing the quality assurance controls for software supporting a safety or waste isolation function that satisfy the above review provisions; and

(m) As applicable, other requirements of the U.S. Department of Energy quality assurance program apply to the control of software supporting a safety or waste isolation function.

(18) Sampling: The basis, including any supporting analyses, for the use of sampling plans for structures, systems, and components and barriers and activities related thereto, such as inspection and commercial-grade item dedication, is required to be documented. The following apply for the use of sampling plans: (i) sampling plans used for high-safety-risk-significant activities are expected to use criteria that provide 95-percent confidence that there are only 5-percent defective items in a lot (95/5); (ii) reduced sampling plans may be used for low-safety-risk-significant activities; and (iii) lots sampled are essentially homogenous;

(19) Design and specification changes, including field changes, are subject to the same design controls that were applicable to the original design;

(20) Measures are provided to assure that responsible repository site personnel are notified of design changes/modifications that may affect performance of their duties;

(21) The applicable change control requirements of 10 CFR 63.44 are described; and

(22) Procedures are established describing methods of reviewing and qualifying data used in design that were collected without a fully implemented 10 CFR Part 63 quality assurance program [NUREG–1298 (U.S. Nuclear Regulatory Commission, 1988)].

(23) Procedures are established describing the use of expert elicitation. The procedure complies with NUREG–1563, “Branch Technical Position on the Use of Expert Elicitation in the High-Level Radioactive Waste Program” (U.S. Nuclear Regulatory Commission, 1996) as addressed in Section 2.5.4 of this review plan; and

(24) Procedures are established describing the use of peer review [NUREG–1297 (U.S. Nuclear Regulatory Commission, 1988)].

Acceptance Criterion 4 The activities related to procurement document control are acceptable provided that:

(1) Procedures are established for the review of procurement documents, to determine that quality requirements are correctly stated, inspectable, and controllable; there are adequate acceptance and rejection criteria; and procurement documents have been prepared, reviewed, and approved in accordance with quality assurance program requirements. To the extent necessary, procurement documents should require contractors and subcontractors to provide an acceptable quality assurance program. The review and documented concurrence of the adequacy of quality requirements stated in procurement documents is performed by independent personnel trained and qualified in quality assurance practices and concepts;

(2) Procedures are established to assure that procurement documents include a statement of work to be performed by the contractor and identify requirements such as: (i) applicable regulatory, design, technical, administrative, and reporting requirements; (ii) drawings; (iii) specifications; (iv) codes and industry standards; (v) test and inspection and acceptance requirements; (vi) access for audit or inspection by the purchaser; (vii) identification of documentation to be submitted to the purchaser or retained by the supplier (including any retention times); (viii) requirements for reporting and disposition of nonconformances; and (ix) special process instructions that should be complied with by suppliers; and

(3) Organizational responsibilities are described for: (i) procurement planning; (ii) the preparation, review, approval, and control of procurement documents; (iii) supplier selection; (iv) bid evaluations; and (v) review and concurrence of supplier quality assurance programs before initiation of activities affected by the program. The involvement of the quality assurance organization is described.

Acceptance Criterion 5 The activities related to instructions, procedures, and drawings are acceptable provided that:

(1) Organizational responsibilities are described for ensuring that activities affecting quality are: (i) prescribed by documented instructions, procedures, and drawings; and (ii) accomplished through implementation of these documents;

(2) Procedures are established to assure that instructions, procedures, and drawings include quantitative (e.g., dimensions, tolerances, operating limits) and qualitative (e.g., workmanship samples) acceptance criteria for determining that important activities have been satisfactorily accomplished; and

(3) Procedures are established for controlling changes to field and laboratory procedures associated with exploratory investigations for site characterization and performance confirmation to assure that such changes are subsequently documented and approved in a timely manner by authorized personnel.

Acceptance Criterion 6 The activities related to document control are acceptable provided that:

(1) The scope of the document control program is described, and the types of controlled documents are identified. Controlled documents are required to include, as a minimum, design documents (e.g., calculations, drawings, specifications, analyses), including documents related to computer software developed to support a safety or waste isolation function; procurement documents; instructions and procedures for such activities as fabrication, construction, modification, installation, testing, and inspection; as-built documents; quality assurance and quality control manuals and quality-affecting procedures; Safety Analysis Reports; nonconformance/deficiency reports; and corrective action reports, including changes thereto;

(2) Procedures for the review, approval, and issuance of documents and changes thereto are established and described to assure technical adequacy and inclusion of appropriate quality requirements before implementation. The quality assurance organization, or an individual other than the one who generated the document, but qualified in quality assurance, reviews and concurs with these documents with respect to quality assurance-related aspects;

(3) Procedures are established to assure that changes to documents are reviewed and approved by the same organizations that performed the initial review and approval or by other qualified responsible organizations delegated by the U.S. Department of Energy;

(4) Procedures are established to assure that documents are available at the location where the activity will be performed before commencing the work;

(5) Procedures are established and described to assure that obsolete or superseded documents are removed and replaced by applicable revisions in work areas in a timely manner;

(6) A master list or equivalent document control system is established to identify the current revision of instructions, procedures, specifications, drawings, and procurement documents. When such a list is used, it should be updated and distributed to predetermined responsible personnel;

(7) Procedures are established and described to provide for the preparation of as-built drawings and related documentation in a timely manner to accurately reflect the actual repository design; and

(8) Maintenance, modification and inspection procedures are reviewed by qualified personnel knowledgeable in the quality assurance discipline (normally the quality assurance organization) to determine: (i) the need for inspection, identification of inspection personnel, and documentation of inspection results; and (ii) that the necessary inspection requirements, methods, and acceptance criteria have been identified.

Acceptance Criterion 7 The activities related to control of purchased material, equipment, and services are acceptable provided that:

(1) Organizational responsibilities are described for the control of purchased material, equipment, software supporting a safety or waste isolation function , and services including interfaces between design, procurement, and quality assurance organizations;

(2) Verification of suppliers' activities during fabrication, inspection, testing, and shipment of material, equipment, and components is planned and performed with quality assurance organization participation in accordance with written procedures to assure conformance to the purchase order requirements. These procedures, as applicable to the method of procurement, provide for:

(a) Specification of the characteristics or processes to be witnessed, inspected, or verified, and accepted; the method of surveillance and the extent of documentation required; and individual positions responsible for implementing these procedures; and

(b) Audits, surveillance, or inspections that assure that the supplier complies with the quality requirements. The quality assurance program requires that the effectiveness of quality control by contractors and subcontractors be assessed.

(3) Selection of suppliers is documented, filed, and maintained as a record;

(4) Procurement of spare or replacement parts for structures, systems, and components and parts thereof important to safety and engineered barriers important to waste isolation are subject to present quality assurance program controls, codes and standards, and technical requirements equal to or better than the original technical requirements, or as required to preclude repetition of defects;

(5) Receiving inspection is performed to assure that:

(a) Material, components, and equipment are properly identified, in correspondence to identification on purchase documents and receiving documentation;

(b) Material, components, equipment, and acceptance records satisfy inspection instructions before installation or use; and

(c) Specified inspection, testing, and other records (such as certificates of conformance attesting that the material, components, and equipment conform to specified requirements) are available at the facility before installation or use.

(6) Items accepted and released are identified as to their inspection status before forwarding them to a controlled storage area or releasing them for installation or further work;

(7) The supplier furnishes the following records to the purchaser:

(a) Documentation that identifies the purchased item and the specific procurement requirements (e.g., codes, standards, and specifications) met by the item;

(b) Documentation identifying any procurement requirements that have not been met; and

(c) A description of nonconformances from the procurement requirements that are dispositioned "accept as is" or "repair."

The review and acceptance of these documents should be described in the purchaser's quality assurance program.

(8) Commercial-Grade Item Dedication: For commercial "off-the-shelf" items, where specific quality assurance controls appropriate for nuclear applications cannot be imposed in a practicable manner, special quality verification requirements must be established and described to provide the necessary assurance of an acceptable item by the purchaser;

For procurement of commercial-grade items, Section 10, "Commercial Grade Items," of Supplement 7S–1 of NQA–1–1983, "Supplementary Requirements for Control of Purchased Items and Services" (American Society of Mechanical Engineers, 1983), does not adequately address commercial-grade item dedication. The guidance provided in this acceptance criteria should be used for commercial-grade item dedication.

Where the U.S. Department of Energy elects to purchase commercial-grade items and dedicate the items for use as basic components, as permitted by the requirements contained in 10 CFR Part 21, the quality assurance program must provide for the following to assure that the dedicated item will perform its intended safety or waste-isolation function:

(a) When applied to facilities licensed pursuant to 10 CFR Part 63, commercial-grade item means an item that is: (i) not subject to design or specification requirements that are unique to that facility or activities; (ii) used in applications other than that facility or activities; and (iii) to be ordered from the manufacturer/supplier on the basis of specifications set forth in the manufacturer's published product description (e.g., catalog). This definition must meet the requirements specified in 10 CFR Part 21;

(b) Important terms having specific meaning that are used in the dedication process, such as "critical characteristics," "dedication," "dedicating entity," "commercial-grade item," etc., are defined. The U.S. Department of Energy should use the following definitions when dedicating commercial-grade items for use as basic components (it is noted that additional definitions such as "commercial-grade survey" may also need to be defined):

   (i) "Critical characteristics" are those important design, material, and performance characteristics of a commercial grade item that, once verified, will provide reasonable assurance that the item will perform its intended safety or waste-isolation function;

   (ii) "Dedicating entity" means the organization that performs the dedication process. Dedication may be performed by the manufacturer of the item, a third-party dedicating entity, or the U.S. Department of Energy itself. The dedicating entity pursuant to 10 CFR 21.21(c) is responsible for identifying and evaluating deviations, reporting defects and failures to comply for the dedicated item, and maintaining auditable records of the dedicating process; and

   (iii) "Dedication" is an acceptance process undertaken to provide reasonable assurance that a commercial-grade item to be used as a basic component will perform its intended safety or waste-isolation function and, in this respect, is deemed equivalent to an item designed and manufactured under a 10 CFR Part 63, Subpart G, quality assurance program. This assurance is achieved by identifying the critical characteristics of the item and verifying their acceptability by inspection, tests, or analyses performed by a purchaser or third-party dedicating entity after delivery, supplemented as necessary by one or more of the following: commercial grade surveys; product inspections or witnessing at hold points at the manufacturer's facilities; and analyses of historical records for acceptable performance. In all cases, the dedication process should be conducted in accordance with the applicable requirements of 10 CFR Part 63, Subpart G. Final dedication of an item occurs after receipt and final acceptance by the U.S. Department of Energy or its contractor, when the item is designated for use as a basic component.

(c) If these definitions are used, the U.S. Department of Energy commits to comply with all the provisions associated with the definitions;

(d) Additional definitions are contained, in 10 CFR 21.3, that are specifically applicable to 10 CFR Part 63 and are required to be applied to U.S. Department of Energy commercial-grade-item dedication activities; and

(e) It is preferred that the above definitions be used. However, additional definitions and guidance for commercial-grade item dedication are provided in Electric Power Research Institute (1988), NP–5652, as endorsed by U.S. Nuclear Regulatory Commission Generic Letter 89-02 (U.S. Nuclear Regulatory Commission, 1989) and Generic Letter 91-05 (U.S. Nuclear Regulatory Commission, 1991). Although these documents are applicable for 10 CFR Part 50 licensees, certain elements of these documents may be appropriate for 10 CFR Part 63 commercial-grade-item-dedication activities.

(9) Sampling plans used for commercial-grade-item dedication activities are required to satisfy the requirements for sampling under Acceptance Criterion 3 of this section;

(10) Suppliers' certificates of conformance are periodically evaluated by audits, independent inspections, or tests to assure that they are valid, and the results are documented;

(11) The quality assurance program describes the responsibilities for, and requires instructions and procedures for, accepting services such as third-party audits and inspections; engineering and consulting services; installation, repair, overhaul, or maintenance work; commercial grade item dedication; and testing. It may be necessary for the acceptance methods to include one or more activities similar to the following: (i) technical verification of data; (ii) surveillance, auditing, or source inspection; and (iii) review of certifications and reports from approved suppliers;

(12) For the purchase of American Society of Mechanical Engineers Section III Code items, the U.S. Nuclear Regulatory Commission considers the referenced edition of NQA–1 in the endorsed versions of the Code to be acceptable only for the construction of American Society of Mechanical Engineers Section III items when the referenced edition of NQA–1 is used in conjunction with the other quality assurance, administrative, and reporting requirements contained in the American Society of Mechanical Engineers Section III Code. Further, applicable provisions contained in the U.S. Department of Energy quality assurance program and requirements contained in the regulations also need to be met and must be used in conjunction with the American Society of Mechanical Engineers Section III Code; and

(13) For audits of American Society of Mechanical Engineers Section III Code suppliers, the U.S. Nuclear Regulatory Commission Information Notice 86-21 and its two supplements discuss the U.S. Nuclear Regulatory Commission recognition of the American Society of Mechanical Engineers accreditation program for N Stamp Holders, and the guidance provided therein should be used by the U.S. Department of Energy. U.S. Department of Energy audits of American Society of Mechanical Engineers Section III Code suppliers shall confirm that the suppliers are satisfactorily implementing: (i) their accredited

American Society of Mechanical Engineers quality assurance program (as approved by the U.S. Department of Energy, (ii) the technical and quality provisions specified in the U.S. Department of Energy purchase order, (iii) the applicable provisions of the U.S. Department of Energy quality assurance program, and (iv) the applicable requirements contained in the regulations.

Acceptance Criterion 8 The activities related to identification and control of materials, parts, and components (including samples) are acceptable provided that:

(1) Controls are established and described to identify and control materials (including consumables), parts, and components, including samples and partially fabricated subassemblies. The description should include organizational responsibilities. Identification requirements for physical samples include the following:

(a) Samples are identified and controlled in a manner consistent with their intended use;

(b) Identification is maintained on the samples or in a manner that assures identification is established and maintained;

(c) Samples are identified from their initial collection through final use;

(d) Sample identification is documented and checked before release of samples for use;

(e) Sample identification methods include use of physical markings; and

(f) If physical markings are either impractical or insufficient, other appropriate means should be employed, such as physical separation, labels or tags attached to bags, containers, or procedural control.

(2) Procedures are established that assure identification is maintained either on the item, software, or sample, or in records traceable to them, to preclude use of incorrect or defective items. Traceability requirements for physical samples include the following:

(a) Sample identification methods assure that traceability is established and maintained from the samples to applicable implementing documents or other specifying documents; and

(b) Sample traceability assures that the sample can be trac4ed at all times from its collection through final use and any post-test retention that may be appropriate.

(3) Identification of materials and parts important to the function of structures, systems, and components important to safety can be traced to the appropriate documentation such as drawings; specifications; purchase orders; technical reports; drilling locations and logs (including well bore and depth); test records; installation and use records; manufacturing and inspection documents; deviation reports; and physical and chemical mill test reports;

(4) Correct identification of materials, parts, and components is verified and documented before release for fabrication, assembling, shipping, and installation;

(5) Correct identification of samples is verified and documented before release for use or analysis. Requirements are established to control the physical markings of samples:

   (a) Physical markings are applied using materials and methods that provide clear and legible identification;

   (b) Physical markings do not detrimentally affect sample content or form;

   (c) Physical markings are transferred to each identified sample portion when the sample is subdivided; and

   (d) Physical markings are not obliterated or hidden by surface treatments or sample preparations, unless other means of identification are substituted.

   Implementing documents specify the representative samples to be archived if the need to archive samples is identified.

(6) Procedures are established for providing traceability of items (when required by codes, standards, or specifications) to: (i) applicable specification and grade of material; (ii) heat, batch, lot, part, or serial number; and (iii) specified inspection, test, or other records such as drawings, purchase orders, deviation reports, or reports of nonconformance and their disposition;

(7) Responsibilities are assigned and procedures or instructions are issued for maintaining identification of items in prolonged storage or storage under adverse conditions by: (i) protecting markings and identification records of items in storage from deterioration caused by environmental exposure or adverse storage conditions; and (ii) restoring or replacing markings or identification records that are damaged because of aging or storage conditions;

(8) Responsibilities are assigned and procedures or instructions are issued for: (i) identifying items with limited calendar or operating life cycles; (ii) establishing records of shelf life or operating life or cycles remaining; (iii) preventing the use of items whose shelf lives have expired; and (iv) preventing further use of items, components, or materials that have reached the ends of their operating lives or cycles;

(9) Handling, storage, and shipping requirements include the following:

   (a) Handling, storage, cleaning, packaging, shipping, and preserving samples are conducted in accordance with established implementing documents or other specified documents;

   (b) Specific measures for handling, storage, cleaning, packaging, shipping, and preserving are identified and use for critical, sensitive, perishable, or high-value samples;

(c) Measures are established for the marking and labeling for packaging, shipping, handling, and storing samples, as necessary, to adequately identify, maintain, and preserve samples;

(d) Markings and labels indicate the presence of special environments or the necessity for special controls;

(e) Special equipment (e.g., containers) and special protective environments (e.g., inert gas, moisture and temperature limits) should be required for particular samples;

(f) Special handling tools and equipment are used and controlled, as necessary, to assure safe and adequate handling;

(g) Special handling tools and equipment are inspected and tested in accordance with implementing documents, and at specified time intervals, to verify that the tools and equipment are adequately maintained; and

(h) Experience and training are specified for operators of special handling and lifting equipment.

(10) Controls are established to preclude the inadvertent use of incorrect or defective items, software supporting a safety or waste isolation function, or samples;

(11) Samples that do not meet requirements specified in work controlling documents (such as job packages, travelers, or work requests) are documented, evaluated, and segregated in accordance with Acceptance Criterion 15 of this section;

(12) the disposition for nonconforming samples is identified and documented and should be limited to "use-as-is," "discard," or, where appropriate, "rework;" and

(13) As applicable, other requirements of the U.S. Department of Energy quality assurance program apply to the identification and control of materials, parts, and components (including samples).

Acceptance Criterion 9 The activities related to control of special processes are acceptable provided that:

(1) The criteria for determining those processes that are controlled as special processes are described. As complete a listing as possible of special processes, which are generally those processes where direct inspection is impossible or disadvantageous, should be provided. Examples of special processes include welding, heat treating, nondestructive examination, and chemical cleaning;

(2) Organizational responsibilities, including those for the quality assurance organization, are described for qualification of special processes, equipment, and personnel;

(3) Procedures, equipment, and personnel associated with special processes are qualified and are in conformance with applicable codes, standards, procedures, and specifications. The quality assurance organization is involved in the qualification activities to assure they are satisfactorily performed;

(4) Procedures are established for recording evidence of acceptable accomplishment of special processes using qualified procedures, equipment, and personnel;

(5) Qualification records of procedures, equipment, and personnel associated with special processes are established, filed, and maintained to be current;

(6) When no applicable codes, standards, or specifications address methods for qualifying special processes associated with scientific investigations, the following methods may be considered: (i) the conducting of a prototype test, if possible, that demonstrates that the process maintains quality or produces a quality product; and (ii) a combination of methods such as peer reviews, technical reviews, models, and testing that provides reasonable assurance that the process maintains quality or produces a quality product;

In all cases, measures are established to assure that special processes associated with scientific investigations are controlled and accomplished by qualified personnel using qualified procedures.

(7) Special processes associated with nondestructive evaluation should be performed in accordance with American Society for Nondestructive Testing–TC–1A (American Society for Nondestructive Testing, 1980). In all cases, the qualification and certification of nondestructive evaluation personnel includes a performance demonstration as part of the practical examination. In lieu of the 3-year recertification interval specified in American Society for Nondestructive Testing–TC–1A, Level III nondestructive examination personnel may be recertified on a 5-year interval.

Acceptance Criterion 10 The activities related to inspection are acceptable, provided that:

(1) The scope of the inspection program is described that indicates that an effective inspection program has been established for verifying conformance of items or activities to specified requirements. Program procedures provide criteria for determining the accuracy requirements of inspection equipment and criteria for determining when inspections are required and defining how and when inspections are performed. The quality assurance organization participates in the above functions;

(2) Organizational responsibilities for inspection are adequately described. Individuals performing inspections are other than those who performed or directly supervised the activity being inspected and do not report directly to the immediate supervisors who are responsible for the activity being inspected. If the individuals performing inspections are not part of the quality assurance organization, the inspection procedures, personnel qualification criteria, and independence from undue pressure such as cost and schedule should be reviewed and found acceptable by the quality assurance organization before the initiation of the activity;

(3) A qualification program for inspectors (including nondestructive examination personnel) is established and documented, and the qualifications and certifications of inspectors are maintained to be current;

(4) Inspection procedures, instructions, or checklists provide for the following: identification of characteristics and activities to be inspected; description of the method of inspection; identification of the individuals or groups responsible for performing the inspection operation in accordance with the provisions of the second bullet under this acceptance criteria; acceptance and rejection criteria; identification of required procedures, drawings, specifications, and revisions thereof; records of the identity of the inspector or data recorder and the results of the inspection operation; and specification of necessary measuring and test equipment, including accuracy requirements;

(5) Procedures are established and described to identify, in pertinent documents, mandatory inspection hold points beyond which work may not proceed until inspected by a designated inspector;

(6) Inspection results are documented and evaluated, and their acceptability is determined by a responsible individual or group;

(7) When inspections associated with normal operations of the site (e.g., routine maintenance, surveillance, tests) are performed by individuals other than those who performed or directly supervised the work, but are within the same group, the following controls are required: (i) the qualification criteria for the inspection personnel are reviewed and found acceptable by the quality assurance organization before initiating the inspection; and (ii) the quality of the work can be objectively demonstrated through a functional test when the activity involves breaching a pressure-retaining item; and

(8) Field surveys conducted under inspection activities are subject to the requirements of this acceptance criterion and other applicable requirements of the U.S. Department of Energy quality assurance program. The field survey system is a permanent system of horizontal and vertical controls; is used in accordance with implementing documents to obtain the accurate location and relocation of designated features, including locations of sample or data collection; and is subject to proper administrative controls and program requirements. Pertinent survey documents are identified, maintained, and verified for completeness as work progresses.

(9) Sampling plans used for inspection activities are required to meet the requirements for sampling under Acceptance Criterion 3 of this section; and

(10) Procedures are established describing methods of reviewing and qualifying data that were collected without a fully implemented 10 CFR Part 63 quality assurance program [NUREG–1298 (U.S. Nuclear Regulatory Commission, 1988)].

Acceptance Criterion 11 The activities related to test control are acceptable provided that:

(1) The scope of the test control program is described that indicates that an effective program has been established for testing activities for verifying conformance of items or activities to specified requirements and demonstrating that items will perform satisfactorily in service. The test control program encompasses, but is not limited to, such testing activities as: acquiring data from samples; prototype qualifications tests; production tests; proof tests before installation; preoperational tests; tests supporting site characterization; tests supporting scientific investigations; tests of software supporting a safety or waste isolation function; construction phase tests; and operational tests. Program procedures provide criteria for determining the accuracy requirements of test equipment and criteria for determining when tests are required and defining how and when testing activities are performed. Tests must be performed in accordance with written test procedures that identify test acceptance criteria and that incorporate, as appropriate, requirements and acceptance limits contained in applicable design documents;

(2) Test procedures or instructions provide, as required, for the following:

(a) The requirements and acceptance limits contained in applicable design and procurement documents;

(b) Instructions for performing the test;

(c) Test prerequisites such as calibrated instrumentation, adequate test equipment, and instrumentation, including their accuracy requirements, completeness of item to be tested, suitable and controlled environmental conditions, and provisions for data collection and storage;

(d) Mandatory inspection hold points for witnessing by the U.S. Department of Energy, contractor, or inspector (as required);

(e) Acceptance and rejection criteria; and

(f) Methods of documenting or recording test data and results, and provisions for ensuring that test prerequisites have been met.

(3) Test results are documented and evaluated, and their acceptability is determined by a responsible individual or group.

Acceptance Criterion 12 The activities related to control of measuring and test equipment are acceptable provided that:

(1) The scope of the program for the control of measuring and test equipment is adequately described and the types of equipment to be controlled are established;

(2) Responsibilities of quality assurance and other organizations are adequately described for establishing, implementing, and ensuring effectiveness of the calibration program;

(3) Procedures are established and described in sufficient detail for calibration (technique and frequency), maintenance, and control of the measuring and test equipment (instruments, tools, gages, fixtures, reference and transfer standards, and nondestructive test equipment) that is used in the measurement, inspection, and monitoring of structures, systems, and components;

(4) The review and documented concurrence of these procedures is described and the organization responsible for these functions is identified;

(5) Measuring and test equipment is identified and traceable to the calibration test data;

(6) Measuring and test equipment is labeled or tagged or "otherwise controlled" to indicate the due date of the next calibration, and such methods of control should be adequately described;

(7) Measuring and test equipment is calibrated at specified intervals, based on the required accuracy, purpose, degree of usage, stability characteristics, and other conditions affecting the measurement. Calibration of this equipment should be against standards that have an accuracy of at least four times the required accuracy of the equipment being calibrated or, when this is not possible, that have an accuracy that assures that the equipment being calibrated will be within required tolerance. The basis of acceptance is documented and authorized by responsible management. The management authorized to perform this function is identified;

(8) Calibration standards have greater accuracy than the standards being calibrated. Calibration standards with the same accuracy may be used if this level of accuracy can be demonstrated to be adequate for the requirements and provided that the basis of acceptance is documented and authorized by responsible management. The management authorized to perform this function is identified;

(9) Reference and transfer standards are traceable to nationally recognized standards. Where national standards do not exist, provisions are established to document the basis for calibration;

(10) When measuring and test equipment is found to be out of calibration, measures are taken and documented to determine the validity of previous inspections performed and the acceptability of items inspected or tested since the last calibration. Inspections or tests are repeated on items determined to be suspect; and

(11) Procedures are established for selecting measuring and test equipment, for use in processes, inspections, and tests, that: (i) is of the type appropriate for measuring specified characteristics of items being processed, inspected, or tested; and (ii) has sufficient range, accuracy, and tolerance to determine conformance to specified requirements.

Acceptance Criterion 13 The activities related to handling, storage, and shipping are acceptable provided that:

(1) Special handling, preservation, storage, cleaning, packaging, and shipping requirements and procedures are established and accomplished by suitably trained and, when appropriate, qualified individuals, in accordance with predetermined work and inspection instructions;

(2) Procedures are established and described to control the cleaning, handling, storage, packaging, and shipping of items, samples, materials, components, and systems, in accordance with design and procurement requirements, to preclude damage, loss, or deterioration from environmental conditions such as temperature or humidity;

(3) Provisions are described for the storage (including the control of shelf life) of chemicals, reagents, lubricants, and other consumable materials;

(4) Provisions are described for identifying special handling tools and equipment that are required for safe handling of items. Provisions are established for inspection and testing of such tools and equipment, including specification of procedures to be implemented at specified intervals to verify that such tools and equipment are adequately maintained; and

(5) Provisions are described for marking or labeling items being shipped, handled, or stored, for the purpose of identifying the items and any special environments or controls required by such items.

Acceptance Criterion 14 The activities related to inspection, test, and operating status are acceptable provided that:

(1) Procedures are established to indicate the inspection, test, and operating status of structures, systems, and components throughout fabrication, installation, testing, and operation;

(2) The status of inspection and test activities should be identified either on the items or in documents traceable to the items where it is necessary to assure that required inspections and tests are performed and to assure that items that have not passed the required inspections and tests are not inadvertently installed, used, or operated;

(3) Inspection, test, and operating status of structures, systems, and components should be identified by status indicators, such as physical location tags, markings, labels, travelers, stamps, inspection records, or other suitable means;

(4) Procedures and authority are established and described to control the application and removal of inspection and welding stamps and status indicators such as those listed in the previous bullet;

(5) Procedures are established and described to control alteration of the sequence of required tests, inspections, and other operations important to waste isolation or

important to safety. Such actions should be subject to the same controls as the original review and approval;

(6) The status of nonconforming, inoperative, or malfunctioning structures, systems, and components is documented and identified to prevent inadvertent use. The organization responsible for this function is clearly identified; and

(7) Procedures are established to prevent inadvertent use or operations of a structure, system, or component that is out of service, by indicating its operating status, by the use of tags or markings on control panels, switches, breakers, and other locations where its use or operation can be initiated.

Acceptance Criterion 15 The activities related to nonconforming materials, parts, or components are acceptable provided that:

(1) Procedures are established and described for identification, documentation, segregation, review, disposition, and notification to affected organizations of nonconforming materials, parts, structures, systems, and components, services and computer software supporting a safety or waste isolation function, if disposition is other than disposal. The procedures provide identification of authorized individuals for independent review of nonconformances, including disposition and closeout;

(2) Procedures are established for preventing the inadvertent use or installation of nonconforming items;

(3) Quality assurance and other organizational responsibilities are described for the definition and implementation of activities related to nonconformance control, including identification of individuals or groups with authority for the disposition of nonconforming items;

(4) Documentation identifies the nonconforming item; describes the nonconformance, the disposition of the nonconformance, and the inspection requirements; and includes signature approval of the disposition. Nonconformances are corrected or resolved before initiation of the preoperational test program on the item;

(5) Reworked, repaired, and replacement items are inspected and tested in accordance with the original inspection and test requirements or acceptable alternatives. Design control measures commensurate with those applied to the original design are applied when dispositioning nonconformance as "use-as-is" or "repair," and the technical bases for such dispositions are documented;

(6) Nonconformance reports are periodically analyzed by the quality assurance organization to show quality trends, and the significant results are reported to upper management for review and assessment; and

(7) Items reworked or repaired are retested or reinspected against the original acceptance criteria unless the disposition of the nonconforming item established alternate

acceptance criteria. (If the latter is the case, then a design change may be required to support the disposition.)

Acceptance Criterion 16 The activities related to corrective action are acceptable provided that:

(1) Procedures are established and described indicating an effective corrective action program has been established. The quality assurance organization reviews and documents concurrence with the procedures;

(2) Corrective action is documented and initiated after the determination of a condition adverse to quality, such as a nonconformance, failure, malfunction, deficiency, deviation, or defect in material, equipment, or samples. Conditions adverse to quality should be identified promptly and corrected as soon as practical. The quality assurance organization is involved in the documented concurrence of the adequacy of the corrective action. Followup action is taken by the quality assurance organization to verify proper implementation of corrective action and to close out the corrective action in a timely manner;

(3) A program for determining adverse quality trends is established and includes: (i) evaluation of nonconformance and other related documents to identify adverse quality trends and assist in identifying root causes; (ii) prompt identification of adverse trends; and (iii) prompt reporting of adverse trends to management;

(4) Significant conditions adverse to quality, the cause of the conditions, and the corrective actions taken to preclude repetition are documented and reported to immediate management and upper levels of management, for review and assessment, including repetitive conditions that are less significant, but when taken collectively:

    (a) Indicate a programmatic failure to properly implement the quality assurance program;

    (b) May be precursors for a significant technical deficiency or problem; or

    (c) May reduce the margin of safety.

    Significant conditions adverse to quality also include, but are not limited to: (i) loss of or potential loss, of a safety or waste-isolation function, to the extent that there is a reduction in the degree of protection provided to the public health and safety; (ii) loss, or potential loss of a safety or waste-isolation function to the extent that there is a reduction in the degree of protection provided for worker safety, (iii) common-cause failures; and (iv) any adverse quality trends.

Acceptance Criterion 17 The activities related to quality assurance records are acceptable provided that:

(1) Quality assurance records that furnish documentary evidence of quality must be specified, prepared, and maintained. These records must be legible, identifiable, and

retrievable. Requirements and responsibilities for quality assurance record transmittal, distribution, retention, maintenance, and disposition must be established and documented.

(2) The scope of the quality assurance records program is described. Quality assurance records include scientific, engineering, and operational data and logs; results of reviews, inspections, tests, audits, and material analyses; monitoring of work performance; maintenance and modification procedures and related inspection results; reportable occurrences; computer software supporting a safety or waste isolation function; qualification of personnel, procedures, and equipment; and other documentation such as design records, drawings, specifications, procurement documents, calibration procedures and reports, design review reports, peer review reports, nonconformance reports, corrective action reports, as-built drawings, and other records required by preclosure and postclosure operating conditions;

(3) Quality assurance and other organizations are identified and their responsibilities are described for the definition and implementation of activities related to quality assurance records, particularly in the retention and duration of record storage;

(4) Criteria are established and described in procedures for determining when a document becomes a quality assurance record, subject to the controls of this section, and the retention period for such records;

(5) Procedures are established describing methods for documenting/recording, reviewing, and confirming the accuracy of quality assurance records, including laboratory and field notebooks and logbooks, data sheets, data-reduction documents, and software supporting a safety or waste isolation function;

(6) Inspection and test records contain the following, where applicable: a description of the type of observation; date and results of the inspection or test; information related to conditions adverse to quality; identification of inspector or data recorder; evidence as to the acceptability of the results; and action taken to resolve any discrepancies noted;

(7) Provisions are made for the disposition of quality assurance records, including: ensuring that disposition of records is governed by the most stringent regulatory requirements that apply to records (this may be an agency other than the U.S. Nuclear Regulatory Commission); ensuring that suppliers' nonpermanent records are properly controlled and retained for required periods; and ensuring that quality assurance records are protected against damage, deterioration, or loss;

(8) Suitable controls are established and described for controlling, protecting, and maintaining quality assurance records before they are entered and stored in a quality assurance record storage area;

(9) Suitable facilities for the storage, preservation, and safekeeping of quality assurance records are described and satisfy the provisions contained in Section 4, "Storage, Preservation, and Safekeeping," of Supplement 17S–1 of NQA–1–1983 (American

Society of Mechanical Engineers, 1983), "Supplementary Requirements for Quality Assurance Records;"

(10) Guidance for storing quality assurance records, using electronic media, is provided in Regulatory Issue Summary 2000-18 (U.S. Nuclear Regulatory Commission, 2000);

(11) The additional records provisions referenced in Section 2.5.1.5 of the Yucca Mountain Review Plan are described;

(12) For quality assurance records, Section 2.8, "Retention of Records," of Supplement 17S–I of NQA–1–1983 (American Society of Mechanical Engineers, 1983), "Supplementary Requirements for Quality Assurance Records," states that the retention period for nonpermanent records is required to be established in writing. Programmatic nonpermanent records should be retained for at least 10 years or the life of the item if less than 10 years. For programmatic nonpermanent records, the retention period should be considered to begin on completion of the activity. For product nonpermanent records generated before facility licensing, the retention period should be considered to begin on completion of delivery. In addition, product and programmatic nonpermanent records should be retained at least until the date of the start of preclosure site operational activities. Table 1 of Regulatory Guide 1.28, Revision 3 (U.S. Nuclear Regulatory Commission, 1985), provides a list of nonpermanent and lifetime records and their respective retention times. Records similar to those identified in Table 1 of Regulatory Guide 1.28 are required to be maintained for the repository for the durations identified. Although Table 1 is intended to be a comprehensive list, it is the U.S. Department of Energy responsibility to assure itself, in accordance with the Records Section of 10 CFR 63.142, that sufficient records are maintained to furnish evidence of activities affecting quality. Table 1 is not applicable for preoperational test or operational phase records at this time because the final design and operating practices have not been developed. Further, Table 1 does not address site characterization records. It should be recognized that the nomenclature of these records may vary. For records not listed in Table 1, the type of record most nearly describing the record in question should be followed with respect to its retention period. The following definitions apply to the records:

(a) Programmatic nonpermanent records are those documents that were used to prescribe activities affecting quality, but that are not considered permanent records. Such records include documents prescribing the planning, execution, and auditing of activities affecting quality. Records such as audit checklists, audit results, and actual examinations used to qualify inspection and test personnel are included in this category; and

(b) Product nonpermanent records document that specific structures, systems, and components of the repository site have been designed and constructed in accordance with applicable requirements, but are such that is it not necessary to retain them as lifetime records. These records include design, verification data, receiving records, calibration records, maintenance records, inspection records, radiographs not associated with in-service inspection, and test records that are not otherwise designated as lifetime records.

(13) This acceptance criterion (i.e., Acceptance Criterion 17 relating to quality assurance records) may be updated to address records for site characterization, preoperational testing, and operations. This update is contingent on the level of detail of records included in the U.S. Department of Energy quality assurance program for these activities. [Note: potential licensing condition.]

Acceptance Criterion 18 The activities related to audits are acceptable provided that:

(1) Responsibilities and procedures are established for audits, for documenting and reviewing audit results, and for designating management levels to review and assess audit results;

(2) Internal and external audits to assure that procedures and activities comply with all aspects of the overall quality assurance program are performed by:

(a) The quality assurance organization, to provide a comprehensive independent verification and evaluation of quality-related procedures and activities; and

(b) The U.S. Department of Energy (and principal contractors), to verify and evaluate the quality assurance programs, procedures, and activities of suppliers. [Note: Internal and external audits are carried out by the U.S. Department of Energy and its contractors to verify that products, services, and activities comply with all aspects of the overall quality assurance program and to determine the effectiveness of the quality assurance program.] The U.S. Department of Energy and its contractors should perform audits of the prime contractor and subcontractors, consultants, vendors, and laboratories.

(3) The audit program should address planning and performance of audits to: (i) verify compliance with drawings, instructions, specifications, and other requirements affecting quality; and (ii) determine the effectiveness of the quality assurance program;

(4) An audit plan is prepared identifying audits to be performed, their frequencies, and schedules. Audits should be regularly scheduled based on the status and safety importance of the activities being performed and should be initiated early enough to assure effective quality assurance during design, procurement, manufacturing, construction, installation, inspection, testing, and performance confirmation. For scheduling audits, Section 2, "Scheduling," of Supplement 18S–1 of NQA–1–1983 (American Society of Mechanical Engineers, 1983), "Supplementary Requirements for Audits," requires audits to be scheduled in a manner that provides coverage and coordination with ongoing quality assurance program activities. The guidelines provided in Regulatory Position C.3.1, "Internal Audits," and C.3.2, "External Audits," of Regulatory Guide 1.28, Revision 3 (U.S. Nuclear Regulatory Commission, 1985), are considered acceptable and should be used for scheduling audits and related audit activities;

(5) Audits include: (i) an objective programmatic and technical evaluation of quality-related practices, procedures, instructions, activities, and items; and (ii) a review of documents and records, including software supporting a safety or waste isolation function and test

data from samples. Audits are conducted to assure that the abovementioned in (i) and (ii) are acceptable and to assure that the quality assurance program is effective and properly implemented;

(6) Provisions are established requiring that audits be performed in all areas where the requirements of 10 CFR Part 63, Subpart G are applicable. Areas that are often neglected but should be included, are activities associated with:

(a) The determination of site features that affect site safety (e.g., site characterization, performance confirmation, core sampling, site and foundation preparation, and methodology);

(b) The preparation, review, approval, and control of early procurements;

(c) Indoctrination and training programs;

(d) Interface control among the U.S. Department of Energy and principal contractors;

(e) Corrective action, calibration, and nonconformance control systems;

(f) Safety Analysis Report commitments;

(g) Development and control of computer software supporting a safety or waste isolation function;

(h) The purchase of American Society of Mechanical Engineers Section III Code items. [Note: For the purchase of such items, the U.S. Nuclear Regulatory Commission has only endorsed certain editions and addenda of the American Society of Mechanical Engineers Section III Code (American Society of Mechanical Engineers, 1998) and in doing so has indirectly endorsed quality assurance standards referenced in the Code. The U.S. Nuclear Regulatory Commission considers the referenced edition of NQA–1 (American Society of Mechanical Engineers, 1983) in the endorsed versions of the Code to be acceptable only for the construction of American Society of Mechanical Engineers Section III items when the referenced edition of NQA–1 is used in conjunction with the other quality assurance, administrative, and reporting requirements contained in the American Society of Mechanical Engineers Section III Code. Applicable provisions contained in the U.S. Department of Energy quality assurance program and requirements contained in the regulations also need to be met]; and

(i) Audits of American Society of Mechanical Engineers Section III Code suppliers. [Note: U.S. Nuclear Regulatory Commission Information Notice 86-21 (U.S. Nuclear Regulatory Commission, 1986) discusses the U.S. Nuclear Regulatory Commission recognition of the American Society of Mechanical Engineers accreditation program for N Stamp Holders, and the guidance provided therein should be used by the U.S. Department of Energy.]

(7) Audit data are analyzed by the quality assurance organization and, as appropriate, the technical staff. The resulting reports describing any quality problems and the effectiveness of the quality assurance program, including the need for an audit of deficient areas, are reported to management for review and assessment;

(8) Audits are performed in accordance with preestablished written procedures or checklists and are conducted by trained, qualified, competent quality assurance and technical personnel having expertise that encompasses the area being audited. Audit team members must not have been directly involved with the work being audited;

(9) Where the on-site quality assurance organization does not report to the off-site organization:

   (a) The off-site quality assurance organization conducts audits sufficient to verify adequacy of activities conducted by the on-site quality assurance organization;

   (b) The off-site quality assurance organization reviews and concurs in the schedule and scope of audits performed by the on-site quality assurance organization; and

   (c) Results of audits performed by the on-site quality assurance organization are provided to the off-site quality assurance organization for review and assessment.

(10) A tracking system for audit findings is established to help assure that all findings are appropriately addressed, prioritized, and trended;

(11) The audited organization describes in a formal report the corrective action to be taken to address findings. This report is submitted to the auditing organization and responsible management of the audited organization; and

(12) Provisions are established and described to assure that the cause of each finding is identified, resulting corrective action is described, and followup action is accomplished to assure proper closeout of deficiencies.

2.5.1.4 Evaluation Findings

If the license application provides sufficient information and the regulatory acceptance criteria in Section 2.5.1.3 are appropriately satisfied, the staff concludes that this portion of the staff evaluation is acceptable. The reviewer writes material suitable for inclusion in the safety evaluation report prepared for the entire application. The report includes a summary statement of what was reviewed and why the reviewer finds the submittal acceptable. The staff can document the review as follows.

The reviewer will prepare evaluation findings based on satisfying the applicable regulatory requirements relating to the U.S. Department of Energy quality assurance program. If the reviewer concludes that information provided with the initial application or a subsequent quality assurance program change submittal shows that the quality assurance program meets the acceptance criteria (or acceptable alternative) provided, the quality assurance program should be considered acceptable. During the review process, clarification may be obtained by the U.S. Department of Energy providing additional information in response to requests by the reviewer. The reviewer will verify that sufficient information has been provided and that the review is sufficiently complete and adequate to support conclusions of the following type to be included in the safety evaluation report.

U.S. Nuclear Regulatory Commission staff has reviewed the Safety Analysis Report and other information submitted in support of the license application and finds, with reasonable assurance, that the requirements at 10 CFR 21.3 are satisfied. Applicable definitions have been appropriately applied to U.S. Department of Energy commercial-grade item dedication.

U.S. Nuclear Regulatory Commission staff has reviewed the Safety Analysis Report and other information submitted in support of the license application and finds, with reasonable assurance, that the requirements at 10 CFR 63.44 are satisfied. Adequate procedures for control of changes, tests, and experiments have been provided.

U.S. Nuclear Regulatory Commission staff has reviewed the Safety Analysis Report and other information submitted in support of the license application and finds, with reasonable assurance, that the requirements at 10 CFR 63.73 are satisfied. Adequate procedures have been established for reporting deficiencies.

U.S. Nuclear Regulatory Commission staff has reviewed the Safety Analysis Report and other information submitted in support of the license application and finds, with reasonable assurance, that the requirements at 10 CFR 63.21(c)(20) are satisfied. Requirements for the content of the license application have been met in that an adequate description of the quality assurance program to be applied to the structures, systems, and components important to safety and to the engineered and natural barriers important to waste isolation has been provided, including a discussion of how the applicable requirements of 10 CFR 63.142 will be satisfied.

U.S. Nuclear Regulatory Commission staff has reviewed the Safety Analysis Report and other information submitted in support of the license application and finds, with reasonable assurance, that the requirements at 10 CFR 63.141 are satisfied. The description of the quality assurance program provided is within the proper scope and includes quality control.

U.S. Nuclear Regulatory Commission staff has reviewed the Safety Analysis Report and other information submitted in support of the license application and finds, with reasonable assurance, that the requirements at 10 CFR 63.142 are satisfied. The quality assurance program described in the license application satisfies requirements of applicability and specified criteria and applies to all structures, systems, and components important to safety, to design and characterization of barriers important to waste isolation, and to activities related thereto.
U.S. Nuclear Regulatory Commission staff has reviewed the Safety Analysis Report and other information submitted in support of the license application and finds, with reasonable assurance, that the requirements at 10 CFR 63.143 are satisfied. The description of the quality assurance program satisfies requirements for the implementation of a program based on the criteria required by 10 CFR 63.142.

U.S. Nuclear Regulatory Commission staff has reviewed the Safety Analysis Report and other information submitted in support of the license application and finds, with reasonable assurance, that the requirements at 10 CFR 63.144 are satisfied. The description of the quality assurance program satisfies requirements and follows procedures for implementation of changes to a previously accepted quality assurance program for cases in which U.S. Nuclear Regulatory Commission approval either is or is not required.

Based on detailed review and evaluation of the quality assurance program description contained in the U.S. Department of Energy license application, the U.S. Nuclear Regulatory Commission staff finds, with reasonable assurance, that:

(1) The organizations and individuals performing quality assurance functions have the required independence and authority to effectively carry out the quality assurance program without undue influence from those directly responsible for costs and schedules;

(2) The quality assurance program describes requirements, procedures, and controls that, when properly implemented, comply with the requirements of 10 CFR Part 63, Subpart G; the requirements of 10 CFR 63.73; the criteria contained in this Section (2.5.1) of the Yucca Mountain Review Plan; and the regulatory requirements, documents, and positions presented in Section 2.5.1 of the Yucca Mountain Review Plan;

A brief description of the U.S. Department of Energy quality assurance program may be provided, along with the more important aspects of the program.

(3) The quality assurance program covers activities affecting structures, systems, and components important to safety and barriers important to waste isolation as identified in the Safety Analysis Report. Accordingly, the staff concludes that the U.S. Department of Energy description of the quality assurance program is in compliance with applicable U.S. Nuclear Regulatory Commission regulations and industry standards and that the quality assurance program can be implemented for the (specify: design, procurement, construction, operation, etc.) phases of the repository life cycle; and

(4) The U.S. Department of Energy quality assurance program description is in compliance with applicable U.S. Nuclear Regulatory Commission regulations.

2.5.1.5 References

Commitments

The U.S. Department of Energy is expected to commit to the use of the staff positions and provisions contained in the following documents in conjunction with any exceptions or clarifications provided in the acceptance criteria. However, as provided for in Section 2.5.1 of the Yucca Mountain Review Plan, exceptions and alternatives to these acceptance criteria and the documents and positions contained in Section 2.5.1.5 of the Yucca Mountain Review Plan may be adopted by the U.S. Department of Energy, provided The U.S. Department of Energy can otherwise demonstrate it satisfies regulatory requirements.

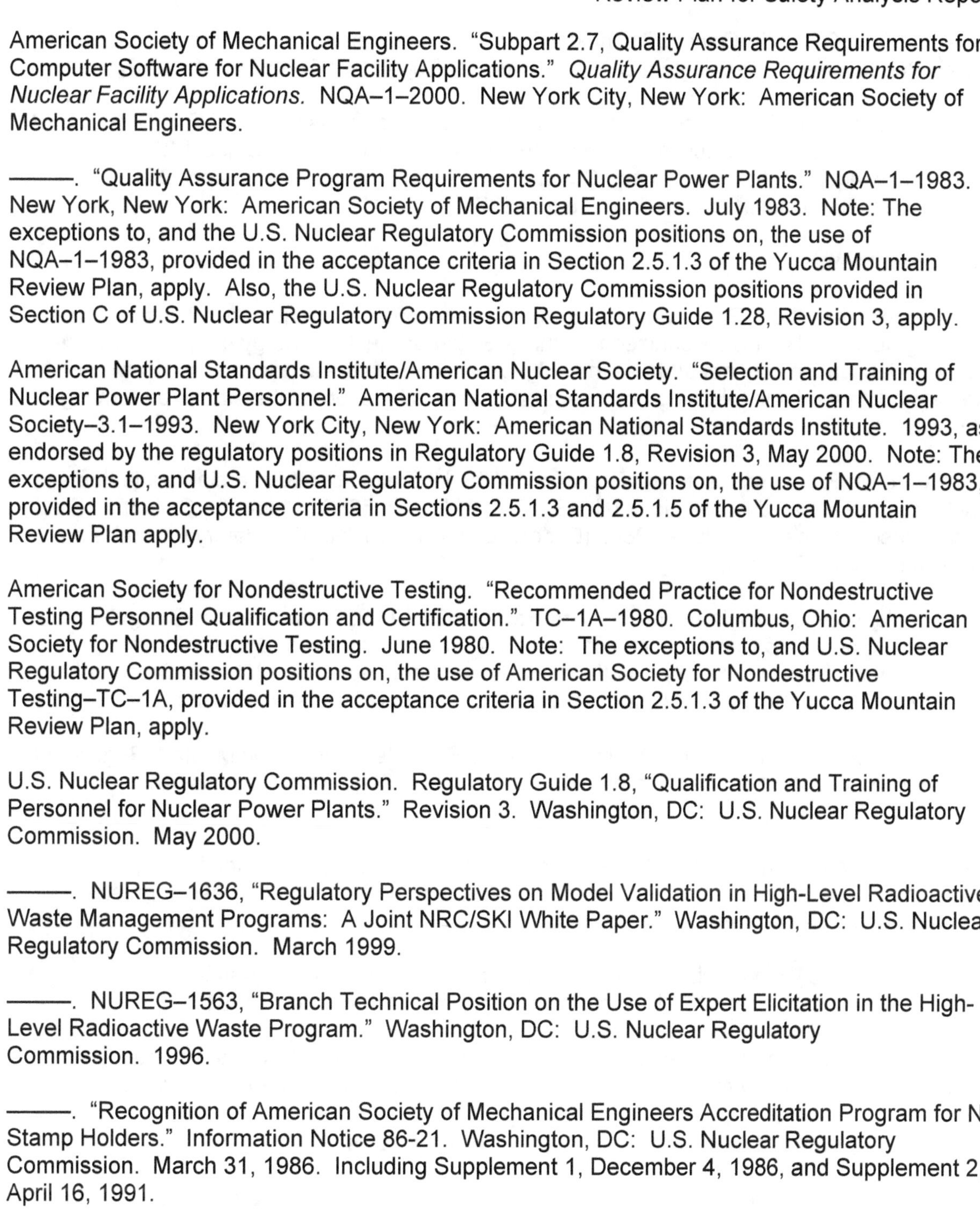

American Society of Mechanical Engineers. "Subpart 2.7, Quality Assurance Requirements for Computer Software for Nuclear Facility Applications." *Quality Assurance Requirements for Nuclear Facility Applications.* NQA–1–2000. New York City, New York: American Society of Mechanical Engineers.

———. "Quality Assurance Program Requirements for Nuclear Power Plants." NQA–1–1983. New York, New York: American Society of Mechanical Engineers. July 1983. Note: The exceptions to, and the U.S. Nuclear Regulatory Commission positions on, the use of NQA–1–1983, provided in the acceptance criteria in Section 2.5.1.3 of the Yucca Mountain Review Plan, apply. Also, the U.S. Nuclear Regulatory Commission positions provided in Section C of U.S. Nuclear Regulatory Commission Regulatory Guide 1.28, Revision 3, apply.

American National Standards Institute/American Nuclear Society. "Selection and Training of Nuclear Power Plant Personnel." American National Standards Institute/American Nuclear Society–3.1–1993. New York City, New York: American National Standards Institute. 1993, as endorsed by the regulatory positions in Regulatory Guide 1.8, Revision 3, May 2000. Note: The exceptions to, and U.S. Nuclear Regulatory Commission positions on, the use of NQA–1–1983, provided in the acceptance criteria in Sections 2.5.1.3 and 2.5.1.5 of the Yucca Mountain Review Plan apply.

American Society for Nondestructive Testing. "Recommended Practice for Nondestructive Testing Personnel Qualification and Certification." TC–1A–1980. Columbus, Ohio: American Society for Nondestructive Testing. June 1980. Note: The exceptions to, and U.S. Nuclear Regulatory Commission positions on, the use of American Society for Nondestructive Testing–TC–1A, provided in the acceptance criteria in Section 2.5.1.3 of the Yucca Mountain Review Plan, apply.

U.S. Nuclear Regulatory Commission. Regulatory Guide 1.8, "Qualification and Training of Personnel for Nuclear Power Plants." Revision 3. Washington, DC: U.S. Nuclear Regulatory Commission. May 2000.

———. NUREG–1636, "Regulatory Perspectives on Model Validation in High-Level Radioactive Waste Management Programs: A Joint NRC/SKI White Paper." Washington, DC: U.S. Nuclear Regulatory Commission. March 1999.

———. NUREG–1563, "Branch Technical Position on the Use of Expert Elicitation in the High-Level Radioactive Waste Program." Washington, DC: U.S. Nuclear Regulatory Commission. 1996.

———. "Recognition of American Society of Mechanical Engineers Accreditation Program for N Stamp Holders." Information Notice 86-21. Washington, DC: U.S. Nuclear Regulatory Commission. March 31, 1986. Including Supplement 1, December 4, 1986, and Supplement 2, April 16, 1991.

———. NUREG–1298, "Qualification of Existing Data for High-Level Nuclear Waste Repositories." Washington, DC: U.S. Nuclear Regulatory Commission. 1988.

———. NUREG–1297, "Generic Technical Position: Peer Review for High-Level Nuclear Waste Repositories." Washington, DC: U.S. Nuclear Regulatory Commission. 1987.

———. U.S. Nuclear Regulatory Commission Regulatory Positions C.1, C.2, C.3, C.3.1, C.3.2 (1, 2, and 3) contained in Section C, "Regulatory Position;" Regulatory Guide 1.28, "Quality Assurance Requirements (Design and Construction)," Revision 3. Washington, DC: U.S. Nuclear Regulatory Commission. August 1985.

Noncommitments

Where applicable, the U.S. Department of Energy may consider the guidance contained in the following documents. It is recognized that the U.S. Department of Energy quality assurance program description may not address the subjects included in these documents. However, if they are addressed, the following documents should also be used by the staff in performing its review.

Electric Power Research Institute. "Guideline for the Utilization of Commercial Grade Items in Nuclear Safety Related Applications (NCIG–07)." EPRI NP–5652. Palo Alto, California: Electric Power Research Institute. June 1988. [Endorsed by U.S. Nuclear Regulatory Commission Generic Letter 89-02 and 91-05.]

"U.S. Nuclear Regulatory Commission Staff Review of the U.S. Department of Energy's Proposed Approach to Risk Significance Categorization of Structures, Systems, and Components Important to Safety." Letter from C. Reamer (September 28) to S. Brocoum, U.S. Department of Energy. Washington, DC: U.S. Nuclear Regulatory Commission. 2001.

———. "Guidance on Managing Quality Assurance Records in Electronic Media." Regulatory Issue Summary 2000-18. Washington, DC: U.S. Nuclear Regulatory Commission. October 13, 2000.

———. Regulatory Guide 1.176, "An Approach for Plant-Specific, Risk-Informed Decisionmaking: Graded Quality Assurance." Washington, DC: U.S. Nuclear Regulatory Commission. August 1998.

———. "Licensee Commercial-Grade Procurement and Dedication Programs." Generic Letter 91-05. Washington, DC: U.S. Nuclear Regulatory Commission. 1991.

———. "Actions to Improve Detection of Counterfeit and Fraudulently Marketed Products." Generic Letter 89-02. Washington, DC: U.S. Nuclear Regulatory Commission. 1989.

2.5.2 Records, Reports, Tests, and Inspections

Although the U.S. Department of Energy is not expected to have prepared procedures and plans for records, reports, tests and inspections at the time of the construction authorization decision, the U.S. Department of Energy should provide a description of the program used to maintain records specified in 10 CFR 63.71 and 63.72.

Review Responsibilities—High-Level Waste Branch and Environmental and Performance Assessment Branch

2.5.2.1 Areas of Review

This section reviews procedures for records, reports, tests, and inspections. Reviewers will evaluate the information required by 10 CFR 63.21(c)(23).

The staff will evaluate the following parts of U.S. Department of Energy procedures for managing records, reports, tests, and inspections using the review methods and acceptance criteria in Sections 2.5.2.2 and 2.5.2.3:

(1) Proposed records of receipt, handling, and disposition of radioactive waste;

(2) Records of construction; and

(3) Ways to ensure use of records by future generations.

2.5.2.2 Review Methods

Review Method 1 Records and Reports

Confirm that the U.S. Department of Energy will maintain records and reports required by conditions of the license or rules, regulations, and orders of the Commission.

Confirm that records of receipt, handling, and disposition of radioactive waste at the geologic repository operations area will contain enough information to provide a complete history of waste movement from the shipper through all phases of storage and disposal.

Verify that records of construction of the geologic repository operations area at the Yucca Mountain site will contain enough information to adequately describe the construction and the resulting as-built configuration. Verify construction records will include the following, as a minimum:

(1) Surveys of the underground facility excavations, shafts, ramps, and boreholes referenced to easily identified surface features or monuments;

(2) A description of the geologic materials and structures encountered;

(3) Geologic maps and geologic cross sections;

(4) Locations and amounts of seepage;

(5) Details of construction equipment, methods, progress, and sequence of work;

(6) Descriptions of construction problems;

(7) Anomalous conditions encountered;

(8) Instrument locations, readings, and analyses;

(9) Locations and descriptions of structural support systems;

(10) Locations and descriptions of dewatering systems;

(11) Details, methods of emplacement, and location of monuments used to identify the site after permanent closure;

(12) Details, methods of emplacement, and location of seals used; and

(13) Geologic repository operations area design records such as specifications and as-built drawings.

Confirm that records of construction of the geologic repository operations area and receipt, handling, and disposition of radioactive waste will be kept in a way to ensure their use by future generations in accordance with 10 CFR 63.51(a)(3).

2.5.2.3 Acceptance Criteria

The following acceptance criteria are based on meeting the requirements of 10 CFR 63.71, 63.72, 63.73, 63.74, and 63.75 relating to records, reports, tests, and inspections.

Acceptance Criterion 1 The U.S. Department of Energy Will Maintain Adequate Records and Reports Required by the Conditions of the License or by Rules, Regulations, and Orders of the Commission.

(1) The U.S. Department of Energy will maintain adequate records and reports that may be required by conditions of the license or rules, regulations, and orders of the Commission;

(2) The records of receipt, handling, and disposition of radioactive waste at the geologic repository operations area will contain enough information to provide a complete history of the movement of the waste from the shipper through all phases of storage and disposal;

(3) The records of construction of the geologic repository operations area at the Yucca Mountain site will contain enough information to give an adequate description of the construction and the resulting as-built configuration. The construction records will include the following, as a minimum:

  (a) Surveys of the underground facility excavations, shafts, ramps, and boreholes referenced to readily identifiable surface features or monuments;

  (b) A description of the geologic materials and structures encountered;

  (c) Geologic maps and geologic cross sections;

  (d) Locations and amounts of seepage;

(e) Details of construction equipment, methods, progress, and sequence of work;

(f) Descriptions of construction problems;

(g) Anomalous conditions encountered;

(h) Instrument locations, readings, and analyses;

(i) Locations and descriptions of structural support systems;

(j) Locations and descriptions of dewatering systems;

(k) Details, methods of emplacement, and location of monuments used to identify the site after permanent closure;

(l) Details, methods of emplacement, and location of seals used; and

(m) Facility design records such as specifications and as-built drawings.

(4) The U.S. Department of Energy will retain the records of construction of the geologic repository operations area and receipt, handling, and disposition of radioactive waste in a way that ensures their use by future generations in accordance with 10 CFR 63.51(a)(3).

2.5.2.4 Evaluation Findings

If the license application provides sufficient information and the regulatory acceptance criteria in Section 2.5.2.3 are appropriately satisfied, the staff concludes that this portion of the staff evaluation is acceptable. The reviewer writes material suitable for inclusion in the safety evaluation report prepared for the entire application. The report includes a summary statement of what was reviewed and why the reviewer finds the submittal acceptable. The staff can document the review as follows.

U.S. Nuclear Regulatory Commission staff has reviewed the Safety Analysis Report and other information submitted in support of the license application and finds, with reasonable assurance, that the requirements at 10 CFR 63.71 are satisfied. The U.S. Department of Energy has provided an adequate description of the record keeping and reporting programs for receipt, handling, and disposal of radioactive waste. These programs also support requirements imposed by license conditions or other rules, records, and orders of the Commission. Therefore, the U.S. Department of Energy meets the requirements for record keeping and reporting of repository operations.

U.S. Nuclear Regulatory Commission staff has reviewed the Safety Analysis Report and other information submitted in support of the license application and finds, with reasonable assurance, that the requirements at 10 CFR 63.72 are satisfied. The U.S. Department of Energy has provided an adequate description of the construction records and record keeping programs. Therefore, the U.S. Department of Energy meets the requirements to maintain records of construction of the geologic repository operations area.

2.5.2.5 References

None.

2.5.3 Training and Certification of Personnel

Although the U.S. Department of Energy is not expected to have prepared procedures and a program for training and certification of personnel at the time of the a construction authorization decision, the U.S. Department of Energy will develop and implement them to meet or exceed the acceptance criteria in this section.

2.5.3.1 U.S. Department of Energy Organizational Structure as it Pertains to Construction and Operation of Geologic Repository Operations Area

Review Responsibilities—High-Level Waste Branch and Environmental and Performance Assessment Branch

2.5.3.1.1 Areas of Review

This section reviews the organizational structure of the U.S. Department of Energy as it pertains to construction and operation of the geologic repository operations area. Reviewers will evaluate the information required by 10 CFR 63.21(c)(22)(i).

The staff will evaluate the following parts of the organizational structure of the U.S. Department of Energy as it pertains to construction and operation of the geologic repository operations area, using the review methods and acceptance criteria in Sections 2.5.3.1.2 and 2.5.3.1.3.

(1) The U.S. Department of Energy delineation of responsibilities and decision-making authority to on-site and Headquarters staff, major contractors, sub-contractors, principal consultants, service organizations, and other affected organizations;

(2) Address of the office of record and the identity of the point of contact of each organizational entity; and

(3) Procedure for delegation of authority.

2.5.3.1.2 Review Methods

Review Method 1 Definition of Responsibilities

Verify that the U.S. Department of Energy provides an adequate delineation of responsibility and decision-making authority during construction and operation of the geologic repository operations area so responsibility for actions can be traced through the management and staff hierarchy (on-site and at Headquarters); contractors; subcontractors; consultants; service organizations; and other affected organizations.

Verify that the address of the office of record for each entity, a point of contact, and a telephone number, fax number, or e-mail address are provided in the license application.

Review Method 2 Procedure for Delegation of Authority

Confirm that an adequate authority delegation procedure is in place for positions having responsibility to act in routine or emergency situations. Confirm that an identified party always has responsibility and sufficient authority to act, and the appropriate qualifications. The development and maintenance of procedures are reviewed using Section 2.5.6 of the Yucca Mountain Review Plan.

2.5.3.1.3 Acceptance Criteria

The following acceptance criteria are based on meeting the requirements of 10 CFR 63.21(c)(22)(i).

Acceptance Criterion 1 Responsibilities Are Adequately Defined.

(1) The U.S. Department of Energy provides an adequate delineation of assignments of responsibility and decision-making authority during construction and operation of the geologic repository operations area, so that responsibility for actions can be traced through the management and staff hierarchy of the U.S. Department of Energy (onsite and at Headquarters); contractors; subcontractors; consultants; service organizations; and other affected organizations; and

(2) The address of the office of record for each entity, a point of contact, and a telephone number, fax number, or e-mail address are provided in the license application.

Acceptance Criterion 2 An Adequate Procedure for Delegation of Authority Situations Is In Place.

(1) There is an adequate authority delegation procedure in place for positions having responsibility to act in routine or emergency situations. An identified party will always have responsibility and sufficient authority to act, along with the appropriate qualifications.

2.5.3.1.4 Evaluation Findings

If the license application provides sufficient information and the regulatory acceptance criteria in Section 2.5.3.1.3 are appropriately satisfied, the staff concludes that this portion of the staff evaluation is acceptable. The reviewer writes material suitable for inclusion in the safety evaluation report prepared for the entire application. The report includes a summary statement of what was reviewed and why the reviewer finds the submittal acceptable. The staff can document the review as follows.

U.S. Nuclear Regulatory Commission staff has reviewed the Safety Analysis Report and other information submitted in support of the license application and finds, with reasonable assurance, that the requirements at 10 CFR 63.21(c)(22)(i) are satisfied. The U.S. Department of Energy has provided an adequate organizational structure as it pertains to the construction and operation of the geologic repository operations area, including the delegation of authority and assignment of responsibilities.

2.5.3.1.5 References

None.

2.5.3.2 Key Positions Assigned Responsibility for Safety and Operations of Geologic Repository Operations Area

Review Responsibilities—High-Level Waste Branch and Environmental and Performance Assessment Branch

At the time of the construction authorization decision, the U.S. Department of Energy is not expected to have identified specific individuals to fill key positions. Therefore, portions of the review defined in this section may be delayed at the time of application for the license. At the time of application to receive, possess, process, store, or dispose of high-level radioactive waste, the U.S. Department of Energy is required to have identified specific individuals to fill key positions.

2.5.3.2.1 Areas of Review

This section reviews key positions assigned responsibility for safety and operations of the geologic repository operations area. Reviewers will evaluate the information required by 10 CFR 63.21(c)(22)(ii).

The staff will evaluate the following parts of key positions assigned responsibility for safety and operations of geologic repository operations area, using the review methods and acceptance criteria in Sections 2.5.3.2.2 and 2.5.3.2.3.

(1) Descriptions of the key positions assigned responsibility for safety at the geologic repository operations area, including minimum skills and experience for each position; and

(2) Identification of alternates for persons in key positions.

2.5.3.2.2 Review Methods

Review Method 1 Descriptions of Key Positions

Verify that the U.S. Department of Energy provides an adequate description of each key position at the geologic repository operations area that includes the minimum skills and experience necessary to hold each position. These positions include, but are not limited to, those with responsibilities in health physics, nuclear criticality safety, training and certification, emergency planning and response, operations, maintenance, engineering, and quality assurance.

Confirm that qualified alternates are identified, to act in the absence of individuals assigned to geologic repository operations area key positions, based on minimum skills and experience necessary to hold each key position.

2.5.3.2.3 Acceptance Criteria

The following acceptance criterion is based on meeting the requirements of 10 CFR 63.21(c)(22)(ii).

Acceptance Criterion 1 Description of Key Positions for Safety at the Geologic Repository Operations Area Are Adequate.

(1) The U.S. Department of Energy provides an adequate description, of each key position at the geologic repository operations area, that includes the minimum skills and experience necessary to hold each position; and

(2) Qualified alternates are identified to act in the absence of individuals assigned to geologic repository operations area key positions, based on minimum skills and experience necessary to hold each key position.

2.5.3.2.4 Evaluation Findings

If the license application provides sufficient information and the regulatory acceptance criteria in Section 2.5.3.2.3 are appropriately satisfied, the staff concludes that this portion of the staff evaluation is acceptable. The reviewer writes material suitable for inclusion in the safety evaluation report prepared for the entire application. The report includes a summary statement of what was reviewed and why the reviewer finds the submittal acceptable. The staff can document the review as follows.

U.S. Nuclear Regulatory Commission staff has reviewed the Safety Analysis Report and other information submitted in support of the license application and finds, with reasonable assurance, that the requirements of 10 CFR 63.21(c)(22)(ii) are satisfied. The U.S. Department of Energy provides an adequate description of the key positions assigned responsibility for safety and operations of the geologic repository operations area and the qualifications of the persons occupying these positions.

2.5.3.2.5 References

None.

2.5.3.3 Personnel Qualifications and Training Requirements

Review Responsibilities—High-Level Waste Branch and Environmental and Performance Assessment Branch

At the time of the construction authorization decision, the U.S. Department of Energy is not required to have a U.S. Nuclear Regulatory Commission-approved personnel training and qualification program in place. The U.S. Department of Energy will have a U.S. Nuclear Regulatory Commission-approved personnel training and qualification program in place before it may receive, possess, process, store, or dispose of high-level radioactive waste.

2.5.3.3.1 Areas of Review

This section reviews personnel qualifications and training requirements. Reviewers will evaluate the information required by 10 CFR 63.21(c)(22)(iii).

The staff will evaluate the following parts of personnel qualifications and training requirements, using the review methods and acceptance criteria in Sections 2.5.3.3.2 and 2.5.3.3.3.

(1) Standards used for selection, training, and certification of personnel;

(2) Program for general training, proficiency testing, and certification of geologic repository operations area personnel;

(3) Procedures for managing and maintaining the training program;

(4) Preoperational and operational radioactive materials training program;

(5) Operator and supervisor training and certification programs and requirements for structures, systems, and components important to safety;

(6) Operator and supervisor requalification program;

(7) Physical requirements for personnel operating equipment and controls that are important to safety;

(8) Methods for selecting and training security guards;

(9) Methods used to evaluate operator testing procedures; and

(10) Qualifications of personnel assigned to key positions important to safety at the geologic repository operations area.

2.5.3.3.2 Review Methods

Review Method 1 Standards for Selection, Training, and Certification of Personnel

Confirm that any standards used for the programs for selection, training, and certification of personnel are adequate. For example, the U.S. Department of Energy may use a systems approach to training such as described at 10 CFR 55.4.

Review Method 2 Programs for General Training, Proficiency Testing, and Certification of Geologic Repository Operations Area Personnel

Additional guidance to support a review of training programs for nuclear facility operators is in Regulatory Guide 1.8, "Qualification and Training of Personnel for Nuclear Power Plants" (U.S. Nuclear Regulatory Commission, 2000).

Verify that the training program establishes the bases for geologic repository operations area personnel qualification and defines the qualification requirements of operators, supervisors, and other staff. The characteristics of this program should be consistent with American National Standards Institute/American Nuclear Society 3.1, Section 5.1, "General Aspects;" Section 5.3, "Training of Personnel Not Requiring U.S. Nuclear Regulatory Commission Licenses;" Section 5.4, "General Employee Training;" and Section 5.5, "Retraining." Confirm that the training program is approved by the U.S. Nuclear Regulatory Commission before receipt of waste at the geologic repository operations area.

Verify the U.S. Department of Energy has procedures to manage and maintain the training program. These procedures should include identification of the personnel responsible for developing training programs, conducting training; retraining employees (including new employee orientations); and maintaining up-to-date records on the status of trained personnel. Development and maintenance of procedures are reviewed using Section 2.5.6 of the Yucca Mountain Review Plan.

Confirm the U.S. Department of Energy specifies training requirements for each job category.

Verify the U.S. Department of Energy will train new hires on a timely schedule.

Review Method 3 Preoperational and Operational Radioactive Materials Training Program

Additional guidance to support a review of the radioactive materials training program for nuclear facility operators is in Regulatory Guide 8.29, "Instructions Concerning Risks from Occupational Radiation Exposure" (U.S. Nuclear Regulatory Commission, 1996); NUREG–0713, "Occupational Radiation Exposure at Commercial Nuclear Power Reactors and Other Facilities" (Raddatz and Hagemayer, 1995); American Society for Testing and Materials E 1168, "Guide for Radiation Protection Training for Nuclear Facility Workers" (American Society for Testing and Materials, 1995); and Regulatory Guide 8.8, "Information Relevant to Ensuring that Occupational Radiation Exposures at Nuclear Power Stations Will Be As Low As Is Reasonably Achievable," paragraph C.1.c (U.S. Nuclear Regulatory Commission, 1984).

Verify the U.S. Department of Energy will implement the radioactive materials training program before conduct of operations involving radioactive material (i.e., preoperational training). Confirm that the U.S. Department of Energy will substantially complete operator training and certification before receipt of radioactive material.

Confirm that operator radiation safety training includes such topics as the nature and sources of radiation, methods for controlling contamination, interactions of radiation with matter, biological effects of radiation, use of monitoring equipment, as low as is reasonably achievable concepts, facility access and visitor controls, decontamination procedures, use of personal monitoring and protective equipment, regulatory and administrative exposure and contamination limits, site-specific hazards, and principles of criticality hazards control.

Confirm that individuals who, in the courses of their employment, are likely to receive yearly occupational doses in excess of 100 mrem (1 mSv), are instructed in the health protection issues associated with exposure to radioactive materials or radiation, in accordance with 10 CFR 19.12.

Verify that individuals involved are informed of estimated doses and associated risks before any special exposures occur, in accordance with 10 CFR 20.1206.

Verify the U.S. Department of Energy will provide training in radiation protection and facility exposure control procedures for all personnel whose duties require: (i) working with radioactive materials; (ii) entering radiation areas; and (iii) directing the activities of others who work with radioactive materials or enter radiation areas.

Confirm that facility personnel whose duties do not require entering radiation areas or working with radioactive materials receive sufficient instructions in radiation protection and facility rules and regulations to understand why they should not enter such areas.

Review Method 4 Operation of Equipment and Controls Important to Safety

Confirm that operators of equipment and controls identified as important to safety are either trained and certified in the operations or will be under the direct visual supervision of an individual who is trained and certified.

Verify that supervisory personnel who personally direct the operation of equipment and controls that are important to safety are trained and certified in such operations.

Verify that operational training includes topics such as installation, design, and operation of structures, systems, and components; decontamination procedures; and emergency procedures.

Review Method 5 Operator and Supervisor Requalification Program for Structures, Systems, and Components Important to Safety

Confirm that the U.S. Department of Energy defines an adequate program for requalification of operators, supervisors, and other staff.

Verify that the frequency of retraining and the nature and duration of training and testing records have been specified. Confirm that retraining will be periodic and conducted at least every 2 years.

Review Method 6 Physical Condition and General Health of Personnel

Additional guidance to support this review is in Regulatory Guide 1.134, “Medical Evaluation of Licensed Personnel for Nuclear Power Plants” (U.S. Nuclear Regulatory Commission, 1998).

Confirm that any condition that might impair judgment or motor coordination, resulting in the inability of an operator to perform activities that are important to safety, has been considered, in selecting personnel to operate such equipment and controls. Such impaired judgment or motor coordination conditions need not categorically disqualify a person from operating equipment and controls important to safety provided appropriate provisions are made to accommodate any such condition.

Review Method 7 Methods for Selecting, Training, and Qualifying Security Guards

Verify that the process by which security guards (including watchmen, armed response persons, etc.) will be selected and qualified is described as required by 10 CFR 73.55(b)(4)(ii). This information will be submitted as part of the physical security plan and reviewed using Section 1.3 of the Yucca Mountain Review Plan. Confirm that selection and training criteria will conform to the general criteria for security personnel contained in 10 CFR Part 73, Appendix B. Regulatory Guide 5.20, "Training, Equipping, and Qualifying of Guards and Watchmen" (U.S. Nuclear Regulatory Commission, 1974) provides additional guidance.

Review Method 8 Methods for Evaluating Operator Testing Procedures

Verify that the methods for evaluating the effectiveness of the training program are described and that program effectiveness is determined by comparison to established objectives and criteria.

Review Method 9 Qualifications of Personnel

Evaluate the qualifications of personnel assigned to geologic repository operations area key positions important to safety, based on the minimum skills and experience necessary to hold each key position.

2.5.3.3.3 Acceptance Criteria

The following acceptance criteria are based on meeting the requirements of 10 CFR 63.151, 63.152, and 63.153.

Acceptance Criterion 1 Adequate Standards Are Used for Selection, Training, and Certification of Personnel.

(1) Any standards used for the programs for selection, training, and certification of personnel are adequate.

Acceptance Criterion 2 Programs for General Training, Proficiency Testing, and Certification of Geologic Repository Operations Area Personnel Are Acceptable.

(1) The training program adequately establishes the bases for geologic repository operations area personnel qualification and defines the qualification requirements of operators, supervisors, and other staff. The characteristics of this program are consistent with American National Standards Institute/American Nuclear Society 3.1, Section 5.1 "General Aspects;" Section 5.3, "Training of Personnel Not Requiring U.S. Nuclear Regulatory Commission Licenses;" Section 5.4, "General Employee Training;" and Section 5.5, "Retraining." The training program will be approved by the U.S. Nuclear Regulatory Commission before receipt of waste at the geologic repository operations area;

(2) The U.S. Department of Energy establishes adequate procedures for managing and maintaining the training program. These procedures include identification of the personnel responsible for developing training programs; conducting training; retraining employees (including new employee orientations); and maintaining up-to-date records on the status of trained personnel;

(3) The U.S. Department of Energy specifies training requirements for each job category; and

(4) The U.S. Department of Energy will train new hires on a timely schedule.

Acceptance Criterion 3 An Acceptable Preoperational and Operational Radioactive Materials Training Program Is Provided.

(1) The U.S. Department of Energy will implement the radioactive materials training program before conduct of operations involving radioactive material (i.e., preoperational training). The U.S. Department of Energy will substantially complete such operator training and certification before receipt of the radioactive material;

(2) The operator radiation safety training includes such topics as the nature and sources of radiation, methods for controlling contamination, interactions of radiation with matter, biological effects of radiation, use of monitoring equipment, as low as is reasonably achievable concepts, facility access and visitor controls, decontamination procedures, use of personal monitoring and protective equipment, regulatory and administrative exposure and contamination limits, site-specific hazards, and principles of criticality hazards control;

(3) The U.S. Department of Energy will instruct all individuals who, in the course of their employment, are likely to receive yearly occupational doses in excess of 100 mrem (1 mSv), in the health protection issues associated with exposure to radioactive materials or radiation per 10 CFR 19.12;

(4) Before any special exposures occur, the U.S. Department of Energy will inform the individuals involved of estimated doses and associated risks, in accordance with 10 CFR 20.1206;

(5) The U.S. Department of Energy will provide adequate training in radiation protection and facility exposure control procedures for personnel whose duties require: (i) working with radioactive materials, (ii) entering radiation areas, and (iii) directing the activities of others who work with radioactive materials or enter radiation areas; and

(6) The facility personnel whose duties do not require entering radiation areas or working with radioactive materials will receive sufficient instructions in radiation protection and facility rules and regulations to understand why they should not enter such areas.

Acceptance Criterion 4 Operation of Equipment and Controls Identified as Important to Safety Is Limited to Trained and Certified Personnel or Is under the Direct Visual Supervision of an Individual with Training and Certification in Their Operation.

(1) Operators of all equipment and controls identified as important to safety are either trained and certified in the operations or will be under the direct visual supervision of an individual who is trained and certified in the operations;

(2) Supervisory personnel who personally direct the operation of equipment and controls important to safety are trained and certified in such operations; and

(3) Operational training includes topics such as installation, design, and operation of structures, systems, and components; decontamination procedures; and emergency procedures.

Acceptance Criterion 5 An Acceptable Operator and Supervisor Requalification Program for Structures, Systems, and Components Important to Safety Is Provided.

(1) The U.S. Department of Energy defines an adequate program for requalification of operators, supervisors, and other staff; and

(2) Frequency of retraining and the nature and duration of training and testing records are specified. Retraining will be periodic and conducted at least every 2 years.

Acceptance Criterion 6 The Physical Condition and the General Health of Personnel Certified for the Operation of Equipment and Controls Important to Safety Are Such That Operational Errors That Could Endanger Other In-Plant Personnel or the Public Health and Safety Will Not Occur.

(1) Conditions that might impair judgment or motor coordination resulting in the inability of an operator to perform activities that are important to safety are adequately considered in the selection of personnel to operate such equipment and controls.

Acceptance Criterion 7 Methods for Selecting, Training, and Qualifying Security Guards Are Acceptable.

(1) The process by which security guards (including watchmen, armed response persons, etc.) will be selected and qualified is adequate as required by 10 CFR 73.55(b)(4)(ii). Selection and training criteria conform to the general criteria for security personnel contained in 10 CFR Part 73, Appendix B.

Acceptance Criterion 8 Methods Used to Evaluate Operator Testing Procedures Are Acceptable.

(1) Methods for evaluating the effectiveness of the training program are described and program effectiveness is determined by comparison to established objectives and criteria.

Acceptance Criterion 9 Qualifications of Personnel Are Adequate.

(1) The U.S. Department of Energy provides an acceptable description of the qualifications of the personnel assigned to geologic repository operations area key positions important to safety based on the minimum skills and experience necessary to hold each key position.

2.5.3.3.4 Evaluation Findings

If the license application provides sufficient information and the regulatory acceptance criteria in Section 2.5.3.3.3 are appropriately satisfied, the staff concludes that this portion of the staff evaluation is acceptable. The reviewer writes material suitable for inclusion in the safety evaluation report prepared for the entire application. The report includes a summary statement of what was reviewed and why the reviewer finds the submittal acceptable. The staff can document the review as follows.

U.S. Nuclear Regulatory Commission staff has reviewed the Safety Analysis Report and other information submitted in support of the license application and finds, with reasonable assurance, that the requirements of 10 CFR 63.151 are satisfied. Operation of systems and components important to safety will be performed only by trained and certified personnel or by personnel under the direct supervision of an individual with training and certification in such operation. Supervisory personnel will also be certified in the operations they supervise.

U.S. Nuclear Regulatory Commission staff has reviewed the Safety Analysis Report and other information submitted in support of the license application and finds, with reasonable assurance, that the requirements of 10 CFR 63.152 are satisfied. The U.S. Department of Energy has established an adequate program for training, proficiency testing, certification, and requalification of operating and supervisory personnel.

U.S. Nuclear Regulatory Commission staff has reviewed the Safety Analysis Report and other information submitted in support of the license application and finds, with reasonable assurance, that the requirements of 10 CFR 63.153 are satisfied. The U.S. Department of Energy has established an adequate program for evaluating the physical condition and general health of personnel certified for operations that are important to safety. Conditions that might cause impaired judgment or motor coordination are adequately considered in the selection of personnel for activities important to safety. The qualifications of personnel holding key positions important to safety are adequate.

2.5.3.3.5 References

American National Standards Institute/American Nuclear Society. "Selection, Qualification and Training of Personnel for Nuclear Power Plants." American National Standards Institute/American Nuclear Society 3.1. 1993.

American Society of Testing and Materials. "Guide for Radiation Protection Training for Nuclear Facility Workers." E 1168. 1995.

U.S. Nuclear Regulatory Commission. Regulatory Guide 1.134, "Medical Evaluation of Licensed Personnel for Nuclear Power Plants." Revision 3. Washington, DC: U.S. Nuclear Regulatory Commission. March 1998.

——. Regulatory Guide 8.29, "Instructions Concerning Risks from Occupational Radiation Exposure." Revision 1. Washington, DC: U.S. Nuclear Regulatory Commission. February 1996.

——. Regulatory Guide 1.8, "Qualification and Training of Personnel for Nuclear Power Plants." Revision 3. Washington, DC: U.S. Nuclear Regulatory Commission. May 2000.

——. Regulatory Guide 8.8, "Information Relevant to Ensuring that Occupational Radiation Exposures at Nuclear Power Stations Will Be As Low As Is Reasonably Achievable (ALARA)." Draft OP–618–4. Second Proposed Revision 4. Washington, DC: U.S. Nuclear Regulatory Commission. May 1984.

——. Regulatory Guide 5.20, "Training, Equipping, and Qualifying of Guards and Watchmen." Washington, DC: U.S. Nuclear Regulatory Commission. January 1974.

Raddatz, C.T. and D. Hagemayer. NUREG–0713, "Occupational Radiation Exposure at Commercial Nuclear Power Reactors and Other Facilities." Vol. 15. Washington, DC: U.S. Nuclear Regulatory Commission. January 1995.

2.5.4 Expert Elicitation

Review Responsibilities—High-Level Waste Branch and Environmental and Performance Assessment Branch

2.5.4.1 Areas of Review

This section reviews expert elicitation. Reviewers will evaluate the information required by 10 CFR 63.21(c)(19).

The U.S. Department of Energy may consider the use of expert elicitation when: (i) empirical data are not reasonably obtainable, or the analyses are not practical to perform; (ii) uncertainties are large and significant to a demonstration of compliance; (iii) more than one conceptual model can explain, and be consistent with, the available data; or (iv) technical judgments are required to assess whether bounding assumptions or calculations are appropriately conservative.

The staff will evaluate the following parts of expert elicitation using the review methods and acceptance criteria in Sections 2.5.4.2 and 2.5.4.3.

(1) Techniques to conduct expert elicitations;

(2) Extent to which guidance in NUREG–1563, "Branch Technical Position on the Use of Expert Elicitation in the High-Level Radioactive Waste Program" (Kotra, et al., 1996) was used to perform expert elicitations; and

(3) Rationales for any discrepancies between staff guidance in NUREG–1563 (Kotra, et al., 1996) and the U.S. Department of Energy conduct of expert elicitations.

2.5.4.2 Review Methods

Review Method 1 Use of NUREG–1563 (Kotra, et al., 1996) or Equivalent Procedures

Verify that expert elicitations either followed the nine-step procedure suggested in NUREG–1563 (Kotra, et al., 1996) or used equivalent procedures. Specifically:

(1) Objectives were defined;

(2) Criteria used to select normative experts and generalists included experts who:

  (a) Possessed the required knowledge and expertise;

  (b) Showed ability to apply their knowledge and expertise;

  (c) In aggregate, represented a broad diversity of independent opinion and approaches to address the topic(s);

  (d) Were willing to be identified publicly with their judgments; and

  (e) Were willing to publicly disclose potential conflicts of interest.

(3) Participants refined the issues and decomposed the problem to clearly and precisely specify more focused and simpler subissues;

(4) Basic information was adequately assembled and was circulated uniformly to the experts;

(5) The experts received preelicitation training that included:

  (a) Familiarization with the subject matter;

  (b) Familiarization with the elicitation process;

  (c) Education in uncertainty and probability encoding and the expression of expert judgment, using subjective probability;

  (d) Practice in formally stating judgments and clearly identifying their associated assumptions and rationales; and

  (e) Identification of biases that could unduly influence judgments.

(6) The conduct of expert elicitations included the following:

  (a) An appropriate setting;

(b) The presence of generalists and normative experts;

(c) A summary of issues, definitions, and assumptions;

(d) Uniform questioning of subject-matter experts; and

(e) Documentation of responses.

(7) Each subject-matter expert got timely feedback from the elicitation team. The rationale for any revisions to elicited judgments was thoroughly documented;

(8) If expert judgments were combined, differing views were treated appropriately as suggested in staff guidance (Kotra, et al., 1996). For combined judgments, the reviewer should confirm that:

(a) The U.S. Department of Energy provided a rationale for the technique used to combine differing views;

(b) The U.S. Department of Energy provided enough documentation to trace the impact of an individual expert's judgment on the consolidated judgment; and

(c) The U.S. Department of Energy discussed effects that the disparate views have had on geologic repository operations area design or health and safety. The U.S. Department of Energy should present significantly different views as individual outputs of the elicitations so that such views may be directly used in the technical assessments or used to condition the extremes in sensitivity analyses.

(9) The U.S. Department of Energy properly documented the expert elicitation, including what was done, why it was done, and who did it.

Verify that the U.S. Department of Energy provided an adequate explanation for any variance from NUREG–1563 (Kotra, et al., 1996) guidance.

Review Method 2 Updating Expert Elicitations

Confirm that any required updating of expert elicitations has been adequately documented to provide a transparent view of the updating process and resulting judgments and uses an appropriate method.

2.5.4.3 Acceptance Criteria

The following acceptance criterion is based on meeting the requirements of 10 CFR 63.21(c)(19).

Acceptance Criterion 1 The U.S. Department of Energy Used NUREG–1563 (Kotra, et al., 1996) or Equivalent Procedures.

(1) Expert elicitations follow the nine-step procedure in NUREG–1563 (Kotra, et al., 1996). Specifically:

(a) Objectives are defined;

(b) Criteria used to select normative experts and generalists include:

(i) Experts possess the required knowledge and expertise;

(ii) Experts demonstrate ability to apply their knowledge and expertise;

(iii) Experts, as a group, represent a broad diversity of independent opinion and approaches to address the topic(s);

(iv) Experts are willing to be identified publicly with their judgments; and

(v) Experts are willing to publicly disclose potential conflicts of interest.

(c) Participants refined the issues and broke down the problem to clearly specify more focused and simpler subissues;

(d) The U.S. Department of Energy adequately assembled and uniformly distributed the basic information to the experts;

(e) The experts received preelicitation training that included:

(i) Familiarization with the subject matter;

(ii) Familiarization with the elicitation process;

(iii) Education in uncertainty and probability encoding and how to express expert judgment, using subjective probability;

(iv) Practice in formally articulating judgments and explicitly identifying their associated assumptions and rationales; and

(v) Identification of biases that could unduly affect judgments.

(f) The conduct of expert elicitations includes the following:

(i) An appropriate setting;

(ii) The presence of generalists and normative experts;

(iii) A summary of issues, definitions, and assumptions;

(iv) Uniform questioning of subject-matter experts; and

(v) Documentation of responses.

(g) Each subject-matter expert received timely feedback from the elicitation team. The rationale for revising elicited judgments is thoroughly documented;

(h) If expert judgments are combined, differing views are treated as suggested in staff guidance (Kotra, et al., 1996). Specifically:

(i) The U.S. Department of Energy provided a rationale for the technique used to combine differing views: the U.S. Department of Energy included enough documentation to trace the impact of an individual expert's judgment on the combined judgment; and

(ii) The U.S. Department of Energy discussed the effects of differing views on facility design or health and safety. The U.S. Department of Energy presented significantly different views as individual outputs of the elicitations so that such views are directly used in the technical assessments or used to condition the extremes in sensitivity analyses.

(i) The U.S. Department of Energy properly documented the expert elicitation including what is done, why it is done, and who did it.

(2) The U.S. Department of Energy adequately explained any variance from the guidance and techniques in NUREG–1563 (Kotra, et al., 1996).

Acceptance Criteria 2 Any Updates to Expert Elicitations Are Adequately Documented and Use Appropriate Methods.

2.5.4.4 Evaluation Findings

If the license application provides sufficient information and the regulatory acceptance criteria in Section 2.5.4.3 are appropriately satisfied, the staff concludes that this portion of the staff evaluation is acceptable. The reviewer writes material suitable for inclusion in the safety evaluation report prepared for the entire application. The report includes a summary statement of what was reviewed and why the reviewer finds the submittal acceptable. The staff can document the review as follows.

U.S. Nuclear Regulatory Commission staff has reviewed the Safety Analysis Report and other information submitted in support of the license application and finds, with reasonable assurance, that the requirements of 10 CFR 63.21(c)(19) are satisfied. The U.S. Department of Energy met the requirements for the contents of the license application. In particular, the Safety Analysis Report explains how and the extent to which expert elicitation was used to characterize: (i) features, events, and processes; (ii) response of geomechanical, hydrogeological, and geochemical systems to thermal loadings; (iii) performance of the geologic repository after permanent closure; (iv) ability of the repository to limit radiological exposures in the event of limited human intrusion into the engineered barrier system; and (v) any other use of expert elicitation to evaluate performance.

2.5.4.5 Reference

Kotra, J.B., et al. NUREG–1563, “Branch Technical Position on the Use Of Expert Elicitation in the High-Level Radioactive Waste Program.” Washington, DC: U.S. Nuclear Regulatory Commission. 1996.

2.5.5 Plans for Startup Activities and Testing

Although the U.S. Department of Energy is not expected to have prepared plans for startup activities and testing at the time of the construction authorization decision, the U.S. Department of Energy will develop and implement these plans that meet the acceptance criteria in this section.

Review Responsibilities—High-Level Waste Branch and Environmental and Performance Assessment Branch

2.5.5.1 Areas of Review

This section reviews plans for startup activities and testing. The reviewers will evaluate the information required by 10 CFR 63.21(c)(22)(iv).

The staff will evaluate the following parts of plans for startup activities and testing, using the review methods and acceptance criteria in Sections 2.5.5.2 and 2.5.5.3.

A review of plans for pre-startup testing and startup activities to be used to evaluate the readiness to receive, possess, process, store, and dispose of high-level radioactive waste should include assessment of planned tests and operations for the structures, systems, and components of the geologic repository operations area. The U.S. Department of Energy is not required to have conducted testing and startup activities or to have detailed procedures in place at the time of construction authorization decision. The U.S. Department of Energy will have an approved testing and startup activities program for structures, systems, and components important to safety in place before receipt of waste. The U.S. Department of Energy is required to have either conducted testing and startup activities or to have detailed procedures in place for such testing and startup activities at the time of application to receive, possess, process, store, or dispose of high-level radioactive waste.

(1) Systems used to develop, review, approve, and execute individual test procedures to evaluate, document, and approve test results;

(2) Pre-startup test program and objectives;

(3) Type and source of design performance information;

(4) Format and content of test procedures and individual test descriptions;

(5) Pre-startup test program compatibility with regulatory guides (if any);

(6) Use of prior experience in developing pre-startup tests;

(7) Assessment of whether initial operating procedures will endanger worker and public health and safety;

(8) Planned user testing for operating, emergency, and surveillance procedures;

(9) Schedules for the testing program relative to the first fuel receipt, repackaging, storage, and disposal, including any overlaps in component and system testing;

(10) Plans for initial startup; and

(11) Evaluation of safety of systematic facility functions and associated activities.

2.5.5.2 Review Methods

Review Method 1 Systems Used to Develop, Review, and Approve Pre-Startup Test Procedures

Verify, based on a summary description, that systems used to develop, review, and approve individual test procedures for each geologic repository operations area component important to safety are acceptable. The summary description should include:

(1) Responsibilities and functions of organizational units for development, review, and approval of test procedures;

(2) Qualification requirements for people assigned responsibilities for test procedure development; and

(3) A description of the general steps for developing, reviewing, approving, and executing tests and for documenting test results.

Review Method 2 Summaries of Pre-Startup Test Programs and Objectives

Verify, based on a summary, that test programs and objectives for each geologic repository operations area structure, system, and component important to safety are acceptable. Evaluate the adequacy of the: (i) type of tests to be performed; (ii) expected response to the tests; (iii) acceptable margin of difference from the expected response; (iv) method of test validation; and (v) appropriateness of proposed corrective action for unexpected or unacceptable test results.

Review Method 3 Incorporation of Design Performance Information in Pre-Startup Testing Plans

Confirm that design information and data from preconstruction performance assessments have been adequately considered in developing pre-startup testing plans. Specifically, functions or parameters of structures, systems, and components important to safety should be tested to the extent feasible.

Review Method 4 Format and Content of Test Procedures

Evaluate the format and content of test procedures for geologic repository operations area structures, systems, and components important to safety and confirm that they are acceptable.

Review Method 5 Test Descriptions

Verify test descriptions are provided for structures, systems, and components that: (i) will be used to establish conformance with safety limits or limiting conditions for operation in the geologic repository operations area technical specifications; (ii) are classified as engineered safety features or will be used to support or ensure the operations of engineered safety features within design limits; (iii) are assumed to function or for which credit is taken in event sequence analyses in the preclosure safety analysis; or (iv) will be used to process, store, control, measure, or limit the release of radioactive materials. Review the conduct of the preclosure safety analysis using Section 2.1.1 of the Yucca Mountain Review Plan.

Confirm that test descriptions contain objectives for each test and a summary of prerequisites, test method(s), and acceptance criteria that will ensure the functional adequacy of structures, systems, and components important to safety and that design features will be demonstrated by the tests.

Verify that test descriptions are consistent with the design requirements. Coordinate with the reviewer of Section 2.1.1.7 ("Design of Structures, Systems, and Components Important to Safety and Safety Controls") of the Yucca Mountain Review Plan to confirm the design requirements.

Confirm that test descriptions contain sufficient information to justify the test method used, particularly if the test method for a structure, system, or component important to safety will not subject the item or system to the range of design operating conditions.

Review Method 6 Compatibility of Test Programs with Applicable Regulatory Guidance

Verify that pre-startup test programs for geologic repository operations area structures, systems, and components are consistent with applicable guidance in Regulatory Guide 3.48 (U.S. Nuclear Regulatory Commission, 1989). Verify that, if the U.S. Department of Energy takes positions inconsistent with guidance, it provides suitable justification for the inconsistencies. For specific components, check for regulatory guidance that may be pertinent.

Review Method 7 Use of Experience from Similar Facilities

Confirm the license application provides an assessment of testing results and operational lessons learned from similar facilities. This assessment should be used to develop testing procedures of adequate scope.

Review Method 8 Protection of Worker and Public

Verify that procedures that will guide initial operation of geologic repository operations area structures, systems, and components important to safety, and any prerequisites and precautionary measures associated with these procedures, are acceptable. Make this assessment based on evaluations of procedures using system diagrams and reviewer experience. Initial operating procedures should include the following:

(1) Purpose and role of test in evaluating performance of structure, system, or component function;

(2) Prerequisites for normal readiness testing, such as:

 (a) Calibrations should be performed or checked;

 (b) Instrumentation should be on hand for necessary performance evaluations;

 (c) Tools and special equipment should be on hand to facilitate evaluations;

 (d) Notifications with lead times necessary to eliminate unnecessary downtime during performance evaluations;

 (e) Checking/setting equipment controls (e.g., physical travel limits for overhead crane);

 (f) Checks of radiation, environmental, or other monitors for acceptable range;

 (g) Identification of subject(s) of tests (e.g., fuel rods to be loaded, cask to be retrieved); and

 (h) Logs and forms to be completed.

(3) Description of preceding function and relationship to function;

(4) Description of series of operations, including expected results, projected times, projected instrument and gauge readings, controls to be used in performance (e.g., torque, time at pressure), and threshold limits requiring contingency actions (such as hold, initiating a contingency sequence, notification);

(5) Requirements for records, including forms to be completed during operation (if any);

(6) Disposition of records and identification of parties to be notified on successful or unsuccessful completion (may be different parties) of function evaluation; and

(7) Identification of following function and relation to function being evaluated.

Review Method 9 Schedules

Verify that the U.S. Department of Energy provides schedules for conducting each phase of the testing program and that these schedules are compatible with schedules for high-level radioactive waste receipt, repackaging, storage, and disposal, including any schedule overlaps. Pay particular attention to start-up-sequence timing and the time available between approval of test procedures and their intended use.

Review Method 10 Testing and Evaluating Functional Adequacy of Structures, Systems, and Components

Verify that new structures, systems, and components important to safety, or untested configurations of such components, will be tested and evaluated before receipt of radioactive waste and that their performance is acceptable. Review schedules and programs for unresolved safety issues, using Section 2.3.2 of the Yucca Mountain Review Plan.

Review Method 11 Plans for Initial Startup of Geologic Repository Operations Area Structures, Systems, and Components and Integrated Operation of the Geologic Repository Operations Area

Verify that the U.S. Department of Energy has acceptable plans for a dry run (cold test) of each operation involving radioactive material to be received, handled, stored, or disposed. Confirm that the U.S. Department of Energy will use the results of these to make necessary changes to equipment and procedures to ensure public and worker health and safety.

Confirm that the U.S. Department of Energy has acceptable plans to conduct routine full-load tests of equipment that is to carry high-level radioactive waste containers, to ensure public and worker health and safety.

For as low as is reasonably achievable considerations, verify that as many operating startup actions as feasible will be performed during preoperational testing before sources of radiation exposure are present.

Confirm that plans for operating start-up of the geologic repository operations area structures, systems, and components and subsequent integrated operation of the entire facility are acceptable. The operating start-up plan should include, but not be limited to, the following elements:

(1) Tests and confirmations of procedures and exposure times involving actual radioactive sources (e.g., radiation monitoring, repackaging operations);

(2) Direct radiation monitoring of casks and shielding for radiation dose rates, streaming, and surface hot-spots;

(3) Verification of effectiveness of heat removal procedures;

(4) Tests of structures, systems, and components important to safety as identified by the preclosure safety analysis (review identification of structures, systems, and components important to safety using Section 2.1.1.6 of the Yucca Mountain Review Plan); and

(5) Documentation of results and test evaluations.

Review Method 12 Overall Geologic Repository Operations Area Safety Supported by Startup and Testing Plans

Confirm that the overall evaluation of geologic repository operations area safety for workers and the public is supported by the aggregate effects of planned start-up activities and associated testing.

2.5.5.3 Acceptance Criteria

The following acceptance criteria are based on meeting the requirements of 10 CFR 63.21(c)(22)(iv).

Acceptance Criterion 1 Systems Used to Develop, Review, and Approve Individual Pre-Startup Test Procedures Are Acceptable.

(1) Based on a summary description, the systems used to develop, review, and approve individual test procedures for each geologic repository operations area component important to safety are acceptable. The summary adequately defines:

  (a) Responsibilities and functions of organizational units for development, review, and approval of test procedures;

  (b) Qualification requirements for people assigned responsibilities for test procedure development; and

  (c) General steps to be followed when developing, reviewing, approving, and executing tests and for documenting test results.

Acceptance Criterion 2 Summaries of Pre-Startup Test Programs and Objectives Are Adequate.

(1) Based on a summary description, the test programs and objectives for each geologic repository operations area structure, system, or component important to safety are acceptable. The summary adequately presents: (i) type of tests to be performed; (ii) expected response to the tests; (iii) acceptable margin of difference from the expected response; (iv) method of test validation; and (v) appropriateness of proposed corrective action for unexpected or unacceptable test results.

Acceptance Criterion 3 Design Performance Information Is Adequately Incorporated in Pre-Startup Testing Plans.

(1) The design information and data from preconstruction performance assessments are adequately considered in the development of pre-startup testing plans. Specifically,

functions or parameters of structures, systems, and components that are important to safety are tested to the extent feasible.

Acceptance Criterion 4 The Format and Content of Test Procedures Are Acceptable.

(1) The format and content of the test procedures for geologic repository operations area structures, systems, and components important to safety are acceptable.

Acceptance Criterion 5 Test Descriptions Are Acceptable.

(1) Adequate test descriptions are provided for those structures, systems, and components that: (i) will be used to establish conformance with safety limits or limiting conditions for operation in the geologic repository operations area technical specifications; (ii) are classified as engineered safety features or will be used to support or ensure the operations of engineered safety features within design limits; (iii) are assumed to function or for which credit is taken in event sequence analyses in the preclosure safety analysis; or (iv) will be used to process, store, control, measure, or limit the release of radioactive materials;

(2) The test descriptions contain acceptable objectives for each test, a summary of prerequisites, test method(s), and specific acceptance criteria for each test that will ensure that both the functional adequacy of structures, systems, and components important to safety and design features are demonstrated by the tests;

(3) The test descriptions are consistent with the design requirements; and

(4) The test descriptions contain sufficient information to justify the test method used, particularly if the test method for a given structure, system, and component important to safety will not subject the item or system under test to the range of design operating conditions.

Acceptance Criterion 6 Test Programs Are Compatible with Applicable Regulatory Guidance.

(1) The pre-startup test programs for geologic repository operations area structures, systems, and components are consistent with applicable regulatory guidance in Regulatory Guide 3.48 (U.S. Nuclear Regulatory Commission, 1989). If the U.S. Department of Energy takes positions inconsistent with guidance, a suitable justification for the inconsistencies is provided.

Acceptance Criterion 7 Adequate Use Is Made of Experience from Similar Facilities.

(1) The license application provides an assessment of testing results and operational lessons learned from similar facilities, and this assessment is used to develop testing procedures of adequate scope.

Acceptance Criterion 8 Initial Operating Procedures Will Protect Workers and the Public.

(1) Procedures that will guide initial operation of the geologic repository operations area structures, systems, and components important to safety and any prerequisites and precautionary measures associated with these procedures are acceptable.

Acceptance Criterion 9 Schedules for Each Phase of the Testing Program Are Acceptable.

(1) The U.S. Department of Energy provides schedules for conducting each phase of the testing program, and these schedules are compatible with schedules for high-level radioactive waste receipt, repackaging, storage, and disposal, including any schedule overlaps.

Acceptance Criterion 10 Structures, Systems, and Components Important to Safety Whose Functional Adequacy Has Not Been Demonstrated by Prior Use or Otherwise Validated Are Tested and Evaluated Before the Receipt of Radioactive Waste.

(1) The new structures, systems, and components important to safety, or untested configurations of such components, are tested and evaluated before receipt of radioactive waste, and their performance is acceptable.

Acceptance Criterion 11 Plans for Initial Start up of Geologic Repository Operations Area Structures, Systems, and Components Important to Safety and Integrated Operation of the Geologic Repository Operations Area Are Acceptable.

(1) The U.S. Department of Energy has acceptable plans to perform a dry run (cold test) of each operation involving radioactive material to be received, handled, stored, or disposed. The results of these tests will be used to make necessary changes to equipment and procedures to ensure public and worker health and safety;

(2) The U.S. Department of Energy has acceptable plans to conduct routine full-load tests of any equipment that is to carry high-level radioactive waste containers, to ensure public and worker health and safety;

(3) For as low as is reasonably achievable considerations, as many of the operating startup actions as feasible are performed during preoperational testing, before sources of radiation exposure are present; and

(4) Plans for operating startup of the geologic repository operations area structures, systems, and components and subsequent integrated operation of the entire facility are acceptable.

Acceptance Criterion 12 Overall Geologic Repository Operations Area Safety Is Adequately Supported by Facility Start-Up and Testing Plans.

(1) The overall evaluation of safety of the facility for workers and the public is supported by the aggregate of planned startup activities and associated testing.

2.5.5.4 Evaluation Findings

If the license application provides sufficient information and the regulatory acceptance criteria in Section 2.5.5.3 are appropriately satisfied, the staff concludes that this portion of the staff evaluation is acceptable. The reviewer writes material suitable for inclusion in the safety evaluation report prepared for the entire application. The report includes a summary statement of what was reviewed and why the reviewer finds the submittal acceptable. The staff can document the review as follows.

U.S. Nuclear Regulatory Commission staff has reviewed the Safety Analysis Report and other information submitted in support of the license application and finds, with reasonable assurance, that the requirements of 10 CFR 63.21(c)(22)(iv) are satisfied. Requirements for the content of the license application have been met. In particular, the plans for testing and startup of structures, systems, and components important to safety of the geologic repository operations area to receive, possess, store, process, and dispose of spent nuclear fuel and high-level radioactive waste are acceptable.

2.5.5.5 Reference

U.S. Nuclear Regulatory Commission. Regulatory Guide 3.48, "Standard Format and Content for the Safety Analysis Report for an Independent Spent Fuel Storage Installation or Monitored Retrievable Storage Installation (Dry Storage)." Revision 1. Washington, DC: U.S. Nuclear Regulatory Commission, Office of Standards Development. August 1989.

2.5.6 Plans for Conduct of Normal Activities Including Maintenance, Surveillance, and Periodic Testing

The U.S. Department of Energy will develop and implement these procedures and plans prior to receipt and possession of waste.

Review Responsibilities—High-Level Waste Branch and Environmental and Performance Assessment Branch

2.5.6.1 Areas of Review

This section reviews plans for conduct of normal activities, including maintenance, surveillance, and periodic testing. Reviewers will evaluate the information required by 10 CFR 63.21(c)(22)(v).

The staff will evaluate the following parts of plans for conduct of normal activities, including maintenance, surveillance, and periodic testing, using the review methods and acceptance criteria in Sections 2.5.6.2 and 2.5.6.3.

Normal operations at the geologic repository operations area may include, among other operations: (i) acceptance of waste; (ii) storage of waste before repackaging; (iii) repackaging of waste; (iv) removal/reuse of transport containers; (v) storage of repackaged waste before disposal; and (vi) disposal of waste. Each activity important to safety should have written procedures in place for normal operations, maintenance, surveillance, and periodic testing:

(1) Procedures and plans;

(2) Descriptions of activities;

(3) Administrative procedures for review, change, and approval; and

(4) Independence of review of procedure development by persons outside the operating management function.

2.5.6.2 Review Methods

Review Method 1 Plans and Procedures for Normal Operations

Verify that the U.S. Department of Energy has provided adequate written procedures for normal operation of structures, systems, and components important to safety, as identified in the preclosure safety analysis and reviewed in Section 2.1 of the Yucca Mountain Review Plan, to include routine and contingency operations and any procedural requirements necessitated by technical specifications. Normal operating procedures should include the following:

(1) Purpose of the procedure;

(2) Responsibilities, training, and qualifications of personnel;

(3) Prerequisites such as:

 (a) Calibrations to be performed or checked;

 (b) Instrumentation;

 (c) Tools and special equipment;

 (d) Notifications to other operations personnel with associated lead times;

 (e) Checks or settings for equipment or controls (e.g., physical travel limits for overhead crane);

 (f) Operational checks of radiation, environmental, or other monitors; and

 (g) Logs and records associated with the test.

(4) Description of the series of operations, including expected results, expected radiation dose, projected times for completion, expected instrument and gauge readings, controls

to be used (e.g., torque, time at pressure); and threshold limits requiring contingency actions (such as hold points, corrective action sequences, and notifications);

(5) Disposition of records and identification of parties to be notified on completion of the operation; and

(6) Identification of any required follow-on actions.

Verify that administrative procedures for the review, change, and approval of normal operating procedures for structures, systems, and components important to safety are adequate and that these procedures have adequate management controls.

Confirm that appropriate industry standards or U.S. Nuclear Regulatory Commission guidance is used as the basis for the operating procedures for structures, systems, and components important to safety.

Verify that normal operations of structures, systems, and components that are important to safety are performed according to written procedures that are reviewed by health, safety, and quality assurance personnel who are independent of the operating management function. Personnel assigned responsibility for these independent reviews should be specified, in both number and technical disciplines, and should collectively have the experience and competence required to review problems in the following areas:

(1) Nuclear engineering;

(2) Chemistry and radiochemistry;

(3) Metallurgy;

(4) Nondestructive testing;

(5) Instrumentation and control;

(6) Radiological safety;

(7) Mechanical, civil, and electrical engineering;

(8) Administrative controls and quality assurance practices; and

(9) Other appropriate fields associated with the characteristics of a repository for high-level radioactive waste.

An individual may possess competence in more than one speciality area.

Review Method 2 Plans and Procedures for Maintenance

Verify that written procedures are provided for maintenance of structures, systems, and components important to safety and include the following:

(1) Purpose of the maintenance procedure;

(2) Responsibilities, training, and qualifications of personnel;

(3) Prerequisites such as:

(a) Calibrations to be performed or checked;

(b) Instrumentation;

(c) Tools and special equipment;

(d) Notifications to other operations or maintenance personnel with associated lead times;

(e) Checks or settings for equipment or controls;

(f) Operational checks of radiation, environmental, or other monitors; and

(g) Logs and records associated with the maintenance.

(4) Description of the maintenance activities, including expected results, expected radiation dose, projected times for completion, expected instrument and gauge readings, controls to be used, and threshold limits requiring contingency actions; and

(5) Disposition of records and identification of parties to be notified on completion.

Verify that administrative procedures for the review, change, and approval of maintenance procedures for structures, systems, and components important to safety are adequate and that these procedures have adequate management controls.

Confirm that appropriate industry standards or U.S. Nuclear Regulatory Commission guidance is used as the basis for the maintenance procedures for structures, systems, and components important to safety.

Verify that maintenance activities on structures, systems, and components that are important to safety are performed according to written procedures that are reviewed by health, safety, and quality assurance personnel who are independent of the operating management function. Personnel assigned responsibility for these independent reviews should be specified, in both number and technical disciplines, and should collectively have the experience and competence required to review problems in the following areas:

(1) Nuclear engineering;

(2) Chemistry and radiochemistry;

(3) Metallurgy;

(4) Nondestructive testing;

(5) Instrumentation and control;

(6) Radiological safety;

(7) Mechanical, civil, and electrical engineering;

(8) Administrative controls and quality assurance practices; and

(9) Other appropriate fields associated with the characteristics of a repository for high-level radioactive waste.

An individual may possess competence in more than one speciality area.

Review Method 3 Plans and Procedures for Surveillance

Verify that written procedures are provided to routinely evaluate, through surveillance, the proper functioning of structures, systems, and components important to safety and include the following:

(1) Purpose of the routine surveillance;

(2) Responsibilities, training, and qualifications of personnel;

(3) Prerequisites such as:

(a) Calibrations to be performed or checked;

(b) Instrumentation;

(c) Tools and special equipment;

(d) Notifications to operations personnel with associated lead times;

(e) Checks or settings for equipment or controls;

(f) Operational checks of radiation, environmental, or other monitors; and

(g) Logs or records associated with the surveillance.

(4) Description of the surveillance activities, including expected results, expected radiation dose, projected times for completion, expected instrument and gauge readings, controls to be assessed; and

(5) Disposition of records and identification of parties to be notified on completion.

Verify that if structures, systems, and components important to safety are found operating outside the tolerance for normal operation during surveillance, adequate procedures are in place to assure they will be restored to normal conditions in a reasonably short time so worker and public health and safety are protected.

Verify that administrative procedures for the review, change, and approval of surveillance procedures for structures, systems, and components important to safety are adequate and that these procedures have adequate management controls.

Confirm that appropriate industry or U.S. Nuclear Regulatory Commission standards, if applicable, are used as the basis for the surveillance procedures for structures, systems, and components important to safety.

Verify that surveillance activities on structures, systems, and components that are important to safety are performed according to written procedures that are reviewed by health, safety, and quality assurance personnel who are independent of the operating management function. Personnel assigned responsibility for these independent reviews should be specified, in both number and technical disciplines, and should collectively have the experience and competence required to review problems in the following areas:

(1) Nuclear engineering;

(2) Chemistry and radiochemistry;

(3) Metallurgy;

(4) Nondestructive testing;

(5) Instrumentation and control;

(6) Radiological safety;

(7) Mechanical, civil, and electrical engineering;

(8) Administrative controls and quality assurance practices; and

(9) Other appropriate fields associated with the characteristics of a repository for high-level radioactive waste.

An individual may possess competence in more than one speciality area.

Review Method 4 Plans and Procedures for Routine Periodic Testing

Verify that written procedures for periodic testing designed to ensure that structures, systems, and components important to safety will perform their design function during normal operations are in place. This testing should be accomplished on a defined schedule and at a frequency sufficient to ensure protection of worker and public safety. The reviewer should verify that procedures for periodic testing of structures, systems, and components important to safety include the following:

(1) Purpose of testing;

(2) Responsibilities, training, and qualifications of personnel;

(3) Prerequisites such as:

   (a) Calibrations to be performed or checked;

   (b) Instrumentation;

   (c) Tools and special equipment;

   (d) Notifications to other operations or testing personnel with associated lead times;

   (e) Checks or settings for equipment or controls;

   (f) Operational checks of radiation, environmental, or other monitors; and

   (g) Logs or records associated with the testing.

(4) Description of the testing activities, including expected results, expected radiation dose, projected times for completion, expected instrument and gauge readings, controls to be used, and threshold limits requiring contingency actions; and

(5) Disposition of records and identification of parties to be notified on completion.

Verify that if structures, systems, and components important to safety are found operating outside the tolerance for normal operation during periodic testing, adequate procedures are in place to assure that they will be restored to normal conditions in a reasonably short time such that worker and public health and safety are protected.

Verify that administrative procedures for the review, change, and approval of periodic testing procedures for structures, systems, and components important to safety are adequate and that these procedures have adequate management controls.

Confirm that appropriate industry or U.S. Nuclear Regulatory Commission standards, if applicable, are used as the basis for the periodic testing procedures for structures, systems, and components important to safety.

Verify that periodic testing activities on structures, systems, and components that are important to safety are performed according to written procedures that are reviewed by health, safety, and quality assurance personnel who are independent of the operating management function. Personnel assigned responsibility for these independent reviews should be specified, in both number and technical disciplines, and should collectively have the experience and competence required to review problems in the following areas:

(1) Nuclear engineering;

(2) Chemistry and radiochemistry;

(3) Metallurgy;

(4) Nondestructive testing;

(5) Instrumentation and control;

(6) Radiological safety;

(7) Mechanical, civil, and electrical engineering;

(8) Administrative controls and quality assurance practices; and

(9) Other appropriate fields associated with the characteristics of a repository for high-level radioactive waste.

An individual may possess competence in more than one speciality area.

2.5.6.3 Acceptance Criteria

The following acceptance criteria are based on meeting the requirements of 10 CFR 63.21(c)(22)(v).

Acceptance Criterion 1 Plans for Normal Operation of Structures, Systems, and Components of the Geologic Repository Operations Area That Are Important to Safety Are Acceptable.

(1) Acceptable written procedures are provided for normal operation of structures, systems, and components important to safety, as identified in the preclosure safety analysis and reviewed in Section 2.1 of the Yucca Mountain Review Plan, to include routine and contingency operations as well as any procedural requirements necessitated by technical specifications. Normal operating procedures include the following:

   (a) Purpose of the procedure;

   (b) Responsibilities, training, and qualifications of personnel;

   (c) Prerequisites such as:

      (i) Calibrations to be performed or checked;

      (ii) Instrumentation;

(iii) Tools and special equipment;

(iv) Notifications to other operations personnel with associated lead times;

(v) Checks or settings for equipment or controls (e.g., physical travel limits for overhead crane);

(vi) Operational checks of radiation, environmental, or other monitors; and

(vii) Logs and records associated with the test.

(d) Description of the series of operations to be performed, including expected results, expected radiation dose, projected times for completion, expected instrument and gauge readings, controls to be used (e.g., torque, time at pressure), and threshold limits requiring contingency actions (such as hold points, corrective-action sequences, and notifications);

(e) Disposition of records and identification of parties to be notified on completion of the operation; and

(f) Identification of any required follow-on actions.

(2) Administrative procedures for the review, change, and approval of normal operating procedures for structures, systems, and components important to safety are adequate, and these procedures have adequate management controls;

(3) Appropriate industry standards or U.S. Nuclear Regulatory Commission guidance is used as the basis for the operating procedures for structures, systems, and components important to safety; and

(4) Normal operations of structures, systems, and components that are important to safety are performed according to written procedures that are reviewed by health, safety, and quality assurance personnel who are independent of the operating management function. Personnel assigned responsibility for these independent reviews are specified, in both number and technical disciplines, and collectively have the experience and competence required to review problems in the following areas:

(a) Nuclear engineering;

(b) Chemistry and radiochemistry;

(c) Metallurgy;

(d) Nondestructive testing;

(e) Instrumentation and control;

(f) Radiological safety;

(g) Mechanical, civil, and electrical engineering;

(h) Administrative controls and quality assurance practices; and

(i) Other appropriate fields associated with the characteristics of a repository for high-level radioactive waste.

Acceptance Criterion 2 Plans and Procedures for Maintenance of Structures, Systems, and Components of the Geologic Repository Operations Area That Are Important to Safety Are Acceptable.

(1) Written procedures are provided for maintenance of structures, systems, and components important to safety and include the following:

(a) Purpose of the maintenance procedure;

(b) Responsibilities, training, and qualifications of personnel;

(c) Prerequisites such as:

(i) Calibrations to be performed or checked;

(ii) Instrumentation;

(iii) Tools and special equipment;

(iv) Notifications to other operations or maintenance personnel with associated lead times;

(v) Checks or settings for equipment or controls;

(vi) Operational checks of radiation, environmental, or other monitors; and

(vii) Logs and records associated with the maintenance.

(d) Description of the maintenance activities, including expected results, expected radiation dose, projected times for completion, expected instrument and gauge readings, controls to be used, and threshold limits requiring contingency actions; and

(e) Disposition of records and identification of parties to be notified on completion.

(2) Administrative procedures for the review, change, and approval of maintenance procedures for structures, systems, and components important to safety are adequate, and these procedures have adequate management controls;

(3) Appropriate industry standards or U.S. Nuclear Regulatory Commission guidance is used as the basis for the maintenance procedures for structures, systems, and components important to safety; and

(4) Maintenance activities on structures, systems, and components that are important to safety are performed according to written procedures that are reviewed by health, safety, and quality assurance personnel who are independent of the operating management function. Personnel assigned responsibility for these independent reviews are specified, in both number and technical disciplines, and collectively have the experience and competence required to review problems in the following areas:

  (a) Nuclear engineering;

  (b) Chemistry and radiochemistry;

  (c) Metallurgy;

  (d) Nondestructive testing;

  (e) Instrumentation and control;

  (f) Radiological safety;

  (g) Mechanical, civil, and electrical engineering;

  (h) Administrative controls and quality assurance practices; and

  (i) Other appropriate fields associated with the characteristics of a repository for high-level radioactive waste.

Acceptance Criterion 3 Plans and Procedures for Surveillance of Structures, Systems, and Components of the Geologic Repository Operations Area That Are Important to Safety Are Acceptable.

(1) Written procedures are provided to routinely evaluate, through surveillance, the proper functioning of structures, systems, and components important to safety and include the following:

  (a) Purpose of the routine surveillance;

  (b) Responsibilities, training, and qualifications of personnel;

  (c) Prerequisites such as:

    (i) Calibrations to be performed or checked;

    (ii) Instrumentation;

(iii) Tools and special equipment;

(iv) Notifications to operations personnel with associated lead times;

(v) Checks or settings for equipment or controls;

(vi) Operational checks of radiation, environmental, or other monitors; and

(vii) Logs or records associated with the surveillance.

(d) Description of the surveillance activities, including expected results, expected radiation dose, projected times for completion, expected instrument and gauge readings, controls to be assessed; and

(e) Disposition of records and identification of parties to be notified on completion.

(2) If structures, systems, and components important to safety are found operating outside the tolerance for normal operation during surveillance, adequate procedures are in place to assure that they will be restored to normal conditions in a reasonably short time such that worker and public health and safety are protected;

(3) Administrative procedures for the review, change, and approval of surveillance procedures for structures, systems, and components important to safety are adequate, and these procedures have adequate management controls;

(4) Appropriate industry or U.S. Nuclear Regulatory Commission standards, if applicable, are used as the basis for the surveillance procedures for structures, systems, and components important to safety; and

(5) Surveillance activities on structures, systems, and components that are important to safety are performed according to written procedures that are reviewed by health, safety, and quality assurance personnel who are independent of the operating management function. Personnel assigned responsibility for these independent reviews are specified, in both number and technical disciplines, and collectively have the experience and competence required to review problems in the following areas:

(a) Nuclear engineering;

(b) Chemistry and radiochemistry;

(c) Metallurgy;

(d) Nondestructive testing;

(e) Instrumentation and control;

(f) Radiological safety;

(g) Mechanical, civil, and electrical engineering;

(h) Administrative controls and quality assurance practices; and

(i) Other appropriate fields associated with the characteristics of a repository for high-level radioactive waste.

Acceptance Criterion 4 Plans and Procedures for Routine Periodic Testing of Structures, Systems, and Components of the Geologic Repository Operations Area That Are Important to Safety Are Acceptable.

(1) Written procedures for periodic testing designed to ensure that structures, systems, and components important to safety will perform their design function during normal operations are in place. This testing will be accomplished on a defined schedule and at a frequency sufficient to ensure protection of worker and public safety. Procedures for periodic testing of structures, systems, and components important to safety include the following:

(a) Purpose of testing;

(b) Responsibilities, training, and qualifications of personnel;

(c) Prerequisites such as:

(i) Calibrations to be performed or checked;

(ii) Instrumentation;

(iii) Tools and special equipment;

(iv) Notifications to other operations or testing personnel with associated lead times;

(v) Checks or settings for equipment or controls;

(vi) Operational checks of radiation, environmental, or other monitors; and

(vii) Logs or records associated with the testing.

(d) Description of the testing activities, including expected results, expected radiation dose, projected times for completion, expected instrument and gauge readings, controls to be used, and threshold limits requiring contingency actions; and

(e) Disposition of records and identification of parties to be notified on completion.

(2) If structures, systems, and components important to safety are found operating outside the tolerance for normal operation during periodic testing, adequate procedures are in

place to assure that they will be restored to normal conditions in a reasonably short time such that worker and public health and safety are protected;

(3) Administrative procedures for the review, change, and approval of periodic testing procedures for structures, systems, and components important to safety are adequate, and these procedures have adequate management controls;

(4) Appropriate industry or U.S. Nuclear Regulatory Commission standards, if applicable, are used as the basis for the periodic testing procedures for structures, systems, and components important to safety; and

(5) Periodic testing activities on structures, systems, and components that are important to safety are performed according to written procedures that are reviewed by health, safety, and quality assurance personnel who are independent of the operating management function. Personnel assigned responsibility for these independent reviews are specified, in both number and technical disciplines, and collectively have the experience and competence required to review problems in the following areas:

    (a) Nuclear engineering;

    (b) Chemistry and radiochemistry;

    (c) Metallurgy;

    (d) Nondestructive testing;

    (e) Instrumentation and control;

    (f) Radiological safety;

    (g) Mechanical, civil, and electrical engineering;

    (h) Administrative controls and quality assurance practices; and

    (i) Other appropriate fields associated with the characteristics of a repository for high-level radioactive waste.

2.5.6.4 Evaluation Findings

If the license application provides sufficient information and the regulatory acceptance criteria in Section 2.5.6.3 are appropriately satisfied, the staff concludes that this portion of the staff evaluation is acceptable. The reviewer writes material suitable for inclusion in the safety evaluation report prepared for the entire application. The report includes a summary statement of what was reviewed and why the reviewer finds the submittal acceptable. The staff can document the review as follows.

U.S. Nuclear Regulatory Commission staff has reviewed the Safety Analysis Report and other information submitted in support of the license application and finds, with reasonable assurance, that the requirements of 10 CFR 63.21(c)(22)(v) are satisfied. The U.S. Department of Energy has provided an adequate plan for conducting normal activities, including operations, maintenance, surveillance, and periodic testing of structures, systems, and components important to safety at the geologic repository operations area.

2.5.6.5 References

None.

2.5.7 Emergency Planning

The review determines with reasonable assurance whether the U.S. Department of Energy has provided an emergency plan that meets with the requirements of 10 CFR Part 63, Subpart I, in the light of information that is reasonably available, and provides reasonable assurance that adequate protection measures can be taken in the event of an emergency.

Review Responsibilities—High-Level Waste Branch, Division of Fuel Cycle Safety and Safeguards, and Environmental and Performance Assessment Branch

2.5.7.1 Areas of Review

This section reviews emergency planning. Reviewers will also evaluate the information required by 10 CFR 63.21(c)(21).

The staff will evaluate the following parts of emergency planning, using the review methods and acceptance criteria in Sections 2.5.7.2 and 2.5.7.3.

(1) Descriptions of the geologic repository operations area and nearby areas;

(2) Types and classifications of potential radioactive materials accidents;

(3) Means for detection of key initiating events and accident conditions;

(4) Actions to mitigate consequences of accidents;

(5) Methods and equipment to assess radioactive materials releases;

(6) Responsibilities of facility personnel during emergencies;

(7) Responsibilities for developing, maintaining, and updating the emergency plan;

(8) Means to notify and coordinate with off-site response organizations;

(9) Information to be communicated to off-site organizations;

(10) Training plans for emergency response;

(11) Means for restoring the facility to a safe condition;

(12) Provisions for quarterly communications checks;

(13) Plans for biennial emergency response exercises;

(14) Plans for semiannual radiological/health physics, medical, and fire drills;

(15) Certification that hazardous chemicals responsibilities are met under the Emergency Planning and Community Right-to-Know Act of 1986;

(16) Comments and their resolution on the emergency plan from off-site emergency response organizations;

(17) Assignments for off-site assistance; and

(18) Arrangements for providing information to the public.

2.5.7.2 Review Methods

Additional guidance for conducting this review is found in NUREG–1567, "Standard Review Plan for Spent Fuel Storage Facilities" (U.S. Nuclear Regulatory Commission, 2000). Criteria for an acceptable emergency plan are in 10 CFR 72.32(b).

Review Method 1 Emergency Plan

Confirm that the U.S. Department of Energy has included a description of the geologic repository operations area and the area near the site sufficient to support an evaluation of the emergency plan.

Verify that the application identifies each plausible type of radioactive materials accident. The radiological emergencies and accidents identified in the emergency plan should be the same as those identified during the review of event sequences conducted using Section 2.1.1.4 of the Yucca Mountain Review Plan.

Verify that the U.S. Department of Energy defines an adequate classification system to identify accidents as "alerts" or "site area emergencies."

Assess the adequacy of the means (instruments, equipment, procedures, etc.) to detect key initiating events and accident conditions. Assess the rationale for the locations and types of detection devices deployed.

Assess the adequacy of planned means to mitigate the consequences of each type of accident, including the means to protect site workers and the program to maintain mitigative equipment.

Verify that methods and equipment planned to be used to assess releases of radioactive materials are adequate to support effective emergency response.

Verify that the U.S. Department of Energy clearly defines the responsibilities of facility personnel during a radiological accident and identifies personnel responsible for prompt notification of off-site response organizations and the U.S. Nuclear Regulatory Commission.

Confirm the adequacy of information provided for off-site response organizations, including the point of contact; address; and phone number, fax, and e-mail addresses.

Verify that the U.S. Department of Energy assigns responsibilities for developing, maintaining, and updating the emergency plan.

Verify that the U.S. Department of Energy provides a brief description of, the means to promptly notify off-site response organizations and request off-site assistance, including medical assistance for the treatment of contaminated injured on-site workers. Confirm that:

(1) A control point will be established;

(2) The unavailability of some personnel, parts of the facility, and some equipment will not prevent the notification of and coordination with off-site response organizations; and

(3) The U.S. Nuclear Regulatory Commission operations center will be notified within 1 hour after an emergency is declared.

Assess the description of the types of information to be provided on geologic repository operations area status, radioactive releases, and recommended protective actions (if necessary). Confirm that this information will be adequate and that it will be provided in a timely manner to off-site response organizations and the U.S. Nuclear Regulatory Commission.

Confirm that emergency response training provided to workers and any special instructions and orientation tours offered for fire, police, medical, and other off-site-based emergency personnel are adequate to support effective actions. Review the geologic repository operations area training program using Section 2.5.3 ("Training and Certification of Personnel") of the Yucca Mountain Review Plan.

Confirm that means to restore the geologic repository operations area to a safe condition after an accident will be adequate.

Confirm that quarterly communications checks with off-site response organizations and biennial on-site exercises to test response to simulated emergencies are planned and include the following:

(1) A check and update of all necessary phone numbers, fax numbers, and e-mail addresses;

(2) An invitation to off-site response organizations to participate in the biennial exercises (participation of off-site organizations in biennial exercises is recommended but not required);

(3) Use of scenarios not known to most exercise participants;

(4) A plan for critiques of each exercise by individuals not having direct implementation responsibility for conducting the exercise. Verify that critiques will evaluate the appropriateness of the plan, emergency procedures, facilities and equipment, training of personnel, and the overall effectiveness of the response; and

(5) Provisions to correct deficiencies identified by the critiques.

Confirm that on-site exercises to test response to simulated emergencies will be conducted biennially.

Confirm that radiological/health physics, medical, and fire drills are planned semiannually.

Verify that the geologic repository operations area operations will satisfy the Emergency Planning and Community Right-to-Know Act of 1986, with respect to hazardous materials at the facility.

Confirm that off-site response organizations were allowed 60 days to comment on the initial submittal of the emergency plan before it was transmitted to the U.S. Nuclear Regulatory Commission. Verify that subsequent plan changes will have a 60-day comment period if the changes affect the off-site response organizations. Confirm that any comments received during the 60-day comment period, and licensee responses, were submitted to the U.S. Nuclear Regulatory Commission with the emergency plan.

Verify that plans for use of off-site assistance include:

(1) Arrangements for requesting and effectively using off-site assistance and provisions for using other organizations that can augment the planned on-site response, as required;

(2) Provisions for prompt communication among principal response organizations to off-site personnel who would be responding onsite;

(3) Provision of adequate emergency facilities and equipment to support the emergency response onsite;

(4) Specification of methods, systems, and equipment for assessing and monitoring consequences of radiological emergency conditions;

(5) Arrangements for medical services for on-site contaminated and injured individuals; and

(6) Training in radiological emergency response for off-site personnel who may be called to assist in an emergency.

Confirm that adequate arrangements for providing timely information to the public exist.

2.5.7.3 Acceptance Criteria

The following acceptance criteria are based on meeting the requirements of 10 CFR 63.161.

Acceptance Criterion 1 An Adequate Emergency Plan for Responding to Potential Radiological Materials and Other Accidents at the Geologic Repository Operations Area Is Provided.

(1) A description of the geologic repository operations area and the area near the site sufficient to support an evaluation of the emergency plan is included;

(2) The U.S. Department of Energy identifies each plausible type of radioactive materials accident. The radiological emergencies and accidents identified in the emergency plan are the same as those identified in event sequences;

(3) The classification system to identify accidents as "alerts" or "site area emergencies" is adequate;

(4) The means (instruments, equipment, procedures, etc.) used to detect key initiating events and accident conditions are adequate. The rationale for the locations and types of detection devices deployed is acceptable;

(5) The planned means for mitigating the consequences of each type of accident, including the means provided to protect site workers and the program to maintain mitigative equipment, are adequate;

(6) The methods and equipment planned to be used to assess releases of radioactive materials to support effective emergency response actions are adequate;

(7) The responsibilities and identities of facility personnel during a radiological accident and of personnel responsible for prompt notification of off-site response organizations and the U.S. Nuclear Regulatory Commission are adequately defined;

(8) Information provided for off-site response organizations, including the point of contact; address; and phone number, fax, and e-mail addresses, is adequate;

(9) Responsibilities for developing, maintaining, and updating the emergency plan are acceptably defined;

(10) A brief description of, the means to promptly notify off-site response organizations and request off-site assistance, including medical assistance for the treatment of contaminated injured on-site workers, are provided. The description also includes sufficient information to verify that:

    (a) A control point will be established;

    (b) The unavailability of some personnel, parts of the facility, and some equipment will not prevent the notification of and coordination with off-site response organizations; and

    (c) The U.S. Nuclear Regulatory Commission operations center will be notified within 1 hour after an emergency is declared.

(11) The types of information to be provided on facility status, radioactive releases, and recommended protective actions (if necessary) are adequate and this information will be provided in a timely manner to off-site response organizations and the U.S. Nuclear Regulatory Commission;

(12) The emergency response training provided to workers, and any special instructions and orientation tours offered for fire, police, medical, and other off-site-based emergency personnel are adequate to support effective actions;

(13) The means to restore the facility to a safe condition after an accident are adequate;

(14) Quarterly communications checks with off-site response organizations and biennial on-site exercises to test response to simulated emergencies are planned and include the following:

   (a) A check and update of all necessary phone numbers, fax numbers, and e-mail addresses;

   (b) An invitation to off-site response organizations to participate in the biennial exercises;

   (c) Use scenarios not known to most exercise participants;

   (d) A plan for critiques of each exercise by individuals not having direct implementation responsibility for conducting the exercise. Critiques will evaluate the appropriateness of the plan, emergency procedures, facilities and equipment, training of personnel, and the overall effectiveness of the response; and

   (e) Provisions to correct deficiencies identified by the critiques.

(15) On-site exercises to test response to simulated emergencies are conducted biennially;

(16) Radiological/health physics, medical, and fire drills are planned semiannually;

(17) The geologic repository operations area operations will satisfy the Emergency Planning and Community Right-to-Know Act of 1986, with respect to hazardous materials at the facility;

(18) The off-site response organizations are allowed 60 days to comment on the initial submittal of the emergency plan before transmittal to the U.S. Nuclear Regulatory Commission. Subsequent plan changes will have a 60-day comment period if the changes affect the off-site response organizations. Comments received during the 60-day comment period and licensee responses are submitted to the U.S. Nuclear Regulatory Commission with the emergency plan;

(19) Plans for use of off-site assistance include:

(a) Arrangements for requesting and effectively using off-site assistance and provisions for using other organizations that can augment the planned on-site response, as required;

(b) Provisions for prompt communication among principal response organizations to off-site personnel who would be responding onsite;

(c) Provision of adequate emergency facilities and equipment to support the emergency response onsite;

(d) Specification of methods, systems, and equipment for assessing and monitoring consequences of radiological emergency conditions;

(e) Arrangements for medical services for on-site contaminated and injured individuals; and

(f) Training in radiological emergency response for off-site personnel who may be called to assist in an emergency.

(20) Adequate arrangements for providing timely information to the public exist.

2.5.7.4 Evaluation Findings

If the license application provides sufficient information and the regulatory acceptance criteria in Section 2.5.7.3 are appropriately satisfied, the staff concludes that this portion of the staff evaluation is acceptable. The reviewer writes material suitable for inclusion in the safety evaluation report prepared for the entire application. The report includes a summary statement of what was reviewed and why the reviewer finds the submittal acceptable. The staff can document the review as follows.

U.S. Nuclear Regulatory Commission staff has reviewed the Safety Analysis Report and other information submitted in support of the license application and finds, with reasonable assurance, that the requirements of 10 CFR 63.161 are satisfied. An acceptable emergency plan for coping with radiological accidents through permanent closure, including dismantlement and decontamination of the surface facilities at the geologic repository operations area, is provided in accordance with 10 CFR 72.32(b). Aspects of this plan include:

(1) Facility and nearby area descriptions;

(2) Types and classifications of radioactive materials accidents;

(3) Means for detection of accident conditions;

(4) Means for mitigation of consequences of accidents;

(5) Adequate assessment of radioactive materials releases;

(6) Definition of responsibilities for facility personnel during an emergency;

(7) Responsibilities for developing, maintaining, and updating the emergency plan;

(8) Identification of off-site response organizations;

(9) Notification and coordination with off-site response organizations;

(10) Information to be communicated to off-site response organizations;

(11) Training of on-site emergency response staff;

(12) Safe condition restoration;

(13) Exercises to demonstrate readiness to act in emergency situations;

(14) Hazardous chemicals responsibilities under the Emergency Planning and Community Right-to-Know Act of 1986;

(15) Comments on the emergency plan from off-site emergency response team members;

(16) Off-site assistance requirements; and

(17) Arrangements for providing information to the public.

2.5.7.5 Reference

U.S. Nuclear Regulatory Commission. NUREG–1567, "Standard Review Plan for Spent Fuel Storage Facilities." Washington, DC: U.S. Nuclear Regulatory Commission, Spent Fuel Project Office. March 2000.

2.5.8 Controls to Restrict Access and Regulate Land Uses

Review Responsibilities—High-Level Waste Branch and Environmental and Performance Assessment Branch

2.5.8.1 Areas of Review

This section reviews controls to restrict access and regulate land uses. Reviewers will also evaluate the information required by 10 CFR 63.21(c)(24).

Controls to restrict access and regulate land uses are implemented to reduce the likelihood of adverse human actions that could reduce the ability of the repository to isolate waste.
The staff will evaluate the following parts of controls to restrict access and regulate land uses, using the review methods and acceptance criteria in Sections 2.5.8.2 and 2.5.8.3.

(1) Extent and adequacy of geologic repository operations area land acquisition or withdrawal;

(2) Compatibility of geologic repository operations area boundaries in the geologic repository operations area design and natural features;

(3) Means used to identify encumbrances or subsurface interests within the geologic repository operations area;

(4) Acceptability of additional controls for permanent closure;

(5) Acceptability of additional controls through permanent closure;

(6) Adequacy of water rights;

(7) Control over surface and subsurface estates;

(8) Means used to identify encumbrances outside the geologic repository operations area; and

(9) Acceptability of monument design.

2.5.8.2 Review Methods

Review Method 1 Ownership of Land

Verify that steps within the U.S. Department of Energy purview to establish effective jurisdiction and control and legislative or other transfer activities underway will be completed before the completion of the U.S. Nuclear Regulatory Commission review and decision on the license application.

Confirm that the land area of the geologic repository operations area is either land acquired by the U.S. Department of Energy, or is permanently withdrawn and is reserved for U.S. Department of Energy use, and is held by the U.S. Department of Energy free and clear of all significant encumbrances including: (i) rights arising under the general mining laws; (ii) easements for right-of-way; and (iii) all other rights arising under lease, rights of entry, deed, patent, mortgage, appropriation, prescription, or otherwise.

Confirm that legal documentation of ownership for the geologic repository operations area includes sufficient indexes of ownership and/or control to satisfy a purchaser-of-record such as: a recorded title search showing any and all interests in the land, or a Bureau of Land Management Master Title Plat that indicates all recorded interests and claims.

If a statutory withdrawal of the geologic repository operations area land has been enacted, verify that the license application includes a copy of the legislation and that the legal descriptions of the land area contained in the statute and the description in the application agree. Since the land area of the proposed repository site would be totally in Federal ownership, the statutory withdrawal would constitute complete ownership documentation, subject to subordinate interests.

Review Method 2 Additional Controls for Permanent Closure

Evaluate whether any controls established over surface and subsurface estates, at the geologic repository operations area or outside the geologic repository operations area, to prevent adverse human actions that could reduce the ability of the geologic repository to isolate the waste, are acceptable and sufficient. Such controls may take the form of: (i) possessory interests; (ii) servitudes; (iii) water rights; (iv) withdrawals from location or patent under the general mining laws; and (v) land use restrictions.

Confirm that the size and boundaries of the geologic repository operations area and the affected area outside the geologic repository operations area are consistent with the design or natural features, to assure the ability of the repository to achieve isolation and to reduce the risk of human activity that could adversely impact waste isolation. Collaborate with the reviewers of the site characteristics completed using Sections 1.1, "General Description" and 2.2, "Repository Safety after Permanent Closure" of the Yucca Mountain Review Plan.

Verify that legal documentation of ownership and/or control of the area outside the geologic repository operations area includes sufficient indexes of ownership and control to satisfy a purchaser-of-record such as: a recorded title search showing any and all interests in the land, or the Bureau of Land Management Master Title Plat, which indicates all recorded interests and claims.

Verify that if a statutory withdrawal has not been enacted for land outside the geologic repository operations area, the U.S. Department of Energy has taken or plans to take appropriate steps within its purview to establish effective jurisdiction and control. Legislative or other transfer activities underway should be completed before the completion of the U.S. Nuclear Regulatory Commission review and decision on the license application.

Confirm that any existing or proposed permissible rights or encumbrances that exist and may be continued, or that should be established outside the geologic repository operations area, are identified, and the nature of any activities that may permissibly occur under these rights are assessed adequately.

Evaluate the U.S. Department of Energy plan for administering and controlling its ownership rights or oversight of land. Verify that the means, such as title search and Bureau of Land Management records search, used to identify any existing or future encumbrances or other surface or subsurface interests of record in the land area outside the geologic repository operations area, were appropriate.

Review Method 3 Additional Controls Through Permanent Closure

Evaluate whether any controls necessary to ensure that the requirements at 10 CFR 63.111(a) and (b) are met, are acceptable and sufficient. Such controls, if necessary, should include land use restrictions and the authority to exclude members of the public.

Confirm that the size and boundaries of the geologic repository operations area, and the affected area outside the geologic repository operations area, are consistent with the design or natural features, to ensure that the requirements at 10 CFR 63.111(a) and (b) are met.

Collaborate with the reviewers of Section 2.1 of the Yucca Mountain Review Plan "Repository Safety Prior to Permanent Closure" and with the reviewers of the site characteristics completed using Sections 1.1, "General Description" and 2.2, "Repository Safety after Permanent Closure," of the Yucca Mountain Review Plan.

Verify that legal documentation of ownership and/or control of the area outside the geologic repository operations area includes sufficient indexes of ownership and/or control to satisfy a purchaser-of-record such as: a recorded title search showing any and all interests in the land, or the Bureau of Land Management Master Title Plat, which indicates all recorded interests and claims.

Verify that if a statutory withdrawal has not been enacted for land outside the geologic repository operations area, the U.S. Department of Energy has taken appropriate steps within its purview to establish effective jurisdiction and control. Legislative or other transfer activities underway should be completed before the completion of the U.S. Nuclear Regulatory Commission review and decision on the license application.

Confirm that any existing or proposed permissible rights or encumbrances that exist and may be continued, or that should be established outside the geologic repository operations are, are identified, and the nature of any activities that may permissibly occur under these rights are assessed adequately.

Evaluate the U.S. Department of Energy plan for administering and controlling its ownership rights or oversight of land. Verify that the means, such as title search and Bureau of Land Management records search, used to identify any existing or future encumbrances or other surface or subsurface interests of record in the land area outside the geologic repository operations area, were appropriate.

Review Method 4 Water Rights

Confirm that the U.S. Department of Energy has obtained such water rights as may be necessary to accomplish the purpose of the geologic repository operations area. Coordinate with the reviewers of the geologic repository operations area design-conducted using Section 2.1, "Repository Safety Prior to Permanent Closure," of the Yucca Mountain Review Plan, to evaluate compliance with the water use requirements.

Review Method 5 Conceptual Design of Monuments

Confirm that the conceptual design of monuments planned to identify the site after permanent closure is adequate. The monuments should accurately identify the location of the repository, be designed to be as permanent as practicable, convey a warning against intrusion into the underground repository, because of risk to public health and safety from radioactive wastes, and have a design life of at least a few hundred years.

2.5.8.3 Acceptance Criteria

The following acceptance criteria are based on meeting the requirements of 10 CFR 63.121 and 63.21(c)(24), regarding controls to restrict access and regulate land use and the conceptual design of monuments.

Acceptance Criterion 1 Ownership of Land Is Adequately Demonstrated.

(1) Steps within U.S. Department of Energy purview to establish effective jurisdiction and control and legislative or other transfer activities underway are complete;

(2) The land area of the geologic repository operations area is either land acquired by the U.S. Department of Energy, or is permanently withdrawn and is reserved for U.S. Department of Energy use, and is held by the U.S. Department of Energy free and clear of all significant encumbrances;

(3) Legal documentation of ownership for the geologic repository operations area includes sufficient indexes of ownership and control to satisfy a purchaser-of-record; and

(4) If a statutory withdrawal of the geologic repository operations area land has been enacted, the license application includes a copy of the legislation, and the legal descriptions of the land area contained in the statute and the description in the application agree.

Acceptance Criterion 2 Additional Controls for Permanent Closure Are Acceptable.

(1) Any additional controls established over surface and subsurface estates, at the geologic repository operations area or outside the geologic repository operations area, to prevent adverse human actions that could reduce the ability of the geologic repository to isolate the waste, are acceptable and sufficient;

(2) The size and boundaries of the geologic repository operations area and the affected area outside the geologic repository operations area are consistent with the design or natural features, to assure the ability of the repository to achieve isolation and to reduce the risk of human activity that could adversely impact waste isolation;

(3) Legal documentation of ownership and/or control of the area outside the geologic repository operations area includes sufficient indexes of ownership and control to satisfy a purchaser-of-record such as: a recorded title search showing any and all interests in the land, or the Bureau of Land Management Master Title Plat;

(4) If a statutory withdrawal has not been enacted for land outside the geologic repository operations area, the U.S. Department of Energy has taken appropriate steps within its purview to establish effective jurisdiction and control. Legislative or other transfer activities are complete;

(5) Any existing or proposed permissible rights or encumbrances that exist and may be continued, or that should be established outside the geologic repository operations area are identified, and the nature of any activities that may permissibly occur under these rights are assessed adequately; and

(6) The means, such as title search and Bureau of Land Management records search, used to identify any existing or future encumbrances or other surface or subsurface interests

of record in the land area outside the geologic repository operations area, were appropriate.

Acceptance Criterion 3 Additional Controls Through Permanent Closure Are Adequate.

(1) Any additional controls necessary to ensure that the requirements, at 10 CFR 63.111(a) and (b), are met, are acceptable and sufficient;

(2) The size and boundaries of the geologic repository operations area, and the affected area outside the geologic repository operations area, are consistent with the design or natural features, to ensure that the requirements at 10 CFR 63.111(a) and (b) are met;

(3) Legal documentation of ownership and/or control of the area outside the geologic repository operations area includes sufficient indexes of ownership and control to satisfy a purchaser-of-record such as a recorded title search showing any and all interests in the land, or the Bureau of Land Management Master Title Plat;

(4) If a statutory withdrawal has not been enacted for land outside the geologic repository operations area, U.S. Department of Energy has taken appropriate steps, within its purview, to establish effective jurisdiction and control. Legislative or other transfer activities are complete;

(5) Any existing or proposed permissible rights or encumbrances that exist and may be continued, or that should be established outside the geologic repository operations area, are identified, and the nature of any activities that may permissibly occur under these rights is assessed adequately; and

(6) The means, such as title search and Bureau of Land Management records search, used to identify any existing or future encumbrances or other surface or subsurface interests of record in the land area outside the geologic repository operations area were appropriate.

Acceptance Criterion 4 The Description of Water Rights Is Adequate.

(1) The U.S. Department of Energy has obtained such water rights as may be necessary to accomplish the purpose of the geologic repository operations area.

Acceptance Criterion 5 The Conceptual Design of Monuments Is Adequate.

(1) The conceptual design of monuments planned to identify the site after permanent closure is adequate.

2.5.8.4 Evaluation Findings

If the license application provides sufficient information and the regulatory acceptance criteria in Section 2.5.8.3 are appropriately satisfied, the staff concludes that this portion of the staff evaluation is acceptable. The reviewer writes material suitable for inclusion in the safety evaluation report prepared for the entire application. The report includes a summary statement

of what was reviewed and why the reviewer finds the submittal acceptable. The staff can document the review as follows.

U.S. Nuclear Regulatory Commission staff has reviewed the Safety Analysis Report and other information submitted in support of the license application and finds, with reasonable assurance, that the requirements of 10 CFR 63.121 and 63.21(c)(24) are met. Requirements for the ownership and control of interests in land and use of permanent monuments to identify the site after permanent closure have been met. In particular:

(1) The geologic repository operations area will be located in and on lands that are either acquired lands under the jurisdiction and control of the U.S. Department of Energy, or are permanently withdrawn and reserved for its use. These lands will be held free and clear of encumbrances such as rights arising under the general mining laws, easements for right-of-way, and other rights arising under lease, rights of entry, deed, patent, mortgage, appropriation, prescription, or otherwise;

(2) Additional controls will be applied for permanent closure to include areas outside the geologic repository operations area. These controls will consist of jurisdiction and control, over surface and subsurface estates, as necessary to prevent adverse human actions that could significantly reduce the repository's ability to achieve isolation;

(3) Additional controls will be applied through permanent closure, including for areas outside the geologic repository operations area. The U.S. Department of Energy will exercise jurisdiction as required to ensure that the preclosure performance objectives in 10 CFR 63.111 are met. The controls include the authority to exclude members of the public;

(4) The U.S. Department of Energy has obtained water rights to accomplish the purposes of the geologic repository operations area; and

(5) The U.S. Department of Energy has provided the conceptual design of monuments to identify the location of the repository after permanent closure.

2.5.8.5 References

None.

2.5.9 Uses of Geologic Repository Operations Area for Purposes Other Than Disposal of Radioactive Wastes

Review Responsibilities—High-Level Waste Branch and Environmental and Performance Assessment Branch

2.5.9.1 Areas of Review

This section reviews the uses of the geologic repository operations area for purposes other than disposal of radioactive wastes. Reviewers will evaluate the information required by 10 CFR 63.21(c)(22)(vii).

The staff will evaluate the following parts of uses of the geologic repository operations area for purposes other than disposal of radioactive wastes, using the review methods and acceptance criteria in Sections 2.5.9.2 and 2.5.9.3.

(1) Proposed activities other than disposal of high-level radioactive waste and their impacts, and

(2) Procedures for conduct and continuing oversight of proposed activities.

2.5.9.2 Review Methods

Review Method 1 Proposed Activities Other Than Disposal

Evaluate whether any proposed activities at the geologic repository operations area, other than the disposal of high-level radioactive waste, will potentially impact structures, systems, and components important to safety and engineered and natural barriers important to waste isolation. Activities to be considered include, but are not limited to:

(1) Long-term interim storage of high-level radioactive waste;

(2) Access for approved purposes unrelated to the disposal of high-level radioactive waste, such as Native American cultural activities, protection of flora and fauna under appropriate regulations, recreation, and resource exploitation (e.g., minerals, geothermal, ground water); and

(3) Performance monitoring or confirmation by groups other than the U.S. Nuclear Regulatory Commission or U.S. Department of Energy.

Review Method 2 Procedures for Proposed Activities that Potentially Affect Structures, Systems, and Components

Assess the adequacy of procedures for the continuing oversight of proposed activities, other than disposal of high-level radioactive waste at the geologic repository operations area, that might affect structures, systems, and components important to safety and engineered and natural barriers important to waste isolation. These procedures should include: (i) purpose of activity; (ii) detailed description of activity; (iii) radiation safety of workers; and (iv) disposition of records and identification of parties to be notified on completion.

2.5.9.3 Acceptance Criteria

The following acceptance criteria are based on meeting the requirements of 10 CFR 63.21(c)(22)(vii), regarding uses of the geologic repository operations area for purposes other than disposal of radioactive wastes.

Acceptance Criterion 1 Proposed Activities Other than Disposal of Radioactive Wastes Are Acceptable.

(1) Proposed activities at the geologic repository operations area, other than the disposal of high-level radioactive waste, are adequately evaluated for their potential impacts on

structures, systems, and components important to safety and engineered and natural barriers important to waste isolation, and the impacts of these activities are acceptable.

Acceptance Criterion 2 Procedures for Proposed Activities Other than Disposal of High-level Radioactive Waste Are Acceptable.

(1) Procedures for the continuing oversight of proposed activities, other than disposal of high-level radioactive waste, at the geologic repository operations area, that might affect structures, systems, and components important to safety, and engineered and natural barriers important to waste isolation, are adequate.

2.5.9.4 Evaluation Findings

If the license application provides sufficient information and the regulatory acceptance criteria in Section 2.5.9.3 are appropriately satisfied, the staff concludes that this portion of the staff evaluation is acceptable. The reviewer writes material suitable for inclusion in the safety evaluation report prepared for the entire application. The report includes a summary statement of what was reviewed and why the reviewer finds the submittal acceptable. The staff can document the review as follows.

U.S. Nuclear Regulatory Commission staff has reviewed the Safety Analysis Report and other information submitted in support of the license application and finds, with reasonable assurance, that the requirements of 10 CFR 63.21(c)(22)(vii) are satisfied. Requirements for the content of the license application have been met in that plans for any uses of the geologic repository operations area for purposes other than disposal of radioactive wastes have been adequately described. These plans include an analysis of the effects, if any, that such uses may have on the operation of the structures, systems, and components important to safety and the engineered and natural barriers important to waste isolation.

2.5.9.5 References

None.

2.5.10 License Specifications

Review Responsibilities—High-Level Waste Branch and Environmental and Performance Assessment Branch

This section reviews the variables, conditions, or other items determined by the U.S. Department of Energy to be probable subjects of license specification. The reviewers will evaluate the information required by 10 CFR 63.21(c)(18).

The review of variables, conditions, or other items that are probable subjects of license specifications, is to be integrated with reviews conducted using other sections of the Yucca Mountain Review Plan. The acceptability of proposed variables, conditions, and other items is assessed in conjunction with a determination that the repository performance objectives will be met, because these specifications define or constrain the operation and construction of the repository. Reviewers should give special attention to items that significantly influence the final design of the geologic repository operations area.

2.5.10.1 Areas of Review

The staff will evaluate the following parts of license specifications, using the review methods and acceptance criteria in Sections 2.5.10.2 and 2.5.10.3. This list is not intended to be comprehensive. The scope of any license specifications will be based on information presented in a license application, not on a predetermined list.

(1) License specifications proposed in the following areas, as appropriate:

(a) Physical and chemical form and radioisotopic content of radioactive waste;

(b) Shape, size, and materials and methods of construction for radioactive waste packaging;

(c) Amount of waste permitted per unit volume of storage space;

(d) Requirements for test, calibration, inspection, surveillance, and monitoring;

(e) Characteristics of drifts, drip shields, backfill, ventilation systems, and other structures, systems, and components;

(f) Controls to restrict access and avoid disturbance; and

(g) Administrative controls.

(2) Technical basis for each proposed variable, condition, or other item, with emphasis given to those items that may significantly influence the final design.

2.5.10.2 Review Methods

Review Method 1 Identification and Technical Bases for Proposed License Specifications

Confirm that proposed license specifications and their technical bases have been identified and justified.

Review Method 2 Plans for Meeting License Specifications

Verify that the U.S. Department of Energy has provided plans for meeting the license specifications and that these plans are consistent with the repository systems designs, based on the results of the reviews conducted using Sections 2.1 and 2.2 of the Yucca Mountain Review Plan.

2.5.10.3 Acceptance Criteria

The following acceptance criteria are based on meeting the requirements of 10 CFR 63.21(c)(18) and 63.43 for license specifications.

Acceptance Criterion 1 Variables, Conditions, and Other Items That Are the Subject of Proposed License Specifications Are Adequately Identified, and Acceptable Technical Bases Have Been Provided.

Acceptance Criterion 2 Plans for Meeting the Proposed License Specifications and Their Technical Bases Are Adequately Defined.

2.5.10.4 Evaluation Findings

If the license application provides sufficient information and the regulatory acceptance criteria in Section 2.5.10.3 are appropriately satisfied, the staff concludes that this portion of the staff evaluation is acceptable. The reviewer writes material suitable for inclusion in the safety evaluation report prepared for the entire application. The report includes a summary statement of what was reviewed and why the reviewer finds the submittal acceptable. The staff can document the review as follows.

The staff has reviewed the Safety Analysis Report and other information submitted in support of the license application and has found, with reasonable assurance, that the requirements of 10 CFR 63.21(c)(18) and 63.43 are satisfied. Requirements for the content of the license application have been met in that those variables, conditions, or other items that are probable subjects of license specifications have been identified and justified. Plans for meeting the license specifications have been defined. Special attention has been given to those items that may significantly influence the final design of the geologic repository operations area.

2.5.10.5 References

None.

3 GLOSSARY

This Glossary is provided for information and is not exhaustive.

absorption: The process of taking up by capillary, osmotic, solvent, or chemical action of molecules (e.g., absorption of gas by water) as distinguished from adsorption.

abstracted model: A model that reproduces, or bounds, the essential elements of a more detailed process model and captures uncertainty and variability in what is often, but not always, a simplified or idealized form. See *abstraction*.

abstraction: Representation of the essential components of a process model into a suitable form for use in a total system performance assessment. Model abstraction is intended to maximize the use of limited computational resources while allowing a sufficient range of sensitivity and uncertainty analyses.

adsorb: To collect a gas, liquid, or dissolved substance on a surface as a condensed layer.

adsorption: The adhesion by chemical or physical forces of molecules or ions (as of gases or liquids) to the surface of solid bodies. For example, the transfer of solute mass, such as radionuclides, in ground water to the solid geologic surfaces with which it comes in contact. The term *sorption* is sometimes used interchangeably with this term.

advection: The process in which solutes, particles, or molecules are transported by the motion of flowing fluid. For example, advection in combination with dispersion controls flux into and out of the elemental volumes of the flow domain in ground-water transport models.

air mass fraction: The mass of air divided by the total mass of gas (typically air plus water vapor) in the gas phase. This expression gives a measure of the "dryness" of the gas phase, which is important in waste package corrosion models.

Alloy 22: A nickel-base corrosion resistant alloy containing approximately 22 weight percent chromium, 13 weight percent molybdenum, and 3 weight percent tungsten as major alloying elements and that may be used as the outer container material in a waste package design (see *outer barrier*).

alluvium: Detrital deposits made by streams on river beds, flood plains, and alluvial fans; especially a deposit of silt or silty clay laid down during time of flood. The term applies to stream deposits of recent time. It does not include subaqueous sediments of seas and lakes.

alternative: Plausible interpretations or designs based on assumptions other than those used in the base case that could also fit or be applicable, based on the available scientific information. When propagated through a quantitative tool such as performance assessment, alternative interpretations can illustrate the significance of the uncertainty in the base case interpretation chosen to represent the repository's probable behavior.

ambient: Undisturbed, natural conditions such as ambient temperature caused by climate or natural subsurface thermal gradients, and other surrounding conditions.

Glossary

anisotropy: The condition that physical properties vary when measured in different directions or along different axes. For example, in layered rock the permeability is often greater within the horizontal layers than across the horizontal layers.

annual frequency: The average number of occurrences of an event in 1 year.

aqueous: Pertaining to water, such as aqueous phase, aqueous species, or aqueous transport.

aquifer: A subsurface, saturated formation of sufficient permeability to transmit ground water and yield water of sufficient quality and quantity for an intended beneficial use.

ash: Bits of volcanic rock that would be broken-up during an eruption to less than 2 mm [0.08 inches] in diameter.

basalt: A type of igneous rock that forms black, rubbly lavas and black-to-red tephras of the type commonly used as lava rocks for barbecues.

borosilicate glass: A predominantly noncrystalline, relatively homogenous glass formed by melting silica and boric oxide together with other constituents such as alkali oxides. A high-level radioactive waste matrix material in which boron takes the place of the lime used in ordinary glass mixtures.

boundary condition: For a model, the establishment of a set condition, often at the geometric edge of the model, for a given variable. An example is using a specified ground-water flux from net infiltration as a boundary condition for an unsaturated flow model.

bound: An analysis or selection of parameter values that yields pessimistic results, such that any actual result is certain to be no worse or could be worse only with an extremely small likelihood.

breach: A penetration in the waste package caused by failure of the outer and inner containers or barriers that allows the spent nuclear fuel or the high-level radioactive waste to be exposed to the external environment and may eventually permit radionuclide release.

burnup: A measure of nuclear reactor fuel consumption expressed either as the percentage of fuel atoms that have undergone fission or as the amount of energy produced per unit weight of fuel.

calibration: (1)The process of comparing the conditions, processes, and parameter values used in a model against actual data points or interpolations (e.g., contour maps) from measurements at or close to the site to ensure that the model is compatible with reality, to the extent feasible. (2) For tools used for field or lab measurements, the process of taking instrument readings on standards known to produce a certain response, to check the accuracy and precision of the instrument. (3) In operations, the process to ensure accuracy of instruments and any setpoints for automation actuations of items important to safety.

canister: A cylindrical metal receptacle that facilitates handling, transportation, storage, and/or disposal of high-level radioactive waste. It may serve as (1) a pour mold and container for vitrified high-level radioactive waste or (2) a container for loose or damaged fuel rods, nonfuel components and assemblies, and other debris containing radionuclides.

carbon steel: A steel made of carbon up to about 2 weight percent and only residual quantities of other elements. Carbon steel is a tough but ductile and malleable material used as baskets to maintain the spent fuel assemblies in fixed positions in a waste package.

Category 1 event sequences: Those event sequences that are expected to occur one or more times before permanent closure of a geologic repository.

Category 2 event sequences: Event sequences other than Category 1 event sequences that have at least one chance in 10,000 of occurring before permanent closure.

Center for Nuclear Waste Regulatory Analyses: A Federally funded research and development center in San Antonio, Texas, sponsored by the U.S. Nuclear Regulatory Commission, to provide the U.S. Nuclear Regulatory Commission with technical assistance for the repository program.

chain reaction: A continuing series of nuclear fission events. Neutrons produced by a split nucleus collide with and split other nuclei causing a chain of fission events.

cladding: The metal outer sheath of a fuel rod generally made of a zirconium alloy, and in the early nuclear power reactors of stainless steel, intended to protect the uranium dioxide pellets, which are the nuclear fuel, from dissolution by exposure to high temperature water under operating conditions in a reactor.

climate: Long-term weather conditions including temperature, wind velocity, precipitation, and other factors, that prevail in a region.

climate states: Representations of climate conditions.

code (computer): The set of commands used to implement a mathematical model on a computer.

colloid: As applied to radionuclide migration, a colloidal system is a group of large molecules or small particles, having at least one dimension with the size range of 10^{-9} to 10^{-6} meters that are suspended in a solvent. Naturally occurring colloids in ground water arise from clay minerals such as smectites and illites. Colloids that are transported in ground water can be filtered out of the water in small pore spaces or very narrow fractures because of the large size of the colloids.

Colloid-Facilitated, Radionuclide Transport Model: A model that represents the enhanced transport of radionuclides by particles that are colloids.

commercial spent nuclear fuel: Nuclear fuel rods, forming a fuel assembly, that have been removed from a nuclear power plant after reaching the specified burnup.

Glossary

common cause failure: Two or more failures that result from a single event or circumstance.

conceptual model: A set of qualitative assumptions used to describe a system or subsystem for a given purpose. Assumptions for the model are compatible with one another and fit the existing data within the context of the given purpose of the model.

consequence: A measurable or calculated outcome of an event or process that, when combined with the probability of occurrence, gives risk.

conservative: A condition of an analysis or a parameter value such that its use provides a pessimistic result, which is worse than the actual result expected.

continuum model: A model that represents fluid flow through numerous individual fractures and matrix blocks by approximating it as continuous flow fields.

corrosion: The deterioration of a material, usually a metal, as a result of a chemical or electrochemical reaction with its environment.

corrosion model: A theoretical representation of a corrosion process based on the application of a combination of fundamental electrochemical (chemical) and thermodynamic principles (or laws) with empirical parameters resulting from experiments, field measurements, or data obtained through industrial experience. Models can describe the penetration of a pit or a crack through a container wall as a function of time.

corrosion resistant alloy: An alloy that exhibits extremely high resistance to general or uniform corrosion in a given environment as a result of the formation of a protective film on its surface. Alloy 22, and other similar nickel-chromium-molybdenum alloys, are considered corrosion resistant alloys because they are extremely resistant to general corrosion in severe aqueous environments (e.g., high temperature brines containing acidic sulfur species).

coupling: The ability to assemble separate analyses or parameters in a performance assessment so that information can be passed among them to develop an overall analysis of system performance. A representation of interrelationship between processes, especially between thermal and hydrologic processes.

crevice corrosion: Localized corrosion of a metal surface at, or immediately adjacent to, an area that is shielded from full exposure to the environment because of close proximity between the metal and the surface of another material.

critical event: See *criticality*.

criticality: (1) A condition that would require the original waste form, which is part of the waste package, to be exposed to degradation, followed by conditions that would allow concentration of sufficient nuclear fuel, the presence of neutron moderators, the absence of neutron absorbers, and favorable geometry. (2) The condition in which a fissile material sustains a chain reaction. It occurs when the number of neutrons present in one generation cycle equals the number generated in the previous cycle. The state is considered critical when a self-sustaining nuclear chain reaction is ongoing.

criticality accident: The release of energy as a result of accidental production of a self-sustaining or divergent neutron chain reaction.

data: Facts or figures measured or derived from site characteristics or standard references from which conclusions may be drawn. Parameters that have been derived from raw data are sometimes, themselves, considered to be data.

U.S. Department of Energy: A Cabinet-level agency of the U.S. federal government charged with the responsibilities of energy security, national security, and environmental quality.

design concept: An idea of how to design and operate the above-ground and below-ground portions of a repository.

diffusion: (1) The spreading or dissemination of a substance caused by concentration gradients. (2) The gradual mixing of the molecules of two or more substances because of random thermal motion.

diffusive transport: Movement of solutes because of their concentration gradient. The process in which substances carried in ground water move through the subsurface by means of diffusion because of a concentration gradient.

dike: A tabular body of igneous rock that cuts across the structure of adjacent rocks or cuts massive rocks.

dimensionality: Modeling in one, two, or three dimensions.

direct exposure: The manner in which an individual receives dose from being in close proximity to a source of radiation. Direct exposures present an external dose pathway.

dispersion (hydrodynamic dispersion): (1) The tendency of a solute (substance dissolved in ground water) to spread out from the path it is expected to follow if only the bulk motion of the flowing fluid were to move it. The tortuous path the solute follows through openings (pores and fractures) causes part of the dispersion effect in the rock. (2) The macroscopic outcome of the actual movement of individual solute particles through a porous medium. Dispersion causes dilution of solutes, including radionuclides, in ground water, and is usually an important mechanism for spreading contaminants in low flow velocities.

disposal container: A cylindrical metal receptacle designed to contain spent nuclear fuel and high-level radioactive waste that will become an integral part of the waste package when loaded with spent nuclear fuel or high-level radioactive waste. In a waste package, the inner container will have spacing structures or baskets to maintain fuel assemblies, shielding components, and neutron absorbing materials in position to control the possibility of criticality.

disruptive event: An off-normal event that, in the case of the potential repository, includes volcanic activity, seismic activity, and nuclear criticality. Disruptive events have two possible effects: (1) direct release of radioactivity to the surface, or (2) alteration of the nominal behavior of the system. For the purposes of screening features, events, and processes for the total system performance assessment, a disruptive event is defined as an event that has a significant effect on the expected annual dose and that has a probability of occurrence during the 10,000-year period of performance less than 1.0, but greater than a cutoff of 0.0001.

disruptive event scenario class: The scenario, or set of related scenarios, that describes the behavior of the system if perturbed by disruptive events. The disruptive scenarios contain all disruptive features, events, and processes that have been retained for analysis.

dissolution: (1) Change from a solid to a liquid state. (2) Dissolving a substance in a solvent.

distribution: The overall scatter of values for a set of observed data. A term used synonymously with frequency distribution or probability distribution function. Distributions have structures that are the probability that a given value occurs in the set.

drift: From mining terminology, a horizontal underground passage. The nearly horizontal underground passageways from the shaft(s) to the alcoves and rooms. Drifts include excavations for emplacement (emplacement drifts) and access (access mains).

drift scale: The scale of an emplacement drift, or approximately 5 meters in diameter.

Drift-Scale Heater Test: A test being conducted in the Exploratory Studies Facility to investigate thermal-hydrologic, thermal-chemical, and thermal-mechanical processes.

drip shield: A metallic structure placed along the extension of the emplacement drifts and above the waste packages to prevent seepage water from directly dripping onto the waste package outer surface.

edge effects: Conditions at the edges of the potential repository that are cooler and wetter because heat dissipates more quickly there than at the center of the repository.

effective porosity: The fraction of a porous medium volume available for fluid flow and/or solute storage, as in the saturated zone. Effective porosity is less than or equal to the total void space (porosity).

empirical: Reliance on experience or experiment rather than on an understanding of the fundamental processes as related to the laws of nature.

emplacement drift: See *drift*.

enrichment: The act of increasing the concentration of fissile isotopes from their value in natural uranium. The enrichment (typically reported in atom percent) is a characteristic of nuclear fuel.

equilibrium: The state of a chemical system in which the phases do not undergo any spontaneous change in properties or proportions with time; a dynamic balance.

events: (1) Occurrences that have a specific starting time and, usually, a duration shorter than the time being simulated in a model. (2) Uncertain occurrences that take place within a short time relative to the time frame of the model. For the purposes of screening features, events, and processes for the total system performance assessment, an event is defined to be a natural or human-caused phenomenon that has a potential to affect disposal system performance and that occurs during an interval that is short compared with the period of performance.

event tree: A modeling tool that illustrates the logical sequence of events that follow an initiating event.

expert elicitation: A formal process through which expert judgment is obtained.

Exploratory Studies Facility: An underground laboratory at Yucca Mountain that includes a 7.9-kilometer [4.9-mile] main loop (tunnel); a 2.8-kilometer [1.75-mile] cross-drift; and a research alcove system constructed for performing underground studies during site characterization. The data collected will contribute toward determining the suitability of the Yucca Mountain site for a repository. Some or all of the Exploratory Studies Facility may eventually be incorporated into the potential repository.

fault (geologic): A planar or gently curved fracture across which there has been displacement parallel to the fracture surface.

fault tree: A graphical logic model that depicts the combinations of events that result in the occurrence of an undesired event.

features: Physical, chemical, thermal, or temporal characteristics of the site or potential repository system. For the purposes of screening features, events, and processes for the total system performance assessment, a feature is defined to be an object, structure, or condition that has a potential to affect disposal system performance.

ferritic steel: A subclass of carbon steels characterized by a relatively low strength but good ductility as a result of the ferrite microstructure. A type of ferritic steel, mild steel, or low-carbon steel containing up to about 0.1 weight percent carbon is the metallic material most commonly used for construction purposes.

film flow: Movement of water as a film along a surface such as a fracture plane.

finite element analysis: A commonly used numerical method for solving mechanical deformation problems. A technique in which algebraic equations are used to approximate the partial differential equations that comprise mathematical models to produce a form of the problem that can be solved on a computer. For this type of approximation, the area being modeled is formed into a grid with irregularly shaped blocks. This method provides an advantage in handling irregularly shaped boundaries, internal features such as faults, and surfaces of engineered materials. Values for parameters are frequently calculated at nodes for convenience, but are defined everywhere in the blocks by means of interpolation functions.

flow: The movement of a fluid such as air, water, or magma. Flow and transport are processes that can move radionuclides from the proposed repository to the receptor group location.

flow pathway: The subsurface course that water or a solute (including radionuclides) would follow in a given ground-water velocity field, governed principally by the hydraulic gradient.

Glossary

fracture: A planar discontinuity in rock along which loss of cohesion has occurred. It is often caused by the stresses that cause folding and faulting. A fracture along which there has been displacement of the sides relative to one another is called a fault. A fracture along which no appreciable movement has occurred is called a joint. Fractures may act as fast paths for ground-water movement.

fracture aperture: The space that separates the sides of a fracture, and the measured width of the space separating the sides of a fracture.

fracture permeability: The capacity of a rock to transmit fluid that is related to fractures in the rock.

frequency: The number of occurrences of an observed or predicted event during a specific time period, or the annual probability of occurrence of an initiating event or an event sequence.

galvanic: Pertains to an electrochemical process in which two dissimilar electronic conductors are in contact with each other and with an electrolyte, or in which two similar electronic conductors are in contact with each other and with dissimilar electrolytes.

galvanic corrosion: Accelerated corrosion of a metal resulting from electrical contact with a more noble metal or nonmetallic conductor in a corrosive electrolyte.

geochemical: The distribution and amounts of the chemical elements in minerals, ores, rocks, soils, water, and the atmosphere; and the movement of the elements in nature on the basis of their properties.

geologic-framework model: A digital, scaled, geometrically congruent , three-dimensional model of the geologic system.

ground water: Water contained in pores or fractures in either the unsaturated or saturated zones below-ground level.

half-life: The time required for a radioactive substance to lose have its activity due to radioactive decay. At the end of one half-life, 50 percent of the original radioactive material has decayed.

heterogeneity: The condition of being composed of parts or elements of different kinds. A condition in which the value of a parameter such as porosity, which is an attribute of an entity of interest such as the tuff rock containing the potential repository, varies over the space an entity occupies, such as the area around the repository, or with the passage of time.

high-level radioactive waste glass: A waste form produced by melting a mixture of high-level radioactive waste and components of borosilicate glass at a high temperature (approximately 1,100 degrees centigrade).

hydrologic: Pertaining to the properties, distribution, and circulation of water on the surface of the land, in the soil and underlying rocks, and in the atmosphere.

igneous: (1) A type of rock that has formed from a molten, or partially molten, material. (2) A type of activity related to the formation and movement of molten rock either in the subsurface (intrusive) or on the surface (volcanic).

infiltration: The process of water entering the soil at the ground surface. Infiltration becomes percolation when water has moved below the depth at which it can be removed (to return to the atmosphere) by evaporation or transpiration. See *net infiltration*.

inner barrier: The inner container in a waste package.

invert: A constructed surface that would provide a level drift floor and enable transport and support of the waste packages.

isothermal: Having a constant temperature.

license application: An application, to the U.S. Nuclear Regulatory Commission for a license to construct and operate a repository.

localized corrosion: Corrosion at discrete sites (e.g., pitting and crevice corrosion).

magma: Molten or partially molten rock that is naturally occurring and is generated within the earth. Magma may contain crystals along with dissolved gasses.

Mathematical Model: A mathematical description of a conceptual model.

matrix: Tuff rock material and its pore space exclusive of fractures. As applied to Yucca Mountain tuff, the ground mass of an igneous rock that contains larger crystals.

matrix diffusion: As used in the Total System Performance Assessment for the Site Recommendation conceptual models, the process by which molecular or ionic solutes, such as radionuclides in ground water, move from areas of higher concentration to areas of lower concentration. This movement is through the pore spaces of the rock material as opposed to movement through the fractures.

matrix permeability: The capability of the matrix to transmit fluid.

mean (arithmetic): For a statistical data set, the sum of the values divided by the number of items in the set. The arithmetic average.

mechanical disruption: Damage to the drip shield or waste package because of external forces.

median: A value such that one-half of the observations are less than that value and one-half are greater than the value, or the value of a cumulative distribution function of a random variable at which the probability is 0.5.

meteorology: The study of climatic conditions such as precipitation, wind, temperature, and relative humidity.

microbe: An organism too small to be viewed with the unaided eye. Examples of microbes are bacteria, protozoa, and some fungi and algae.

microbial influenced corrosion: Deterioration of metals as a result of the metabolic activity of microorganisms.

migration: Radionuclide movement from one location to another within the engineered barrier system or the environment.

mineral model: A description of the kinds and relative abundances of minerals that is used to approximate the true mineralogical system.

mineralogical: Of or relating to the chemical and physical properties of minerals, their occurrence, and their classification.

model: A depiction of a system, phenomenon, or process, including any hypotheses required to describe the system or explain the phenomenon or process.

near field: The area and conditions within the potential repository including the drifts and waste packages and the rock immediately surrounding the drifts. The region around the potential repository where the natural hydrogeologic system has been significantly impacted by the excavation of the repository and the emplacement of waste.

net infiltration: The amount of infiltration that escapes the zone of evapotranspiration, which is generally the zone below the zone of plant roots. See *infiltration*.

nominal behavior: (1) Expected behavior of the system as perturbed only by the presence of the potential repository. (2) Behavior of the system in the absence of disruptive events.

nominal features, events, and processes: Those features, events, and processes expected, given the site conditions as described from current site characterization information.

nominal scenario class: The scenario, or set of related scenarios, that describes the expected or nominal behavior of the system as perturbed only by the presence of the potential repository. The nominal scenarios contain all expected features, events, and processes that have been retained for analysis.

nuclear criticality safety: Protection against the consequences of a criticality accident, preferably by prevention of the accident.

U.S. Nuclear Regulatory Commission: An independent agency, established by the U.S. Congress under the Energy Reorganization Act of 1974, to ensure adequate protection of the public health and safety, the common defense and security, and the environment, in the use of nuclear materials in the United States. The U.S. Nuclear Regulatory Commission scope of responsibility includes regulation of the transport, storage, and disposal of nuclear materials and waste.

Nuclear Waste Policy Act (42 U.S.C. 10101 et seq.): The Federal statute enacted in 1982 that established the Office of Civilian Radioactive Waste Management and defined its mission to develop a federal system for the management, and geologic disposal, of commercial spent nuclear fuel and other high-level radioactive wastes. The Act also: (1) specified other federal responsibilities for nuclear waste management; (2) established the Nuclear Waste Fund to cover the cost of geologic disposal; (3) authorized interim storage under certain circumstances; and (4) defined interactions between federal agencies and the states, local governments, and Indian tribes. The act was substantially amended in 1987.

Nuclear Waste Policy Amendments Act of 1987: Legislation that amended the Nuclear Waste Policy Act to: (1) limit repository site characterization activities to Yucca Mountain, Nevada; (2) establish the Office of the Nuclear Waste Negotiator to seek a state or Indian tribe willing to host a repository or monitored retrievable storage facility; (3) create the Nuclear Waste Technical Review Board; and (4) increase state and local government participation in the waste management program.

numerical model: An approximate representation of a mathematical model that is constructed using a numerical description method such as finite volumes, finite differences, or finite elements. A numerical model is typically represented by a series of program statements that are executed on a computer.

Office of Civilian Radioactive Waste Management: A U.S. Department of Energy office created by the Nuclear Waste Policy Act of 1982 to implement the responsibilities assigned by the Act.

outer barrier: The outer container in a waste package.

oxidation: (1) A corrosion reaction in which the corroded metal forms an oxide, usually applied to reaction with a gas containing elemental oxygen, such as air. (2) An electrochemical reaction in which there is an increase in the valence of an element resulting from the loss of electrons.

parameter: Data, or values, such as those that are input to computer codes for a total system performance assessment calculation.

patch: A circumscribed area of a surface. In the DOE modeling of waste package corrosion, it is the minimal surface area of the outer container over which uniform corrosion occurs, as opposed to localized corrosion in pits.

pathway: A potential route by which radionuclides might reach the accessible environment and pose a threat to humans. For example, direct exposure is an external pathway, and inhalation and ingestion are internal pathways.

permeability: The ability of a material to transmit fluid through its pores when subjected to a difference in head (pressure gradient). Permeability depends on the substance transmitted (oil, air, water, etc.) and on the size and shape of the pores, joints, and fractures in the medium and the manner in which they are interconnected.

phase: A physically homogeneous and distinct portion of a material system, such as the gaseous, liquid, and solid phases of a substance. In liquids and solids, single phases may coexist.

phase stability: A measure of the ability of a particular phase to remain without transformation.

pit: A small cavity formed in a solid as a result of localized dissolution.

pitting corrosion: Localized corrosion of a metal surface, confined to a small area, that takes the form of cavities named pits.

porosity: The ratio of openings, or voids, to the total volume of a soil or rock expressed as a decimal fraction or as a percentage. See also *effective porosity*.

pre-startup and startup testing: Activities to evaluate the readiness to receive, possess, process, store, and dispose of high-level radioactive waste.

probabilistic: (1) Based on or subject to probability. (2) Involving a variate, such as temperature or porosity. At each instance of time, the variate may take on any of the values of a specified set with a certain probability. Data from a probabilistic process are an ordered set of observations, each of which is one item from a probability distribution.

probability: The chance that an outcome will occur from the set of possible outcomes. Statistical probability examines actual events and can be verified by observation or sampling. Knowledge of the exact probability of an event is usually limited by the inability to know, or compile, the complete set of possible outcomes over time or space, a degree of belief.

probability distribution: The set of outcomes (values) and their corresponding probabilities for a random variable.

processes: Phenomena and activities that have gradual, continuous interactions with the system being modeled. For the purposes of screening features, events, and processes for the total system performance assessment, a process is defined as a natural or human-caused phenomenon that has a potential to affect disposal system performance and that operates during all or a significant part of the period of performance.

process model: A depiction or representation of a process, along with any hypotheses required to describe or to explain the process.

radioactive decay: The process in which one radionuclide spontaneously transforms into one or more different radionuclides, which are called daughter radionuclides.

radioactivity: The property possessed by some elements (i.e., uranium) of spontaneously emitting radiation (e.g., alpha particles, beta particles, or gamma rays) by the disintegration of atomic nuclei.

radiolysis: Chemical decomposition by the action of radiation.

radionuclide: Radioactive type of atom with an unstable nucleus that spontaneously decays, usually emitting ionizing radiation in the process. Radioactive elements are characterized by their atomic mass and atomic number.

range (statistics): The numerical difference between the highest and lowest value in any set.

receptor: An individual for whom radiological doses are calculated or measured.

relative permeability: The ability of a material to transmit fluid through its pores when subjected to a pressure gradient under unsaturated conditions. Relative permeability is a function of permeability (has a value between 0 and 1).

repository footprint: The areal extent of the underground repository facility.

retardation: Slowing or stopping radionuclide movement in ground water by mechanisms that include sorption of radionuclides, diffusion into rock matrix pores and microfractures, and trapping of large colloidal molecules in small pore spaces or dead ends of microfractures.

risk: The probability that an undesirable event will occur, multiplied by the consequences of the undesirable event.

risk assessment: An evaluation of potential consequences or hazards that might be the outcome of an action. This assessment focuses on potential negative impacts on human health or the environment.

risk-informed, performance-based: A regulatory approach in which risk insights, engineering analysis and judgments, and performance history are used to: (1) focus attention on the most important activities; (2) establish objective criteria based on risk insights for evaluating performance; (3) develop measurable or calculable parameters for monitoring system and licensee performance; and (4) focus on the results as the primary basis fo regulatory decision making.

rock matrix: See *matrix*.

runoff: Lateral movement of water at the ground surface, such as down steep hillslopes or along channels, that is not able to infiltrate at a specified location. See *runon*.

runon: Lateral movement of water along the ground surface from some upstream location that becomes available for infiltration. See *runoff*.

safety question: A question regarding the adequacy of structures, systems, and components important to safety and engineered or natural barriers important to waste isolation.

scenario: A well-defined, connected sequence of features, events, and processes that can be thought of as an outline of a possible future condition of the potential repository system. Scenarios can be undisturbed, in which case the performance would be the expected, or nominal, behavior for the system. Scenarios can also be disturbed, if altered by disruptive events such as human intrusion or natural phenomena such as volcanism or nuclear criticality.

scenario class: A set of related scenarios sharing sufficient similarities that they can usefully be aggregated for the purposes of screening or analysis. The number and breadth of scenario classes depend on the resolution at which scenarios have been defined. Coarsely defined scenarios result in fewer, broad scenario classes, whereas narrowly defined scenarios result in many narrow scenario classes. Scenario classes (and scenarios) should be aggregated at the coarsest level at which a technically sound argument can be made while still retaining adequate detail for the purposes of the analysis.

seepage: The inflow of ground water moving in fractures or pore spaces of permeable rock to an open space in the rock such as a drift. Seepage rate is the percolation flux that enters the drift. Seepage is an important factor in waste package degradation and mobilization and migration of radionuclides out of the potential repository.

seismic: Pertaining to, characteristic of, or produced by earthquakes or earth vibrations.

shallow infiltration: The amount of infiltration that escapes the root zone and percolates downward into the unsaturated zone. See *net infiltration*.

site recommendation: A recommendation by the Secretary of Energy to the President that the Yucca Mountain site is suitable for development as the Nation's first high-level radioactive waste repository.

sorb: To undergo a process of sorption.

sorption: The binding, on a microscopic scale, of one substance to another. A term that includes both adsorption and absorption. The sorption of dissolved radionuclides onto aquifer solids or waste package materials by means of close-range chemical or physical forces is potentially an important process in a repository. Sorption is a function of the chemistry of the radioisotopes, the fluid in which they are carried, and the mineral material they encounter along the flow path.

sorption coefficient (K_d): Coefficient for a term for the various processes by which one substance binds to another.

source term: Types and amounts of radionuclides that are the source of a potential release.

spatial variability: A measure of how a property, such as rock permeability, varies at different locations in an object such as a rock formation.

speciation: The existence of the elements, such as radionuclides, in different molecular forms in the aqueous phase.

spent nuclear fuel: Fuel that has been withdrawn from a nuclear reactor following irradiation, the constituent elements of which have not been separated by reprocessing. Spent fuel that has been burned (irradiated) in a reactor to the extent that it no longer makes an efficient contribution to a nuclear chain reaction. This fuel is more radioactive than it was before irradiation, and releases significant amounts of heat from the decay of its fission product radionuclides. See *burnup*.

stainless steel: A class of iron-base alloys containing a minimum of approximately 10-percent chromium to provide corrosion resistance in a wide variety of environments.

stratigraphy: The science of rock strata. It is concerned with all characters and attributes of rocks as *strata* and their interpretation in terms of mode of origin and geologic history.

stress corrosion cracking: A cracking process that requires the simultaneous action of a corrodent and sustained (residual or applied) tensile stress. Stress corrosion cracking excludes both the fracture of already corroded sections and the localized corrosion processes that can disintegrate an alloy without the action of residual or applied stress.

structure: In geology, the arrangement of the parts of the geologic feature or area of interest such as folds or faults. This includes features such as fractures created by faulting and joints caused by the heating of rock.

tectonic: Pertaining to geologic forms or effects created by deformation of the earth's crust.

tephra: A collective term for all clastic materials ejected from a volcano and transported through the air. It includes volcanic dust, ash, cinders, lapilli, scoria, pumice, bombs, and blocks.

thermal-chemical: Of or pertaining to the effect of heat on chemical conditions and reactions.

thermal-hydrologic: Of or pertaining to changes in ground-water movement due to the effects of changes in temperature.

thermal-hydrologic processes: Processes that are driven by a combination of thermal and hydrologic factors. These processes include evaporation of water near the potential repository when it is hot and subsequent redistribution of fluids by convection, condensation, and drainage.

thermal hydrology: The study of a system that has both thermal and hydrologic processes. A thermal-hydrologic condition, or system, is expected to occur if heat-generating waste packages are placed in the potential repository at Yucca Mountain.

thermal-mechanical: Of or pertaining to changes in mechanical properties of rocks from effects of changes in temperature.

thermodynamics: A branch of physics that deals with the relationship and transformations between work as a mechanical action and heat.

total system performance assessment: A risk assessment that quantitatively estimates how the potential Yucca Mountain repository system will perform in the future under the influence of specific features, events, and processes, incorporating uncertainty in the models and uncertainty and variability of the data.

transparency: The ease of understanding the process by which a study was carried out, which assumptions are driving the results, how they were arrived at, and the rigor of the analyses leading to the results. A logical structure ensures completeness and facilitates in-depth review of the relevant issues. Transparency is achieved when a reader or reviewer has a clear picture of what was done in the analysis, what the outcome was, and why.

transpiration: The removal of water from the ground by vegetation (roots).

transport: A process that allows substances such as contaminants or radionuclides to be carried in a fluid through (1) the physical mechanisms of convection, diffusion, and dispersion; and (2) the chemical mechanisms of sorption, leaching, precipitation, dissolution, and complexation. Types of transport include advective, diffusive, and colloidal.

tuff: A general term for all consolidated pyroclastic rocks. The most abundant type of rock at the Yucca Mountain site.

uncertainty: How much a calculated or measured value varies from the unknown true value.

uniform corrosion: A type of corrosion attack (deterioration) more or less uniformly distributed over a metal surface. Corrosion that proceeds at approximately the same rate over a metal surface. Also called general corrosion.

unsaturated zone flow: The movement of water in the unsaturated zone driven by capillary, viscous, gravitational, inertial, and evaporative forces.

variable: A nonunique property or attribute.

variability (statistical): A measure of how a quantity varies over time or space.

volcanism: Pertaining to volcanic activity.

watershed: The area drained by a river system including the adjacent ridges and hillslopes.

APPENDIX A

LICENSING REVIEW AND THE YUCCA MOUNTAIN REVIEW PLAN

A1 LICENSING REVIEW AND THE YUCCA MOUNTAIN REVIEW PLAN

A U.S. Nuclear Regulatory Commission license is required, under the provisions of the U.S. Code of Federal Regulations (CFR) Title 10, Part 63 (Part 63), "Disposal of High-Level Radioactive Wastes in a Proposed Geologic Repository at Yucca Mountain, Nevada," for disposal of high-level radioactive waste. U.S. Nuclear Regulatory Commission authority to regulate a high-level radioactive waste repository comes from the Atomic Energy Act of 1954, as amended; the Energy Reorganization Act of 1974, as amended; and the Nuclear Waste Policy Act of 1982, as amended.

The Energy Policy Act of 1992 directed the U.S. Environmental Protection Agency to contract with the National Academy of Sciences, to provide advice on the appropriate technical bases for public health and safety standards governing a Yucca Mountain, Nevada, repository. In its report, "Technical Bases for Yucca Mountain Standards" (National Research Council, 1995), the National Academy of Sciences recommended that an individual protection standard, expressed as a limit on individual risk rather than on dose, would provide a reasonable basis for protecting the health and safety of the general public. The Energy Policy Act of 1992 also directed the U.S. Environmental Protection Agency to issue public health and safety standards for Yucca Mountain that "... prescribe the maximum annual effective dose equivalent to individual members of the public and that are consistent with the findings and recommendations that would be made by the National Academy of Sciences. This approach is different from that contained in the U.S. Environmental Protection Agency disposal standards at 40 CFR Part 191, "Environmental Radiation Protection Standards for Management and Disposal of Spent Nuclear Fuel, High-Level, and Transuranic Wastes," that were applied at the Waste Isolation Pilot Plant in New Mexico. In addition, the Energy Policy Act of 1992 directs the U.S. Nuclear Regulatory Commission to modify the technical requirements and criteria contained in original U.S. Nuclear Regulatory Commission generic regulations for disposal of high-level radioactive waste in 10 CFR Part 60, "Disposal of High-Level Radioactive Wastes in Geologic Repositories," to be consistent with the U.S. Environmental Protection Agency standards applicable to Yucca Mountain. The U.S. Environmental Protection agency published public health and environmental radiation protection standards for Yucca Mountain, Nevada, at 40 CFR Part 197 on June 13, 2001. The Commission has incorporated these standards into its final 10 CFR Part 63. Any license application for a geologic repository at Yucca Mountain, submitted under 10 CFR Part 63, is to contain "General Information," a "Safety Analysis Report," and is to be accompanied by a final environmental impact statement. Any Restricted Data or National Security Information must be separated from unclassified information in any license application. In light of the terrorist attacks of September 11, 2001, the Commission has directed the staff to conduct a comprehensive reevaluation of U.S. Nuclear Regulatory Commission physical protection requirements. If this effort indicates that U.S. Nuclear Regulatory Commission regulations or requirements warrant revision, such changes would occur through appropriate methods and, if necessary, the Yucca Mountain Review Plan would be revised accordingly.

Although the National Environmental Policy Act of 1969 requires an environmental evaluation for major federal actions that significantly affect the human environment, the Nuclear Waste Policy Act of 1982 requires that the U.S. Nuclear Regulatory Commission adopt, to the extent practicable, the final environmental impact statement prepared by the U.S. Department of Energy in connection with the issuance of a construction authorization and license for the Yucca Mountain repository. Thus, the U.S. Nuclear Regulatory Commission would prepare an environmental evaluation only for any areas where it cannot adopt the U.S. Department of

Energy final environmental impact statement. U.S. Nuclear Regulatory Commission regulations at 10 CFR 51.109 contain the criteria the U.S. Nuclear Regulatory Commission will use to determine if the final environmental impact statement published by the U.S. Department of Energy can be adopted. The Yucca Mountain Review Plan is not a staff guidance document for an environmental evaluation. The Commission has previously provided its comments on the draft environmental impact statement, the supplemental draft environmental impact statement, and the final environmental impact statement for a potential high-level waste repository at Yucca Mountain to the U.S. Department of Energy.

The U.S. Department of Energy, as the applicant for a license to construct, operate, and eventually close a geologic repository, and to receive, and possess source, special nuclear, or byproduct material, and dispose of such material at a geologic repository operations area at the Yucca Mountain site is required to provide detailed information on the facilities, equipment, and procedures to be used, and to discuss the effect of proposed operations on public health and safety. Environmental impacts of the proposed repository are evaluated in the final environmental impact statement for the Yucca Mountain site prepared in accordance with the Nuclear Waste Policy Act, and submitted with the license application. The U.S. Nuclear Regulatory Commission staff reviews information submitted by the applicant to determine whether the proposed activities will meet the applicable regulatory requirements, and thus be protective of public health and safety and the environment. General procedures for the issuance of a construction authorization or license to receive and posses waste, as well as an amendment to a construction authorization or license, are described in 10 CFR Part 2, Subpart A. 10 CFR Part 2, Subpart J, contains the procedures applicable to proceedings for the issuance of a construction authorization or licenses for the receipt of high-level radioactive waste at a geologic repository.

10 CFR Part 63, a site-specific rule, will be implemented using this site-specific Yucca Mountain Review Plan. The Yucca Mountain Review Plan provides the staff with guidance on the review of a license application for a high-level radioactive waste disposal facility at Yucca Mountain. The staff will also use the Yucca Mountain Review Plan to review requested amendments to the license application and, potentially, applications to amend a construction authorization or license. The principal purpose of the review plan is to ensure quality and uniformity in the U.S. Nuclear Regulatory Commission staff reviews. Use of published drafts of this Yucca Mountain Review Plan began in the prelicensing consultations conducted consistent with 10 CFR 2.101(a)(1) and 10 CFR 63.16. (Any information provided during the prelicense application consultations, but not included in the license application, will not be considered during the acceptance or technical review under the Yucca Mountain Review Plan.) Each Yucca Mountain Review Plan section provides guidance on what is to be reviewed, the review basis, how the staff review is to be accomplished, what the staff would find acceptable in a demonstration of compliance with the regulations, and the conclusions that are sought regarding the applicable sections in 10 CFR Part 63.

This Yucca Mountain Review Plan covers only those aspects of the U.S. Nuclear Regulatory Commission regulatory mission that are related to the licensing of a high-level radioactive waste disposal facility. As such, the Yucca Mountain Review Plan helps focus the staff review on determining if a facility can be constructed and operated, and waste received and possessed, in compliance with the applicable U.S. Nuclear Regulatory Commission regulations. The Yucca Mountain Review Plan also makes information about regulatory matters widely available and improves communications and understanding of the staff review process by the U.S. Department of Energy, interested members of the public, the State of Nevada, affected units of local governments and Indian tribes, and other stakeholders. For review of any amendments, the focus of the review should be on the changes proposed in the amendment. Reviewers should not reevaluate previously approved actions if they are not part of the amendment, unless the review of the amendment package identifies problems with other aspects of facility operation.

The Yucca Mountain Review Plan is not a regulation. The acceptance criteria laid out in this Yucca Mountain Review Plan guidance for the U.S. Nuclear Regulatory Commission staff responsible for the review of an application for a high-level radioactive waste repository at Yucca Mountain. Review plans do not set forth regulatory requirements, and compliance with the Yucca Mountain Review Plan is not required. Methods and solutions different from those set out in the Yucca Mountain Review Plan will be acceptable if the U.S. Department of Energy demonstrates it has satisfied regulatory requirements, including the findings requisite to the issuance of a construction authorization, license, or amendment of a construction authorization or license by the U.S. Nuclear Regulatory Commission. To the extent practical, the staff has made the Yucca Mountain Review Plan risk-informed and performance-based. This, coupled with the performance-based regulations, will ensure that the U.S. Nuclear Regulatory Commission review is focused on those aspects most important to health and safety and will afford the U.S. Department of Energy flexibility in methods chosen to meet the performance-based regulations. The Yucca Mountain Review Plan, however, contains sufficient detail to support conclusions on regulatory compliance and provide guidance for the proposed first-of-a-kind facility to be reviewed within the 3-year decision period mandated by statute.

It is the responsibility of the U.S. Department of Energy to identify structures, systems, and components important to safety and natural and engineered barriers important to waste isolation. The U.S. Nuclear Regulatory Commission staff review using the guidance in the Yucca Mountain Review Plan will consider the safety strategy of the U.S. Department of Energy. This approach is consistent with the U.S. Nuclear Regulatory Commission policy regarding risk-informed, performance-based regulations (U.S. Nuclear Regulatory Commission, 1999) in which risk insights, engineering analysis, expert judgment, the principle of defense-in-depth, and safety margins, are incorporated in licensing decisions.

A1.1 Conduct of The Yucca Mountain Licensing Review

A1.1.1 Licensing Review Philosophy

Since passage of the Atomic Energy Act of 1954, the U.S. Nuclear Regulatory Commission has been engaged in a continuing process of interpreting and applying the Agency's basic responsibilities to protect public health and safety, assure the common defense and security, minimize danger to life or property, and provide adequate protection from the risks involved in the commercial use of Atomic Energy Act radioactive materials. These terms are not defined in the Atomic Energy Act of 1954, nor are they self-explanatory. The underlying regulatory philosophy used by the U.S. Nuclear Regulatory Commission in fulfilling its regulatory mission can be found in the "U.S. Nuclear Regulatory Commission Strategic Plan" (U.S. Nuclear

Regulatory Commission, 2000) which embodies the principle that the licensee is responsible for the safe operation of a nuclear facility.

The following three principles are important in implementing the U.S. Nuclear Regulatory Commission regulatory mission:

(1) The U.S. Nuclear Regulatory Commission does not select sites or designs, or participate with licensees or applicants in selecting proposed sites or designs;

(2) The U.S. Nuclear Regulatory Commission role is not to monitor all licensee activities, but to oversee and audit them. The U.S. Nuclear Regulatory Commission should evaluate whether a license application meets the applicable regulations based on a review of what is in the application and supporting materials. Reviews using staff audit calculations should be performed in limited situations, such as where there are unique proposals involving new methods or assumptions. Otherwise, the U.S. Nuclear Regulatory Commission staff should review the application to verify that assumptions are justified, methods used are acceptable and applicable over the range presented, models are properly applied, and results are acceptable. Staff may do quick, bounding calculations and performance assessments, and confirmatory analyses using process-level models; however, in-depth, detailed analyses may be limited to a few applications. Figure A1–1 shows the relationship of the level of detail to licensing reviews and inspections during the preclosure period; and

(3) The three outcomes available to the U.S. Nuclear Regulatory Commission at the conclusion of a licensing review are: (i) grant the license; (ii) grant the license subject to conditions; or (iii) deny the license. Other than rejecting an applicant or licensee proposal, the U.S. Nuclear Regulatory Commission has no power to compel a licensee to come forward with, or prepare, a different proposal.

The U.S. Nuclear Regulatory Commission regulatory role in any licensing action is to apply the applicable regulations and guidance, and to review applications for proposed actions to determine if compliance with regulations has been achieved. The burden of proof is on the applicant or licensee to show that the proposed action is safe, to demonstrate that regulations are met, and to ensure continued compliance with the regulations. License conditions should be discussed with the licensee and imposed, as necessary to meet the reasonable assurance and reasonable expectation determinations for issuance of a construction authorization, a license, or any amendment thereto. Failure of the applicant to satisfy regulatory requirements can provide a basis to deny the requested licensing action.

In conducting its reviews, the U.S. Nuclear Regulatory Commission evaluates whether an applicant or licensee has demonstrated that its proposed approach is adequate to meet the codified requirements. The applicant or licensee is not required to present a complete understanding and answers for all issues that could be raised concerning a proposal, including those not related to health and safety. U.S. Nuclear Regulatory Commission staff should examine whether applicant or licensee proposals are acceptable. If a proposal meets the applicable regulations, no additional showing should be required. To do so would be imposing a requirement on an applicant beyond what is required in the regulations.

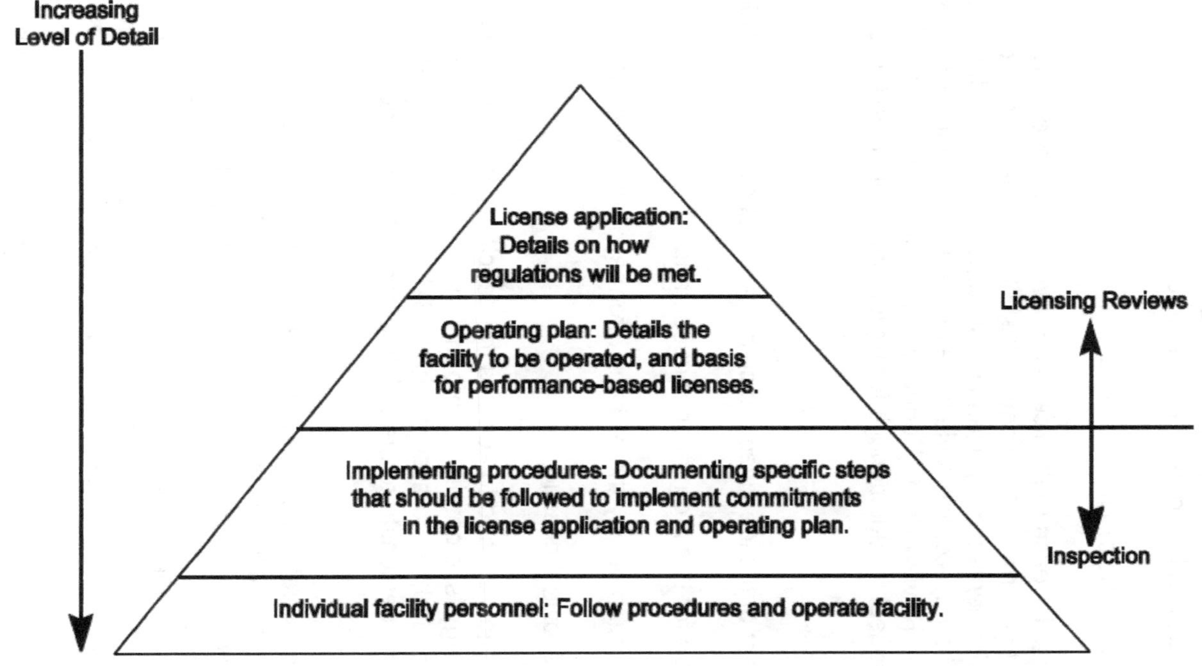

Figure A1–1. Schematic of U.S. Nuclear Regulatory Commission Licensing and Inspection Process and Applicability to Licensing Documents

In no instance should a reviewer determine that alternatives that are less protective than those proposed by the applicant are acceptable. In addition, U.S. Nuclear Regulatory Commission staff should submit requests for additional information when more information is needed to determine whether regulatory requirements are met.

10 CFR 63.21(a) specifies that an application "...must be as complete as possible in the light of information that is reasonably available at the time of docketing." While knowledge of the site and design will evolve over time, at each licensing step, the U.S. Department of Energy must demonstrate compliance with applicable regulatory requirements.

A1.1.2 Format and Content of Documents

Correspondence and documents for each of the licensing review milestones should be logically organized and contain adequate information to convey the U.S. Nuclear Regulatory Commission position and requirements simply, clearly, and concisely. Procedures for conducting and documenting the acceptance review are presented in Appendix B ("Acceptance Review") of the Yucca Mountain Review Plan. Requests for additional information should be focused, brief, and clear. A request for additional information should include three parts:

(1) A statement of the issue;

This presents a summary of the identified deficiency and the regulatory requirement.

(2) A discussion of the basis for the information request; and

This provides an explanation of why the existing information is inadequate.

(3) The action needed to resolve the issue.

This defines the information needed to address the deficiency without specifying how the U.S. Department of Energy is to obtain the information. The staff must be careful, when describing the action needed, not to assume the U.S. Department of Energy responsibility to state and demonstrate the safety of the actions proposed in the license application.

Requests for additional information related to the technical adequacy of the license application should state all relevant problems and issues to be resolved before approval in a manner that is clear, concise, and consistent with the regulations and good engineering practice. A request for additional information is considered primarily an exchange through which the U.S. Nuclear Regulatory Commission staff elicits the information necessary for it to determine if the applicant has demonstrated compliance with the regulations. U.S. Nuclear Regulatory Commission staff may provide further explanatory information, depending on the complexity of the request.

During the technical review, some requests for additional information may be related to an apparent failure to meet regulatory requirements. In this case, the request for additional information should identify the specific section of the regulations, or other supporting documents, (e.g., regulatory guides, standard review plans, U.S. Nuclear Regulatory Commission technical reports, American Society of Mechanical Engineers/American Society for

Testing and Materials codes, or techniques accepted by the scientific community) that relate to the issue. This type of supporting information may provide both a technical and a regulatory perspective for the request.

Requests for additional information should be numbered sequentially, with the numbering for an individual request for additional information remaining constant through the course of the licensing review. The cover letter transmitting the requests for additional information will include a schedule for the applicant to provide responses and the dates of remaining milestones. The letter will also reiterate the statement from the acceptance review that failure to respond within the specified time frame may be grounds for denial of the application, in accordance with 10 CFR 2.108(a).

The content of the safety evaluation report will be based on the guidance provided in the Yucca Mountain Review Plan. Any limits and restrictions imposed as a condition of approval of the construction authorization or license will be specified in the safety evaluation report and the license. The technical reviewer should notify the licensing project manager as soon as practical when potential license conditions or license specifications are identified. The format for the safety evaluation report will follow the structure of the Yucca Mountain Review Plan. The safety evaluation report will describe the information the staff reviewed, provide the technical basis for the staff conclusion regarding compliance, and state an evaluation finding. Information from the U.S. Nuclear Regulatory Commission prelicensing issue resolution process that has not been submitted or referenced in support of the license application may not be relied on to reach a determination of whether regulatory requirements are met. The findings made as a result of the staff's detailed review will be stated in the safety evaluation report at the conclusion of each section.

A1.2 General Review Procedure

A licensing review is not intended to be a detailed evaluation of all aspects of facility operations. Specific information about implementation of the program outlined in an application is obtained through the U.S. Nuclear Regulatory Commission review of procedures and operations done as part of the inspection function. A definition of the differences between licensing reviews and inspections is shown in Figure A1–1. If a construction authorization or license is issued, certain changes to the authorized activities may require the issuance of an amendment as provided by 10 CFR 63.44, 63.45, and 63.46. An application for an amendment should describe the proposed changes in detail, and should discuss any related health and safety, as well as environmental issues. The health and safety aspects of amendment requests should be reviewed using the applicable sections of this Yucca Mountain Review Plan.

In conducting any review, the staff will rely on the approach described in Section A1.1 to ensure the efficient and effective use of resources. This approach will involve drafting a safety evaluation report that identifies where the U.S. Department of Energy has not provided sufficient information to make a regulatory conclusion. These gaps will then serve as the basis for staff requests for additional information. As needed, the U.S. Nuclear Regulatory Commission staff and the U.S. Department of Energy will interact on the responses to the questions either through conference calls or public meetings. These interactions should help ensure that the U.S. Department of Energy fully responds to the requests for additional information, and that the responses do not result in additional requests for additional information. A goal of this process is to prepare only one round of requests for additional information. While the U.S. Department of Energy is addressing requests for additional information, the staff may publish portions of the safety evaluation report

The license application review is described in the following sections.

A1.2.1 Acceptance Review Objectives

The U.S. Nuclear Regulatory Commission staff shall conduct an acceptance review of the tendered application to determine the completeness of the information submitted. This review requires a comparison of the submitted information with the information specified in 10 CFR 63.21. The application will be considered complete for docketing if the information provided is complete, reflects an adequate investigation and physical examination of the regional and site conditions, and provides appropriate analyses and design information to demonstrate that the applicable acceptance criteria will be met. The staff shall complete the acceptance review and inform the applicant as to the results of the review within 3 months of the receipt of the application, along with a projected schedule for the remainder of the review. In this transmittal, the staff may identify any deficiencies in the application, including information needed to make the application complete. Detailed technical questions, although not required, can be included if they are identified during the acceptance review.

A1.2.2 Detailed Review Objectives

After completion of the acceptance review, the staff shall conduct a detailed technical review of the application. The results of this review (i.e., the basis for acceptance or denial of the requested licensing action) are documented by the U.S. Nuclear Regulatory Commission staff in its safety evaluation report. During the course of this review, the staff will publish its safety evaluation report, and possibly one or more supplements. The safety evaluation report and supplements will contain evaluation findings and conclusions reached during the review and any license conditions.

Open items are items that remain outstanding at the time of publication of the safety evaluation report, and which will be addressed in a later supplement. Because the staff has not completed its review of or reached a final position on, these items they are considered open.

Items that are resolved to the staff's satisfaction during the review, but for which certain confirmatory information has not yet been received, are called confirmatory items. In these instances, the U.S. Department of Energy may have committed to provide confirmatory information. The staff would need such information before it could close the item. Not all confirmatory items[1] will need to be resolved before licensing. Some may require information from construction activities before they can be closed. The staff will track these items through its inspection process or address them in licensing.

The staff review may identify license conditions which will be incorporated into any license issued. These conditions are needed to ensure that the applicable requirements are met, for example, during facility operation. A license condition may be in the form of a condition in the body of the license, or aa license specification that outlines the operational limits of the facility (derived from analyses and evaluations in the license application), which is appended to any license issued. It is important to note that any license commitment made by the U.S. Department of Energy in its application that is relied on to make regulatory findings should be included as a license condition. U.S. Department of Energy license conditions identified by the U.S. Nuclear Regulatory Commission staff could be a matter addressed in the hearing on the license application.

A1.2.3 Licensing Review Process

The licensing process for a high-level waste repository at Yucca Mountain is depicted in Figure A1–2. Prelicensing activities, consistent with the Nuclear Waste Policy Act and 10 CFR 2.101(a) and 63.16, have been underway for a number of years. Figure A1–2 portrays the licensing process from the time at which the U.S. Department of Energy submits an application to the U.S. Nuclear Regulatory Commission.

Upon receipt of an application, the U.S. Nuclear Regulatory Commission staff would treat the application as tendered and begin an acceptance review to determine whether the application is sufficiently complete for docketing and to begin detailed technical review. The staff could reach three conclusions as a result of this acceptance review. First, the license application could be determined to be substantially incomplete, in which case it would be rejected and returned to the U.S. Department of Energy with an identification of the deficiencies. Second, the staff could find that the license application is sufficiently complete that the detailed technical review could begin, but that additional information is needed in limited areas. In this case, the staff would docket the application, proceed with the detailed technical review in other areas and prepare requests for additional information regarding the deficient areas. The U.S. Department of Energy would need to provide the needed information within a specified period, to enable the staff to complete the acceptance review. Then, the staff could determine that the license application is complete in all respects. In this case, the application would be docketed, and the detailed technical review of the entire application would begin.

---

[1]Confirmatory items are used during a licensing process to identify items for which a licensee needs to provide additional confirmatory information but which do not prevent the licensing action from proceeding. Closed pending issues, which were defined during the formal high-level waste prelicensing issue resolution process established between the U.S. Nuclear Regulatory Commission and the U.S. Department of Energy, were those issues for which the U.S. Nuclear Regulatory Commission staff had confidence that the U.S. Department of Energy proposed approach, together with any U.S. Department of Energy agreements to provide additional information (through specified testing, analysis, etc.) acceptably addressed the staff questions such that no information beyond that provided, or agreed to, would likely be required at the time of initial license application. Closed pending items do not presuppose whether the U.S. Nuclear Regulatory Commission considers the U.S. Department of Energy license application to meet the acceptance criteria provided in this review plan. The closed pending terminology will not apply during licensing.

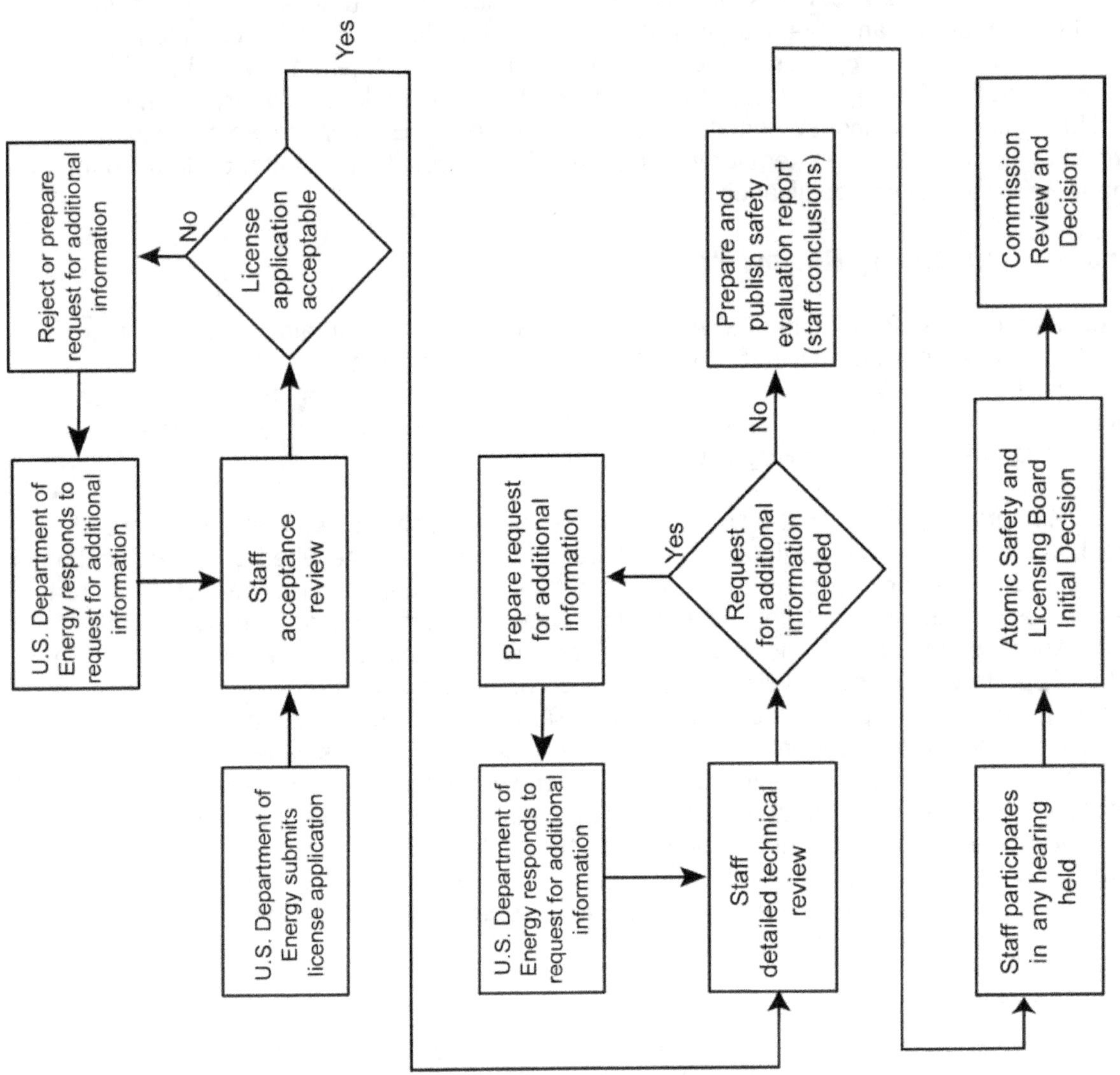

Figure A1–2. Licensing Process

If the tendered license application is found to be acceptable for docketing, the staff would begin its detailed technical review using the review methods and acceptance criteria presented in the Yucca Mountain Review Plan. If the detailed technical review identifies areas where additional information is needed, the staff would prepare requests for additional information and transmit them to the U.S. Department of Energy. The failure of U.S. Department of Energy to provide the requested information within a specified time period would result in a notice of denial of the application, issued pursuant to 10 CFR 2.108. If the staff receives the requested information on a timely basis, the staff would continue the detailed technical review until it has reached conclusions regarding regulatory compliance.

The staff conclusions and evaluation findings on the license application would be documented in a safety evaluation report issued on the application. This safety evaluation report provides the bases for a staff recommendation as to whether a construction authorization or license should be granted and would identify any license conditions or license specifications required to ensure regulatory compliance. The safety evaluation report would be published and made electronically available via the Licensing Support Network. For illustration purposes, this process is generally depicted in Figure A1–3 and is described in detail in Section A1.2.4.

In ruling on contentions admitted in the proceeding regarding the issuance of a construction authorization, the Atomic Safety and Licensing Board would consider evidence admitted in the proceeding, including the staff safety evaluation report, and issue an initial decision. The Commission would rule on any appeals to the initial decision, and review the staff findings, before determining whether to authorize construction.

A1.2.4 Hypothetical Review of a License Application Section

As an example, for illustration purposes only, Figure A1–3 shows how the review of a license application section would proceed and how the U.S. Nuclear Regulatory Commission staff would conduct the review of a section of a license application for a high-level waste repository at Yucca Mountain. The details of this hypothetical review are as follows.

First a project manager would examine the nature of a specific section of the license application and identify the staff and contractor disciplines needed to conduct the review. Once assigned, this team would study the associated Yucca Mountain Review Plan section, applicable regulatory requirements, and other relevant technical background information.

At this point, the team members would conduct the detailed technical review of the specific section of the application using the review methods in the Yucca Mountain Review Plan. If the reviewers find that the approach used in the application is not consistent with the review methods in the Yucca Mountain Review Plan, they would notify the project manager and then establish review methods and acceptance criteria appropriate to the material in the application.

As the team reviews the section of the application using the review methods in the Yucca Mountain Review Plan, it would compare the review results with the acceptance criteria in the review plan. At this stage, the reviewers may discover that insufficient information has been provided in the application to support conclusions as to whether the acceptance criteria have been met. If more information is needed, the reviewer would inform the project manager and prepare a request for additional information to be sent to the U.S. Department of Energy.

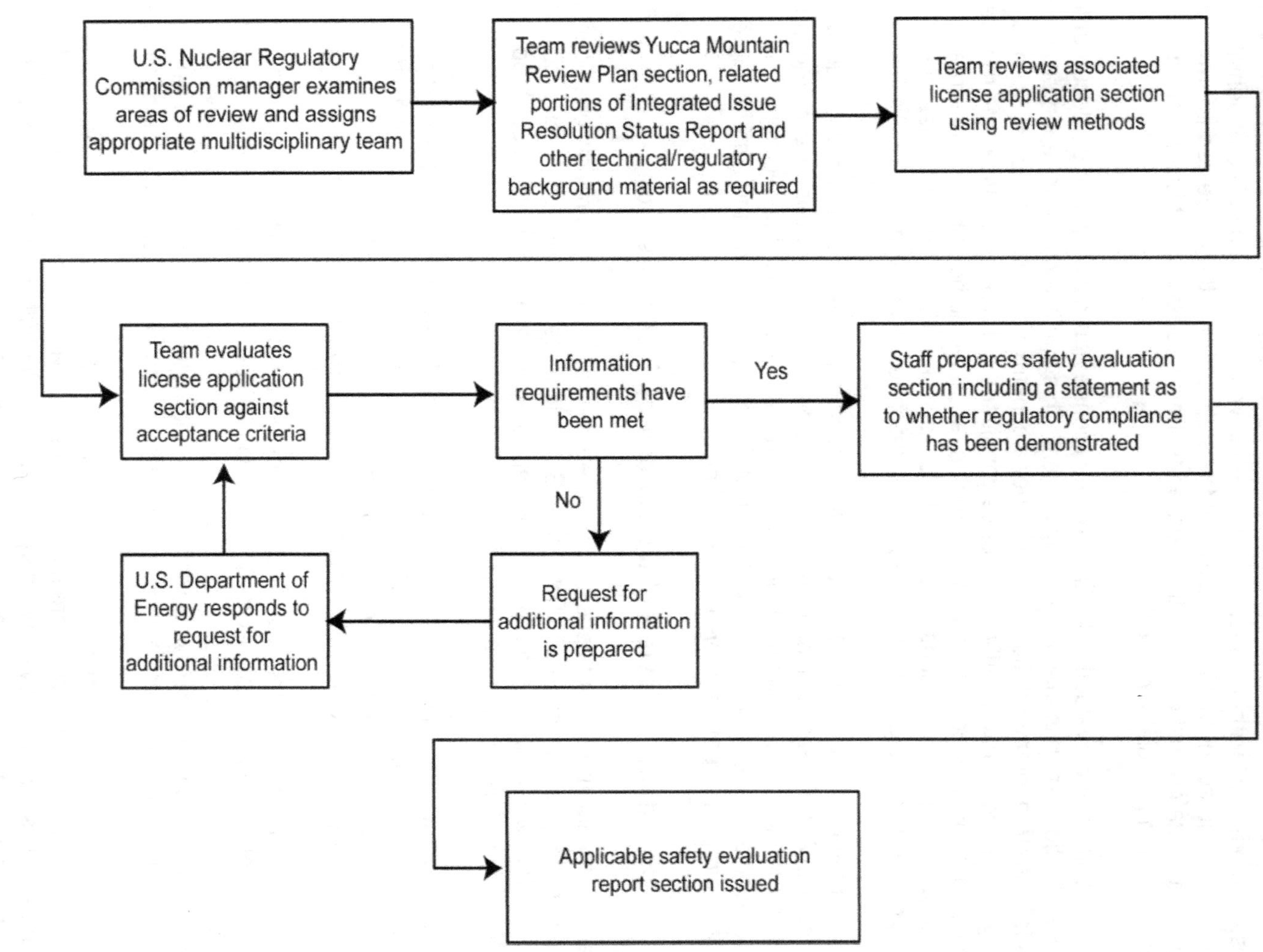

Figure A1–3. Hypothetical Review of a Typical License Application Section

If the U.S. Department of Energy responds adequately to any request for additional information, the reviewers would prepare the final draft of the related portion of the safety evaluation report. The safety evaluation report would describe what was reviewed and document the reviewers' conclusions as to whether the regulatory requirements have been met. The example evaluation findings located at the end of the specific Yucca Mountain Review Plan section would be used as the foundation for preparing these conclusions. The safety evaluation report section would then be combined with similar safety evaluations from reviews of other sections of the application.

The following text is a hypothetical example of how review results and evaluation findings might be prepared. The hypothetical technical information and evaluation findings statements do not reflect any prejudgment of matters related to a license application for Yucca Mountain. The information presented does not reflect U.S. Department of Energy analyses or U.S. Nuclear Regulatory Commission staff assessments or any agreements made by the U.S. Department of Energy to provide additional information to the U.S. Nuclear Regulatory Commission staff. The example presents a hypothetical documentation of a review using Review Method 3 and Acceptance Criterion 3 of Yucca Mountain Review Plan Section 2.2.1.3.9, Radionuclide Transport in the Saturated Zone.

Example: "Safety Evaluation Report Section for Radionuclide Transport in the Saturated Zone" (prepared using Section 2.2.1.3.9, Review Method 3 and Acceptance Criterion 3, of the Yucca Mountain Review Plan).

The Radionuclide Transport in the Saturated Zone model abstraction addresses features and processes that would affect movement of radionuclides in the saturated zone from the area beneath the proposed repository site at Yucca Mountain to the proposed 18-km [11-mi] compliance boundary. Figure A1–4 illustrates the relationship between the radionuclide transport in the saturated zone model abstraction and the flow paths in the saturated zone model abstraction. The overall organization and identification of all the model abstractions are depicted in Figure A1–5. The U.S. Department of Energy description and technical basis for abstraction of radionuclide transport in the saturated zone are described in several supporting analysis and model reports.

Relationship to Other Model Abstractions

This Radionuclide Transport in the Saturated Zone model abstraction incorporates subject matter related to the following additional key technical issues and model abstractions

- Unsaturated and Saturated Flow Under Isothermal Conditions;
- Structural Deformation and Seismicity;
- Container Life and Source Term; and
- Total System Performance Assessment and Integration.

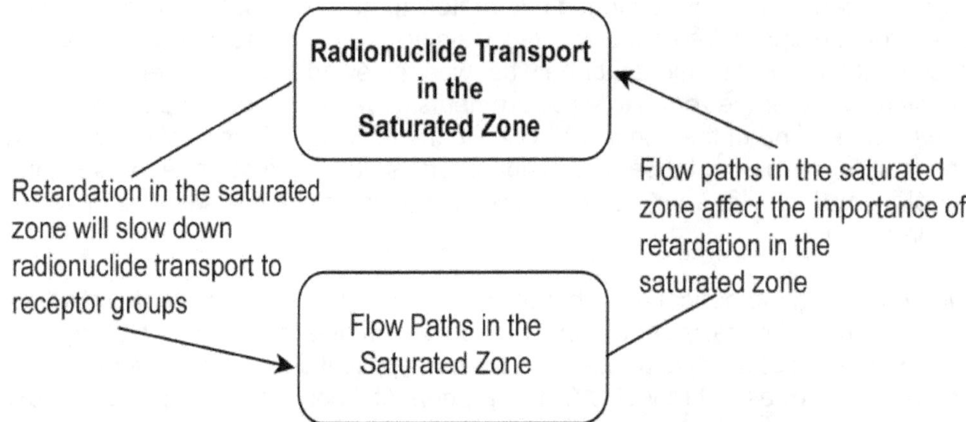

Figure A1–4. Diagram Illustrating the Relationship Between the Radionuclide Transport in the Saturated Zone and Flow Paths in the Saturated Zone Integrated Subissues

Importance to Postclosure Performance

The U.S. Department of Energy identifies radionuclide delay through the saturated zone at Yucca Mountain as a principal factor of the current postclosure safety case (Civilian Radioactive Waste Management System Management and Operation, 2000a). The degree of radionuclide sorption on mineral surfaces within the rock matrix of the tuff aquifer system and in the alluvial aquifer system is the most important process affecting the ability of the saturated zone to act as a natural barrier by attenuating and delaying potentially released radionuclides. In the U.S. Department of Energy approach, sorption of radionuclides in the tuff aquifer system is assumed to occur only within the relatively stagnant rock matrix, whereas flow occurs primarily in fracture networks. Matrix diffusion, a process whereby aqueous radionuclides diffuse from actively flowing pore spaces into the relatively stagnant pore space within the rock matrix, is thus also another important process to be considered because the majority of saturated pore volume in the saturated tuff aquifer system comprises relatively stagnant water within rock matrix.

The U.S. Department of Energy investigated the importance of saturated zone transport through robustness and neutralization analyses (Civilian Radioactive Waste Management System Management and Operation, 2000a,b). The degraded barrier analysis, in which 5th percentile values are used for parameters that positively promote delay of radionuclides in the saturated zone and 95h percentile values are used for parameters that positively promote transport in the saturated zone, suggests modest sensitivity of dose (Civilian Radioactive Waste Management System Management and Operation, 2000a) to the saturated zone transport barrier. The similarity of the peak mean dose in the degraded and basecases is attributed to the dominance in the basecase average dose of the high-dose realizations (Civilian Radioactive Waste Management System Management and Operation, 2000b). A saturated zone transport barrier neutralization analysis, in which the unsaturated zone output is fed directly to the biosphere, yields a curve nearly identical to the robustness analysis (Civilian Radioactive Waste Management System Management and Operation, 2000a). It is apparent that the modeled

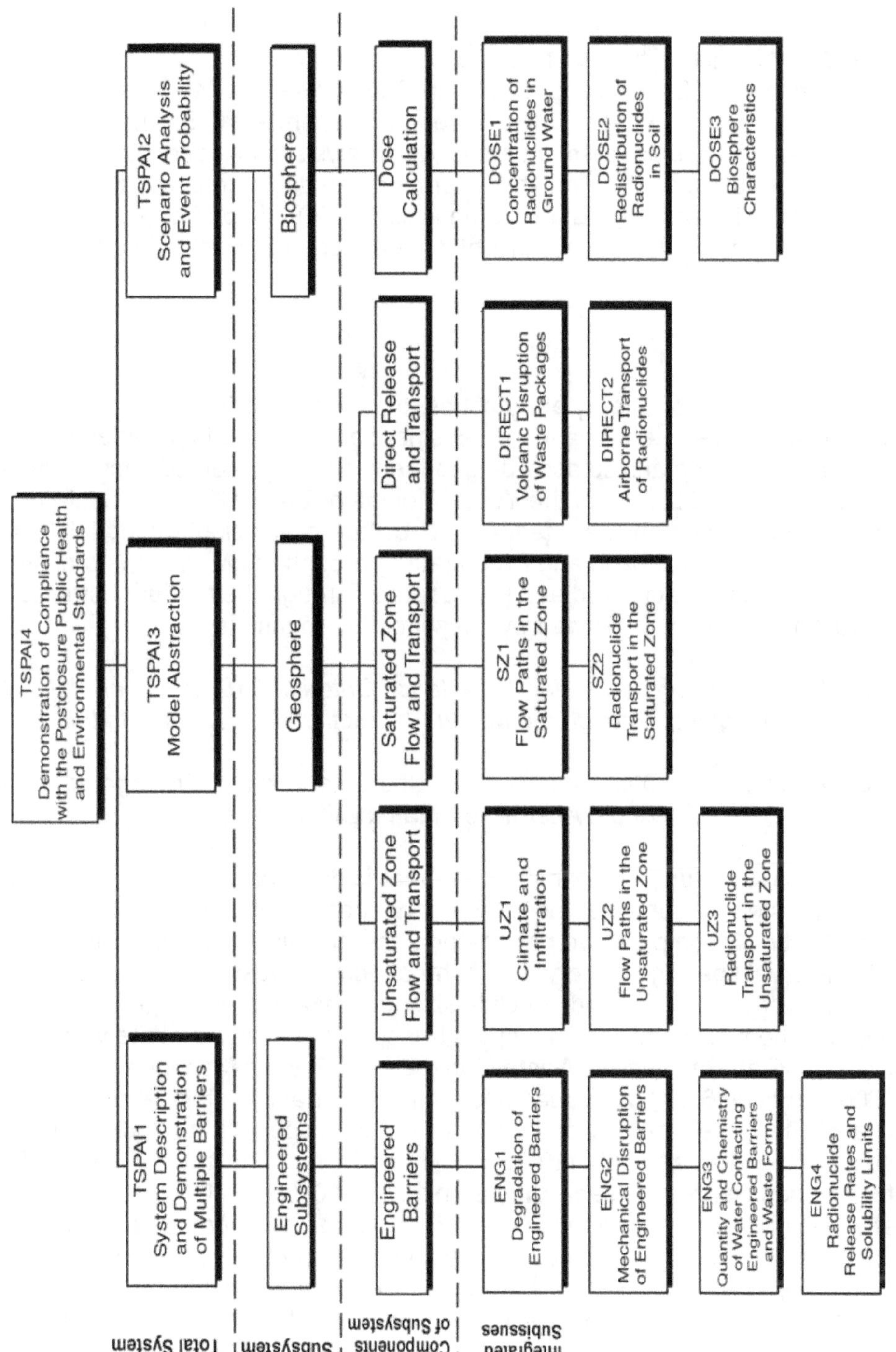

Figure A1–5. Components of Performance Assessment Review

unsaturated zone barrier in the U.S. Department of Energy total system performance assessment is the more important barrier, and this may mask the potential importance of the saturated zone barrier.

Nevertheless, the importance of the saturated zone is reflected in its status as a principal factor, chiefly as a component of defense in depth (Civilian Radioactive Waste Management System Management and Operation, 2000a). Furthermore, an independent U.S. Nuclear Regulatory Commission staff performance assessment sensitivity analysis concluded that retardation in the saturated zone is important, based on much higher modeled doses that result from its removal from the analysis (U.S. Nuclear Regulatory Commission, 1999). In particular, neptunium retardation has been shown to have a significant dose effect (U.S. Nuclear Regulatory Commission, 1999, 2001).

Technical Basis

A review of U.S. Department of Energy approaches for including radionuclide transport in the saturated zone in total system performance assessment abstractions is provided in the following subsections. The review is organized according to the five review methods and acceptance criteria identified in Section 2.2.1.3.9 of the Yucca Mountain Review Plan: (i) System Description and Model Integration Are Adequate, (ii) Data Are Sufficient for Model Justification, (iii) Data Uncertainty Is Characterized and Propagated Through the Model Abstraction, (iv) Model Uncertainty Is Characterized and Propagated Through the Model Abstraction, and (v) Model Abstraction Output Is Supported by Objective Comparisons.

Only the hypothetical review conducted for Acceptance Criterion 3 (Data Uncertainty Is Characterized and Propagated Through the Model Abstraction) is presented in this example.

Acceptance Criterion 3 Data Uncertainty Is Characterized and Propagated Through the Model Abstraction Matrix Diffusion.

Uncertainty in data used to support the inclusion of matrix diffusion in the transport model is treated in the total system performance assessment abstraction of saturated zone radionuclide transport by stochastically sampling two parameters: the effective diffusion coefficient and the effective flowing interval spacing. Uncertainty in the effective diffusion coefficient is a function of the uncertainty and variability in the radionuclide size, temperature, heterogeneity of rock properties, and geochemical conditions along the transport pathway. U.S. Department of Energy analyses (Civilian Radioactive Waste Management System Management and Operation, 2000c, Section 6.8.4) show that most of the uncertainty in this parameter can be attributed to variability in the tortuosity of the connected pore space in the rock matrix. Based on its analyses, the U.S. Department of Energy estimated a range of possible values for effective diffusion coefficients in volcanic tuffs from 10^{-9} to 10^{-6} cm^2/s [10^{-10} to 10^{-7} in^2/s]. To ensure the effective diffusion coefficient is not overestimated, the upper bound of this range is set to below the smallest observed molecular diffusion coefficient. A log-uniform distribution is assumed for this range because it is considered unbiased with respect to the order of magnitude of the sampled parameter value and skewed toward lower values. This approach reasonably encompasses the uncertainty of this parameter.

Another important uncertainty is that of flowing interval spacing. Smaller values for effective flowing interval spacing would result in predictions of more rapid matrix diffusion. Analyses were performed to estimate a lognormally distributed range of flowing interval spacing with a mean $\log_{10}$ value of 1.29 and a standard deviation of 0.43 (Civilian Radioactive Waste Management System Management and Operation, 2000d). This estimate results in a range of approximately 2–200 meters [7–700 feet] with a median flowing interval spacing of approximately 20 meters [70 feet]. This wide range of values reasonably encompasses the uncertainty of flowing interval spacing and, given the highly fractured nature of the volcanic tuffs beneath Yucca Mountain, does not seem overly optimistic. The effective flowing interval spacing is used only as a transport parameter that affects the rate of matrix diffusion; it does not affect modeled ground water fluxes or flow velocities.

Sorption Coefficients

Although a significant amount of laboratory work and literature research is evident in the U.S. Department of Energy process model report (Civilian Radioactive Waste Management System Management and Operation, 2000e) and supporting analysis and model reports (Civilian Radioactive Waste Management System Management and Operation, 2000c,f), the process used in conducting the expert elicitation (or expert judgment) for transport parameter distributions, particularly K_d values, is not described in sufficient detail. Many of the methods normally used in expert elicitation (e.g., panel selection, training, bias, consensus building, dissenting opinions, aggregation, and documentation) are not discussed. In addition, the information used by the expert panel is not described in a way that demonstrates how the strengths and weaknesses of different data sets were evaluated and considered to derive the K_d probability distribution functions. Also, subsequent changes from the initial elicitation are not documented in a transparent manner. This type of information is important to allow a reviewer to trace the process used to develop parameter distributions from the original data and assumptions to the results and conclusions (U.S. Nuclear Regulatory Commission, 1996). Although the parameter distributions used may be appropriate, without the underlying basis for the expert judgments, the radionuclide transport in the saturated zone model abstraction does not provide a sufficient treatment of data uncertainty.

In discussions of geochemical effects on saturated zone transport outlined in Civilian Radioactive Waste Management System Management and Operation (2001), the U.S. Department of Energy states that the specific effects are included because uncertainty distributions of sorption coefficients are broad enough to encompass them. In each case, staff conclude that the U.S. Department of Energy has not provided sufficient technical basis that the uncertainty distributions account for the effects. Specific comments on the included features, events, and processes follow.

Ground Water Chemistry/Composition in Unsaturated and Saturated Zone: This feature, event, and process is included for the saturated zone on the basis that K_d uncertainty ranges bound possible variations because of chemistry variations (Civilian Radioactive Waste Management System Management and Operation, 2001). The discussion of total system performance assessment disposition, however, does not address the potential for correlation among radioelement K_ds and possible performance effects. Furthermore, Civilian Radioactive Waste Management System Management and Operation (2000f) states that K_d values derived from experiments are not considered to be influenced by microbial and precipitation/dissolution processes—the effects of which are asserted to be included.

Radionuclide Transport in a Carrier Plume: This feature, event, and process is included in the saturated zone, based on the assertion that no credit is taken for chemical changes within the plume that would decrease the transport rate (Civilian Radioactive Waste Management System Management and Operation, 2001, p. 56). However, the feature, event, and process discussion does not state how potentially adverse plume effects are accounted for; it is apparent that the U.S. Department of Energy is relying on K_d distributions. This argument appears to ignore the aspects of retardation that suggest sorption is dominated by solution chemistry rather than rock type. Because the U.S. Department of Energy does not explicitly model evolving water chemistry in the migrating carrier plume, including transport effects, it should provide a technical basis that states that ignoring this process is conservative or has negligible consequences.

Geochemical Interactions in the Geosphere: This feature, event, and process, which addresses processes such as dissolution and precipitation, is included (Civilian Radioactive Waste Management System Management and Operation, 2001). There is an inconsistency in that, while the U.S. Department of Energy claims its K_d uncertainty distributions account for variations from possible interactions along the transport path, it is not clear that these processes were considered in deriving the distributions (Civilian Radioactive Waste Management System Management and Operation, 2000f).

Complexation in the Geosphere: This feature, event, and process is stated to be included because the effects of complexation agents in the existing ground water system are included implicitly in the distribution for the K_d value for each element (Civilian Radioactive Waste Management System Management and Operation, 2001, p. 59). Parameter distributions and current U.S. Department of Energy process models do not appear to address adequately the effects of organic complexation on transport parameters (Civilian Radioactive Waste Management System Management and Operation, 2000f).

Microbial Activity in Geosphere: This feature, event, and process is said to be included (Civilian Radioactive Waste Management System Management and Operation, 2001) based on the argument that K_d uncertainty ranges account for effects of microbial activity. The analysis and model report (Civilian Radioactive Waste Management System Management and Operation, 2000f), however, states that K_d values derived from experiments are not considered to be influenced by microbial processes.

The issue common to these five included features, events, and processes is that the U.S. Department of Energy has not adequately demonstrated that uncertainty distributions include all the possible variations in K_d in the saturated zone below Yucca Mountain. Documentation is necessary to determine how the U.S. Department of Energy developed the total system performance assessment transport parameter distributions and the type of information used to support the expert elicitation.

In summary, the U.S. Department of Energy has not provided experimental and field information to constrain data uncertainty for all transport parameters.

Fault Zones

Faults can provide fast pathways for radionuclide transport in the saturated zone. Furthermore, the flow and transport characteristics of fault zone pathways can differ widely from those elsewhere in the tuff aquifer. The U.S. Department of Energy has not adequately accounted for the possible effects of these differences in formulating transport parameter distributions (Civilian Radioactive Waste Management System Management and Operation, 2000c,f).

Colloidal Transport

The U.S. Department of Energy has improved its capability to model saturated zone colloid transport in recent total system performance assessment efforts (Civilian Radioactive Waste Management System Management and Operation, 2000b,e), but many of the parameters (e.g., the colloid partitioning coefficient, K_c) used in the models are not supported by site characterization or laboratory data. The U.S. Department of Energy has addressed this problem, to some extent, by using bounding analyses and sensitivity analyses, but there are insufficient radioelement specific data to determine whether the uncertainty in colloid transport has been constrained in the radionuclide transport in the saturated zone model abstraction. The two key parameters that affect saturated zone colloid transport are colloid partition coefficient K_c and colloid retardation factor R_c; colloid matrix diffusion is neglected (Civilian Radioactive Waste Management System Management and Operation, 2000g). In the saturated zone, R_c is defined for the tuff aquifer on the basis of one field test, and no site-specific data are available for the alluvial aquifer (Civilian Radioactive Waste Management System Management and Operation, 2000c,h). The microspheres used in the tests had diameters between 280 nm [1.1×10^{-5} in] and 640 nm [2.5×10^{-5} in] (Civilian Radioactive Waste Management System Management and Operation, 2000h); this value is large compared with a typical size range in colloids from 1 nm to 450 nm [4×10^{-8} in to 2×10^{-5} in]. Smaller colloids will have a much higher specific surface area and perhaps be greater contributors to the potential colloid load. Conversely, these smaller colloids may be small enough to diffuse into the matrix and be physically filtered, reducing their impact on repository performance. The U.S. Department of Energy discusses these limitations in Section 6.1.5 of Civilian Radioactive Waste Management System Management and Operation (2000h), but does not provide sensitivity analyses to test their effects on repository performance. Finally, in calculating R_c from the field data, assigning equal weight to results from the lower Prow Pass Tuff and the lower Bullfrog Tuff may not be conservative because the lower Bullfrog Tuff is the most transmissive interval at the C-Wells (Civilian Radioactive Waste Management System Management and Operation, 2000e, p. 3-29).

The K_c parameter, used to simulate reversible colloid attachment by lowering the radioelement K_d, is based on data for americium sorption to colloids and is applied to the K_d values for all reversibly attached radionuclides (Civilian Radioactive Waste Management System Management and Operation, 2000c). Calculation of K_c also involves a term for colloid concentration in the water. The colloid concentration adopted is 0.03 mg/L [0.03 ppm]. This value is claimed to be conservative because it corresponds to the highest observed or expected colloid concentration (Civilian Radioactive Waste Management System Management and Operation, 2000g). This concentration, however, is well below the maximum values used in release models for waste form 5 mg/L [5 ppm] and iron (hydr)oxide 1 mg/L [1 ppm] colloids derived from the engineered barrier system (Civilian Radioactive Waste Management System Management and Operation, 2000i). The U.S. Department of Energy has not used any data, site-specific or not, to demonstrate that the reversible colloid attachment parameter will bound the range of possible effects of this process, nor have sensitivity analyses been employed to investigate the effects of parameter uncertainty on modeled repository performance.

Alluvium

Characterization of the alluvial transport path is incomplete and uncertain. It is, therefore, important that parameter distributions used in total system performance assessment reflect those uncertainties. However, K_d distributions for alluvium are based on a limited number of site-specific tests that do not allow strong conclusions to be drawn (Civilian Radioactive Waste Management System Management and Operation, 2000f, p. 92). Furthermore, the U.S. Department of Energy states in the discussion of assumptions in Civilian Radioactive Waste Management System Management and Operation (2000f, p. 36) that it has not confirmed that sorption data are adequate for the alluvium. Parameter uncertainty could be particularly important for relatively poorly sorbing radioelements such as neptunium, iodine, and technetium. The distribution for alluvial effective porosity (Civilian Radioactive Waste Management System Management and Operation, 2000c) uses no site-specific data and rests on unconfirmed assumptions. The alluvial aquifer is subject to possibly large spatial and stratigraphic variations in transport parameters (U.S. Nuclear Regulatory Commission, 2000e), which the U.S. Department of Energy has not demonstrated that uncertainty distributions accommodate.

Evaluation Findings

(These example evaluation findings are based on the ones in Section 2.2.1.3.9, "Radionuclide Transport in the Saturated Zone" of the Yucca Mountain Review Plan. They are hypothetical and do not present any licensing conclusion for a high-level waste repository at Yucca Mountain.)

The U.S. Nuclear Regulatory Commission staff has reviewed the safety analysis report and other information submitted in support of the license application relevant to radionuclide transport in the saturated zone, and has found, with reasonable expectation, that they do not satisfy the requirements of 10 CFR Part 63.114 for model abstraction in this section. Technical requirements for conducting a performance assessment in the area of radionuclide transport in the saturated zone have not been met. In particular, the U.S. Nuclear Regulatory Commission staff found that:

- Appropriate data from the site and surrounding region, uncertainties, and variabilities in parameter values, and alternative conceptual models have not been used in the analyses so as to comply with 10 CFR 63.114(a)–(c);

- Specific features, events, and processes have not been adequately included in the analyses, and appropriate technical bases have not been provided, for inclusion or exclusion, so as to comply with 10 CFR 63.114(e);

- Specific degradation, deterioration, and alteration processes have not been included in the analyses, taking into consideration their effects on annual dose, and appropriate technical bases have not been provided for inclusion or exclusion, so as to comply with 10 CFR63.114(f); and

- Adequate technical bases have not been provided for models used in the performance assessment, as required by 10 CFR 63.114(g).

References

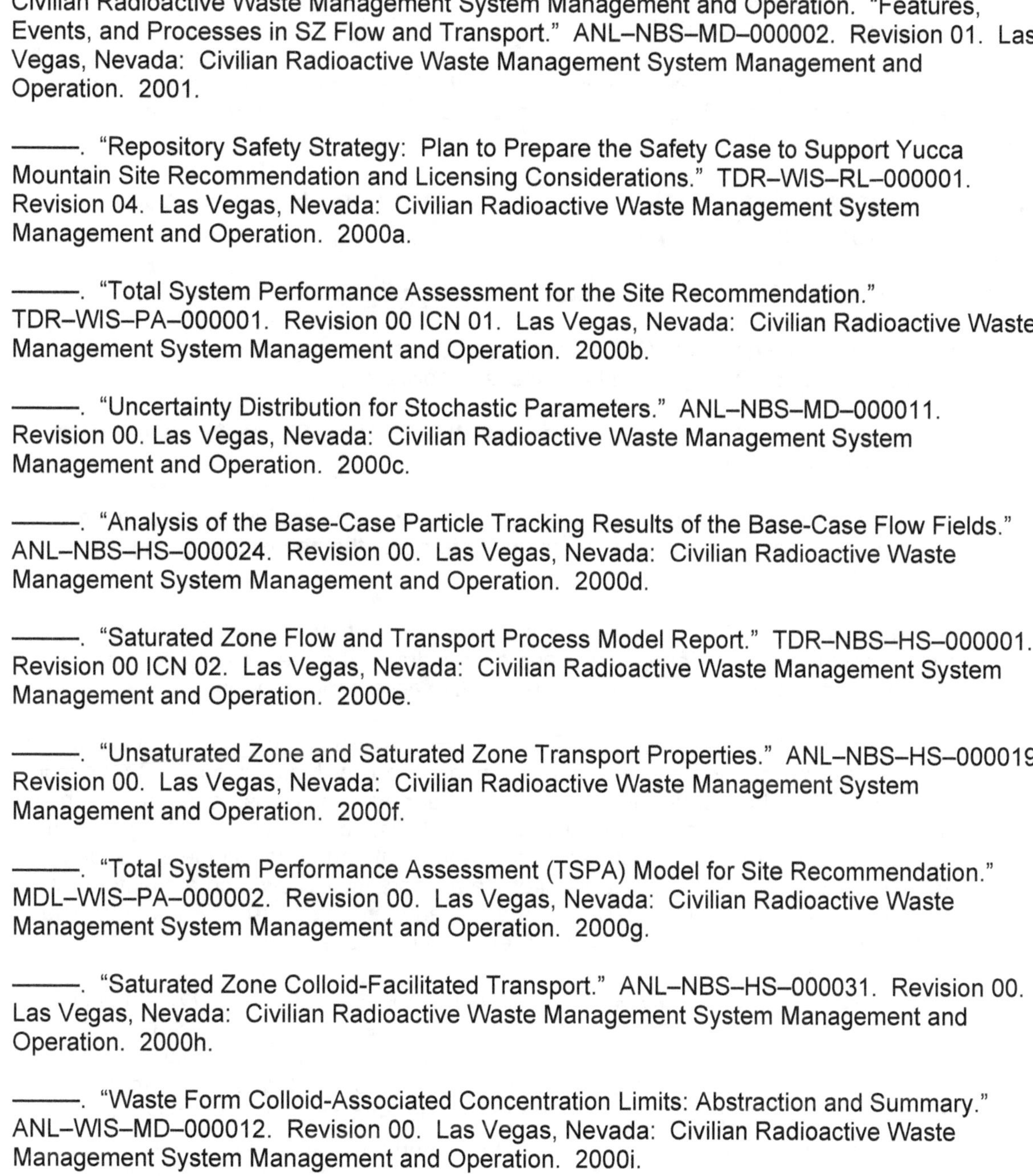

Civilian Radioactive Waste Management System Management and Operation. “Features, Events, and Processes in SZ Flow and Transport.” ANL–NBS–MD–000002. Revision 01. Las Vegas, Nevada: Civilian Radioactive Waste Management System Management and Operation. 2001.

———. “Repository Safety Strategy: Plan to Prepare the Safety Case to Support Yucca Mountain Site Recommendation and Licensing Considerations.” TDR–WIS–RL–000001. Revision 04. Las Vegas, Nevada: Civilian Radioactive Waste Management System Management and Operation. 2000a.

———. “Total System Performance Assessment for the Site Recommendation.” TDR–WIS–PA–000001. Revision 00 ICN 01. Las Vegas, Nevada: Civilian Radioactive Waste Management System Management and Operation. 2000b.

———. “Uncertainty Distribution for Stochastic Parameters.” ANL–NBS–MD–000011. Revision 00. Las Vegas, Nevada: Civilian Radioactive Waste Management System Management and Operation. 2000c.

———. “Analysis of the Base-Case Particle Tracking Results of the Base-Case Flow Fields.” ANL–NBS–HS–000024. Revision 00. Las Vegas, Nevada: Civilian Radioactive Waste Management System Management and Operation. 2000d.

———. “Saturated Zone Flow and Transport Process Model Report.” TDR–NBS–HS–000001. Revision 00 ICN 02. Las Vegas, Nevada: Civilian Radioactive Waste Management System Management and Operation. 2000e.

———. “Unsaturated Zone and Saturated Zone Transport Properties.” ANL–NBS–HS–000019. Revision 00. Las Vegas, Nevada: Civilian Radioactive Waste Management System Management and Operation. 2000f.

———. “Total System Performance Assessment (TSPA) Model for Site Recommendation.” MDL–WIS–PA–000002. Revision 00. Las Vegas, Nevada: Civilian Radioactive Waste Management System Management and Operation. 2000g.

———. “Saturated Zone Colloid-Facilitated Transport.” ANL–NBS–HS–000031. Revision 00. Las Vegas, Nevada: Civilian Radioactive Waste Management System Management and Operation. 2000h.

———. “Waste Form Colloid-Associated Concentration Limits: Abstraction and Summary.” ANL–WIS–MD–000012. Revision 00. Las Vegas, Nevada: Civilian Radioactive Waste Management System Management and Operation. 2000i.

U.S. Nuclear Regulatory Commission. NUREG–1746, "System-Level Repository Sensitivity Analyses Using TPA Version 3.2 Code." Washington, DC: U.S. Nuclear Regulatory Commission. August 2001.

———. NUREG–1668, "NRC Sensitivity and Uncertainty Analyses for a Proposed HLW Repository at Yucca Mountain, Nevada, Using TPA 3.1." Washington, DC: U.S. Nuclear Regulatory Commission. March 1999.

———. NUREG–1563, "Branch Technical Position on the Use of Expert Elicitation in the High-Level Radioactive Waste Program." Washington, DC: U.S. Nuclear Regulatory Commission. November 1996.

A1.3 Developing a Risk-Informed, Performance-Based Yucca Mountain Review Plan

The Yucca Mountain Review Plan incorporates the following four principles:

(1) The U.S. Nuclear Regulatory Commission defends its licensing decision, while the U.S. Department of Energy defends its license application;

(2) The Yucca Mountain Review Plan implements 10 CFR Part 63, a performance-based and site-specific rule;

(3) The Yucca Mountain Review Plan will be consistent with the applicable regulations and the review that the staff needs to complete to make the necessary findings on safety; and

(4) The Yucca Mountain Review Plan incorporates the more than 15 years of knowledge gained about the Yucca Mountain site and design during the prelicensing period and avoids the imposition of unnecessarily prescriptive acceptance criteria.

To support review of the U.S. Department of Energy safety analysis report, these principles are reflected in five major Yucca Mountain Review Plan subsections within Section 2: (i) repository safety before permanent closure; (ii) repository safety after permanent closure; (iii) research and development program to resolve safety questions; (iv) performance confirmation program; and (v) administrative and programmatic requirements. Subordinate portions include this appendix, an appendix providing guidance for the conduct of the acceptance review, and a chapter that supports review of compliance with general information requirements in 10 CFR Part 63. The structure of the Yucca Mountain Review Plan is presented in Figure A1–6.

The preclosure and postclosure safety reviews will focus on whether the U.S. Department of Energy safety analysis report demonstrates, with reasonable assurance for the preclosure period and reasonable expectation for the postclosure period, that the corresponding performance objectives at 10 CFR Part 63 will be met. U.S. Nuclear Regulatory Commission staff is using a total system approach for both the preclosure and postclosure safety reviews that takes advantage of the knowledge of the site and design that has accumulated during the prelicensing period and the rapid growth in preclosure safety analysis and performance assessment capabilities. These improvements in capability include the results of performance assessment work by the U.S. Nuclear Regulatory Commission and industry, and reviews of the

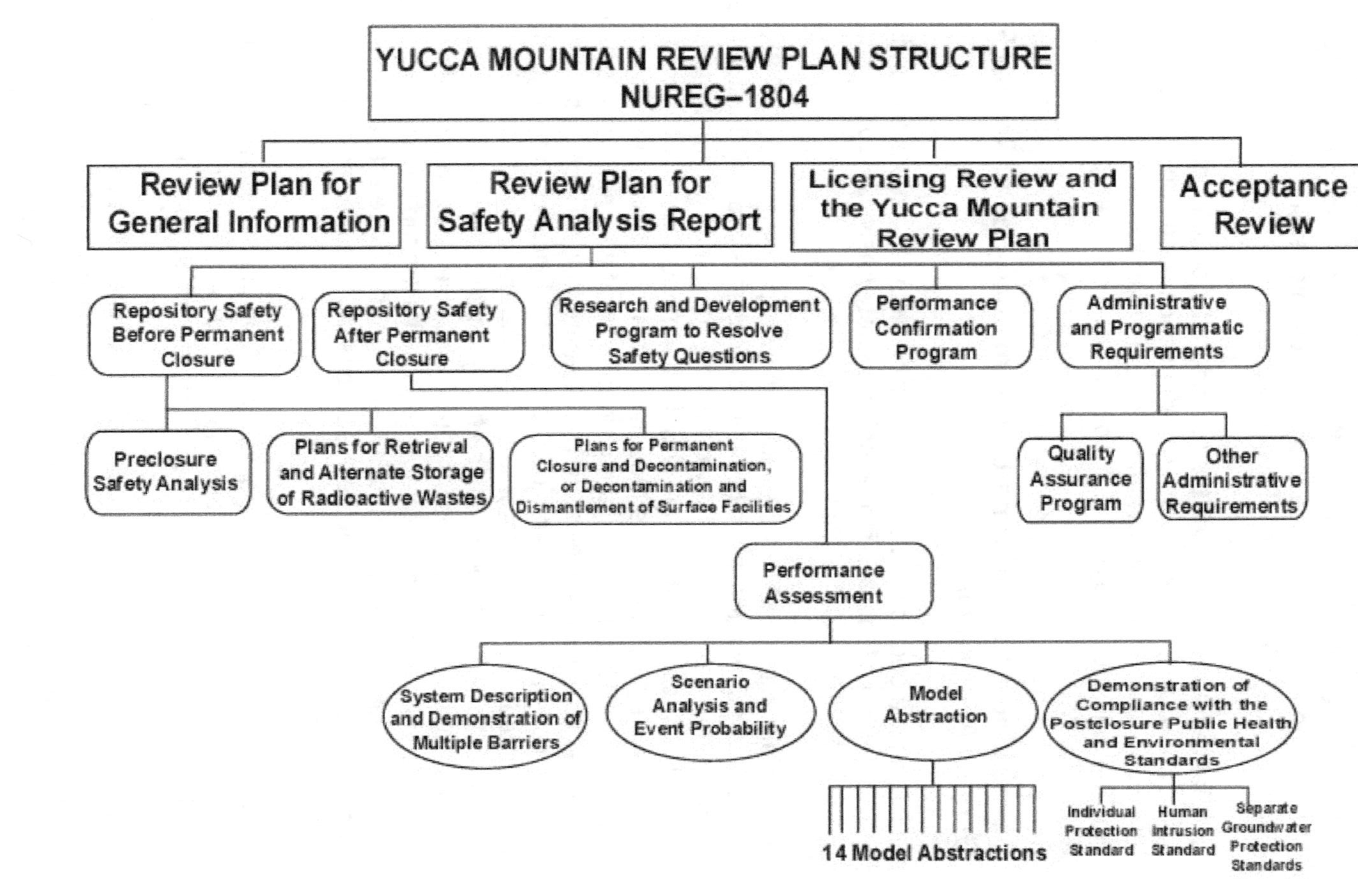

A1–6. Structure of the Yucca Mountain Review Plan

U.S. Department of Energy performance assessments for Yucca Mountain. This total system approach facilitates integration of the technical disciplines required to review a Yucca Mountain license application. The Yucca Mountain Review Plan uses existing U.S. Nuclear Regulatory Commission guidance from other regulatory programs that is applicable to the construction and operation of a geologic repository, modifying it as necessary for consistency with the risk-informed, performance-based philosophy and with the nature and risk of proposed repository operations. The approaches used to develop each of the major Yucca Mountain Review Plan sections are described in the following six subsections.

Appendix B, "Acceptance Review," provides the procedure for conducting the acceptance review of the license application. Risk-informed, performance-based principles are not incorporated in this appendix. The review verifies only that the information in the license application is complete, and therefore does not require a risk-informed, performance-based approach.

A1.3.1 Developing a Risk-Informed, Performance-Based Review Plan for General Information

Any license application for a geologic repository at Yucca Mountain, submitted under 10 CFR Part 63, is to contain "General Information," a "Safety Analysis Report," and is to be accompanied by a final environmental impact statement.

Chapter 1, "General Information," reviews the requirements specified in 10 CFR 63.21(b). The intent of including general information in the license application is twofold. First, it allows the U.S. Department of Energy to provide an overview of its engineering design concept for the repository (Section 1.1). Second, it allows the U.S. Department of Energy to demonstrate its understanding of what aspects of the Yucca Mountain site and its environs (Section 1.5) influence health and safety. Understanding the performance of the design in the context of the Yucca Mountain site and its environs allows the U.S. Department of Energy to make risk-informed, performance-based judgments regarding compliance with the regulations, which are subsequently evaluated by the staff elsewhere in the Safety Analysis Report (Chapter 2). Accordingly, the material to be reviewed by the staff is generally informational in nature, with the more detailed technical discussions and descriptions found elsewhere in the Safety Analysis Report section of the license application. Notable exceptions are the information found in Sections 1.2, 1.3, and 1.4 of the Yucca Mountain Review Plan. There are five sections in Chapter 1, and the extent to which each of these sections incorporates risk-informed, performance-based principles varies.

Section 1.1, "General Description," provides for review of a general description of the geologic repository operations area, including its major structures, systems, and components. The material in this section is generally informational in nature, comparable to that typically found in an executive summary, and no detailed technical analysis is required by the reviewer. The detailed review of information covered by these subjects will be conducted in other sections of the Yucca Mountain Review Plan. Because the geologic repository operations area design is generally presented in the context of how compliance with the performance objectives will be achieved, this section of the license application is implicitly risk informed and performance based.

Section 1.2, "Proposed Schedules for Construction, Receipt, and Emplacement of Waste," provides for review of general schedules for various phases of repository construction and operation. Again, the material to be reviewed is informational in nature, and no detailed technical analysis is required by the reviewer. Because the geologic repository operations area design is generally presented in the context of how compliance with the performance objectives will be achieved, this section of the license application is implicitly risk informed and performance based, and the staff's subsequent review will be likewise.

Section 1.3, "Physical Protection Plan," provides for a review to determine with reasonable assurance that the U.S. Department of Energy will have an adequate physical protection system. The system must provide assurance that activities at a high-level waste repository do not constitute an unreasonable risk to the public health and safety. General and specific performance objectives for the U.S. Department of Energy to meet are listed in this section. The physical protection system should be designed to protect against a loss of control of the geologic repository operations area that could be sufficient to cause radiation exposure exceeding the dose defined in 10 CFR 72.106. Physical protection requirements for high-level radioactive waste at a geologic repository operations area are codified under 10 CFR 73.51. This section is risk informed and performance based.

Section 1.4 provides for a review of the Material Control and Accounting Program plan submitted by the U.S. Department of Energy. The plan describes how the system will be established, implemented, and maintained to ensure that it is adequate to protect against, detect, and respond to the loss of high-level radioactive waste. Material control and accounting for high-level radioactive waste are required by 10 CFR 63.21(b)(4) and stipulated in 10 CFR 63.78. This section provides for a risk-informed, performance-based review of the U.S. Department of Energy program and its capability to meet the requirements in 10 CFR 63.78. High priority will be given to the overall system detection and resolution capabilities at an implementation level.

Section 1.5, "Description of Site Characterization Work," provides for review of an overview description of the site characterization work conducted up to the time of license application, and the results of that work, necessary to support the license application. The material to be reviewed is generally informational in nature and is intended to place the geologic repository operations area in the context of the Yucca Mountain site and environs. Although there are no performance objectives addressed in this section of the license application, the information summaries provided in this section support the detailed safety reviews conducted in the preclosure and postclosure safety evaluation sections of the Yucca Mountain Review Plan. Therefore, the adequacy and sufficiency of site characterization activities and the resulting information will be judged, not in this section of the license application, but in the context of the compliance demonstrations and supporting technical bases provided in the "Safety Analysis Report" section of the license application. Therefore, the information contained in this section of the license application is implicitly risk-informed and performance-based, and the staff's subsequent review will be likewise.

A1.3.2 Developing a Risk-Informed, Performance-Based Review Plan for Repository Safety Before Permanent Closure

Section 2.1.1, "Preclosure Safety Analysis," provides for review of compliance with the performance objectives, in 10 CFR Part 63, which are based on permissible levels of doses to workers and the public established on the basis of acceptable levels of risk. 10 CFR 63.21(c)(5) requires a preclosure safety analysis of the geologic repository operations area for the period before permanent closure, to verify compliance with the performance objectives. Preclosure safety analysis is a systematic examination of the site, the design, the potential hazards and initiating events and their consequences, and the potential dose consequences to workers and the public. Preclosure safety analysis considers the probability of potential hazards, taking into account the range of uncertainty associated with the data that support the probability calculations. Event sequences are defined based on well-established (discipline-specific) methodologies that allow a combination of probabilistic and deterministic estimates. Sequences of human-induced and natural events are used as inputs to calculate consequences of potential failures of structures, systems, and components in terms of doses to workers and the public. These calculated doses are compared to allowable doses in establishing the importance of structures, systems, and components. The structures, systems, and components that must be functional to comply with the performance objective dose limits are identified as structures, systems, and components important to safety. It is the responsibility of the U.S. Department of Energy to specify these structures, systems, and components. Preclosure safety analysis also identifies and describes the controls that are relied on to prevent potential event sequences from occurring or to mitigate their consequences, and identifies measures taken to ensure the availability of the safety systems. The end products of the preclosure safety analysis are a list of structures, systems, and components important to safety (also known as the Q-List) and the associated design criteria and technical specifications necessary to keep them functional and to meet the performance objectives. The structures, systems, and components important to safety may also be further categorized, based on relative safety significance, using risk information from the preclosure safety analysis. This distinction may be used to focus the requirements of design details and the application of quality assurance controls through a graded quality assurance program. The Yucca Mountain Review Plan provides criteria appropriate to evaluate the U.S. Department of Energy technical basis for categorizing structures, systems, and components and grading quality assurance requirements.

The staff review is focused on items that preclosure safety analysis determines to be important to safety. The rigor of review for the design items on the Q-List and the level of attention to detail depend on relative safety significance. No prescriptive design criteria are imposed in the Yucca Mountain Review Plan, because 10 CFR Part 63 allows the U.S. Department of Energy to develop the design criteria and demonstrate their appropriateness. Thus, the U.S. Department of Energy has flexibility to use any codes, standards, and methodologies it demonstrates to be applicable and appropriate. The risk-informed review process in the Yucca Mountain Review Plan focuses on determining compliance with performance objectives, as demonstrated by the U.S. Department of Energy preclosure safety analysis. In summary, the review philosophy is based on the following premises: (i) the U.S. Department of Energy must demonstrate, through its preclosure safety analysis, that the repository will be designed, constructed, and operated to meet the specified exposure limits (performance objectives) throughout the preclosure period; (ii) the staff must focus the review on the design of the structures, systems, and components important to safety, in the context of the design's ability to meet the performance objectives; and finally (iii) the staff resources will be focused proportionately on the inspection and review of high-risk significant structures, systems, and components important to safety.

Section 2.1.2, "Plans for Retrieval and Alternate Storage of Radioactive Wastes," contains the performance objectives specified at 10 CFR Part 63. Review methods and acceptance criteria were developed from the associated regulatory requirements in 10 CFR Part 63. Specific emphasis is placed on allowing the U.S. Department of Energy flexibility in demonstrating compliance, which is a performance-based approach. This section is risk informed because the option is preserved to retrieve waste throughout the period that wastes are being emplaced, until the completion of a performance confirmation program.

Section 2.1.3, "Plans for Permanent Closure and Decontamination and Dismantlement of Surface Facilities," identifies two areas of review: (i) the description of design considerations intended to facilitate closure and decontamination, and (ii) the plans for permanent closure and decontamination. The acceptance criteria emphasize a description of the features incorporated into the design that may facilitate closure. The section makes reference to the Nuclear Material Safety and Safeguards decommissioning plans, which are also consistent with risk-informed, performance-based regulation, only to the extent that the U.S. Department of Energy may have information related to closure and decontamination available at the time of license application submittal. The Yucca Mountain Review Plan explicitly acknowledges that information submitted by the U.S. Department of Energy in the license application, regarding closure and decontamination, will be preliminary.

A1.3.3 Developing a Risk-Informed, Performance-Based Review Plan for Repository Safety After Permanent Closure

Section 2.2, "Repository Safety after Permanent Closure," provides for a risk-informed, performance-based review of the U.S. Department of Energy performance assessment. The performance assessment quantifies repository performance to demonstrate reasonable expectation of compliance with the postclosure public health and environmental standards at 10 CFR Part 63, Subpart L. The U.S. Department of Energy performance assessment is a systematic analysis that answers the three risk questions: what can happen?; how likely is it to happen?; and what are the consequences? The performance assessment and analyses must focus on reasonable parameter distributions, as stated in 10 CFR 63.304(4). The Yucca Mountain performance assessment is a sophisticated analysis that involves various complex considerations and evaluations. Examples include evolution of the natural environment, degradation of engineered barriers over a 10,000-year period, and disruptive events such as seismicity and volcanism. The staff will also consider the technical support for models and parameters of the performance assessment, based on detailed process models, laboratory and field experiments, and natural analogs. Because the performance assessment encompasses such a broad range of issues, the staff will use risk information throughout the review process. Using risk information will ensure that the review focuses on those items most important to health and safety.

Section 2.2.1 requires the staff to apply risk information throughout the review of the performance assessment. First, the staff reviews the barriers important to waste isolation in Section 2.2.1.1. The U.S. Department of Energy must identify the important barriers (engineered and natural) for the performance assessment, describe each barrier's capability, and provide the technical basis for that capability. This risk information includes the U.S. Department of Energy understanding of each barrier's importance. Staff review of the U.S. Department of Energy barrier analysis considers risk insights from previous performance assessments conducted for the Yucca Mountain site, detailed process-level modeling efforts, laboratory and field experiments, and natural analog studies. The result of this review is a staff understanding of each barrier's importance to waste isolation, which focuses the reviews conducted in Sections 2.2.1.2, "Scenario Analysis and Event Probability" and 2.2.1.3, "Model Abstraction." However, it is the responsibility of the U.S. Department of Energy to specify the natural and engineered barriers important to waste isolation.

Scenario analysis and model abstraction are key aspects of the performance assessment. The risk information drawn from the review of the multiple barriers section will direct the staff review to those topics, within scenario analysis and model abstraction, that are important to waste isolation. Section 2.2.1.2 provides the review methods and acceptance criteria for scenarios for both nominal and disruptive events. An acceptable scenario selection method includes identification and screening of features, events, and processes; and construction of scenarios from the retained features, events, and processes considered at the Yucca Mountain site. Then, abstracted models used in the performance assessment for the retained scenarios will be reviewed. The performance assessment review focuses on the 14 model abstractions in Section 2.2.1.3. These model abstractions are derived from those aspects of the engineered, geosphere, and biosphere subsystems shown to be most important to waste isolation, based on prior performance assessments, knowledge of site characteristics, and repository design. Figure A1–5 presents these model abstractions and their relation to subsystem components. The staff developed each of the 14 sections in substantial detail, allowing for a detailed review. However, it is unlikely that each of the abstractions will have the same risk significance. The staff will review the abstractions according to their risk significance as determined in the multiple barrier review. Nevertheless, until the U.S. Department of Energy submits a license application, the review plan sections associated with model abstractions must remain flexible, and in substantial detail, so that the U.S. Department of Energy will understand how the U.S. Nuclear Regulatory Commission will review the abstractions. After the staff completes the review of scenarios and model abstractions, it will update, as necessary, its assessment of the U.S. Department of Energy barrier analysis.

The staff will use the 14 model abstractions in Section 2.2.1.3 to determine compliance with 10 CFR 63.113 and 63.114. The abstractions consider the engineered, geosphere, and biosphere subsystems that may be important to waste isolation. Important to waste isolation means important to meeting the performance objectives specified in 10 CFR 63.113. The staff will focus its review to understand the importance to health and safety of the various assumptions, models, and data in the performance assessment. The staff will also focus its review to verify that the degree of technical support for models and data abstractions is appropriate for their contribution to risk. This means the staff will review each model abstraction to a detail level suitable for the degree to which the U.S. Department of Energy relies on it to demonstrate compliance with the performance objectives. In the multiple barrier review, the staff will evaluate the capability of the barriers. For example, if the U.S. Department of Energy relies on the unsaturated zone to provide significant delay in the transport of radionuclides and/or reduce radionuclide concentrations to the reasonably maximally exposed individual, then the staff will perform a detailed review of this abstraction. However, if the U.S. Department of

Energy shows that this abstraction has a minor impact on the delay of radionuclide transport to the reasonably maximally exposed individual, then the staff will conduct a simplified review focusing on the bounding assumptions. The staff will use the review methods and acceptance criteria in these sections to decide whether the U.S. Department of Energy properly characterized and factored the features, events, and processes into the performance assessment. This is necessary to decide whether the U.S. Department of Energy performance assessment is acceptable and complies with 10 CFR 63.114. The review methods and acceptance criteria the staff will use to evaluate compliance with the postclosure public health and environmental standards are in Section 2.2.1.4 of the Yucca Mountain Review Plan.

Section 2.2.1.4, "Demonstration of Compliance with the Postclosure Public Health and Environmental Standards," focuses on the role of the performance assessment to demonstrate that the performance objectives have been met with reasonable expectation. This is where the probability estimates from Section 2.2.1.2, "Scenario Analysis and Event Probability," and consequence estimates from model abstractions are combined to form the risk estimate for the repository. It includes reasonable expectation of compliance with the postclosure individual protection standard, the human intrusion standard, and the ground-water protection standards. Consideration is given to parameter uncertainty and alternate conceptual models.

A1.3.4 Developing a Risk-Informed, Performance-Based Review Plan for the Research and Development Program to Resolve Safety Questions

Section 2.3 provides for a review of the "Research and Development Program to Resolve Safety Questions." The program applies to structures, systems, and components important to safety and engineered or natural barriers important to waste isolation. The program identifies, describes, and discusses those safety features or components for which further technical information is required to confirm the adequacy of design. This section is performance-based because it focuses on those items most important to safety and waste isolation.

A1.3.5 Developing a Risk-Informed, Performance-Based Review Plan for the Performance Confirmation Program

Section 2.4 provides for a review of the "Performance Confirmation Program." The program is comprised of tests, experiments, and analyses conducted to evaluate the adequacy of the information used to demonstrate compliance with the performance objectives. The need for a performance confirmation program is unique to the high-level radioactive waste program. This uniqueness reflects the uncertainties in estimating geologic repository performance over 10,000 years. The bases for the acceptance criteria are the requirements for performance confirmation, in 10 CFR Part 63, that are performance based. Where suitable, the acceptance criteria are also risk-informed because performance confirmation focuses on those parameters and natural and engineered barriers already identified to be important to health and safety.

A1.3.6 Development of the Administrative and Programmatic Requirements Section

This portion of the Yucca Mountain Review Plan is the most difficult for which to implement a risk-informed, performance-based approach. No performance objectives are provided in 10 CFR Part 63 for this section. Existing U.S. Nuclear Regulatory Commission regulatory guidance and standard review plans were examined for examples of appropriate review methods and acceptance criteria that could be incorporated in the Yucca Mountain Review Plan. However, some of these examples were greatly prescriptive, while others seemed inadequate, based on our knowledge of expected repository operations and administrative programs. This situation is complicated by the unique nature of the high-level radioactive waste regulatory program and the lack of an operational history, or historical performance results, such as are available for most other types of nuclear facilities. To the extent possible, acceptance criteria and review methods for this section of the Yucca Mountain Review Plan are based on similar existing and successful U.S. Nuclear Regulatory Commission regulatory programs, considering expected operations and associated risks, while taking advantage of opportunities to omit prescriptive requirements, when appropriate.

The quality assurance section of the Yucca Mountain Review Plan is risk informed, explicitly as a result of the application of a graded quality assurance program. The review methods and acceptance criteria are written either to accommodate such a graded program or to support review of a nongraded program. The quality assurance section provides for quality assurance controls to be applied in a graded manner, based on the safety-risk-significance of the structures, systems, and components and the barriers important to safety or waste isolation. These quality assurance control provisions are intended to be applied to high-safety-risk-significant structures, systems, and components, and barriers and their related activities. The U.S. Department of Energy may propose reduced quality assurance requirements for low-safety-risk- significant structures, systems, and components, barriers, and their related activities. The quality assurance section also contains many review provisions for areas such as quality assurance for scientific investigations, software, and commercial-grade item dedication. The quality assurance section of the Yucca Mountain Review Plan is performance-based as a result of allowing the U.S. Department of Energy to concentrate its quality assurance activities on high-safety-risk-significant items and activities. 10 CFR Part 63 specifically requires that the quality assurance program be prescriptive by describing how the quality assurance requirements will be satisfied. The prescriptive requirements for the quality assurance program contained in this regulation are similar to regulatory requirements contained in 10 CFR Parts 50, 70, 71, and 72. Thus, the quality assurance section of the Yucca Mountain Review Plan contains prescriptive review provisions that are intended to be applied to high-safety-risk-significant items and activities.

The other administrative and programmatic sections in the Yucca Mountain Review Plan are nonprescriptive, providing flexible acceptance criteria and review methods and referring the reviewer to other U.S. Nuclear Regulatory Commission guidance documents, but not specifying the standards or practices the U.S. Department of Energy must use for compliance demonstration. Rather, these sections require the U.S. Department of Energy to: (i) identify any standards, programs, and procedures that will be used; (ii) demonstrate that those standards, programs, and procedures are appropriate; and (iii) implement them properly. The acceptance criteria and review methods require the staff to evaluate the administrative and programmatic sections of the U.S. Department of Energy license application, based on the validity and adequacy of the basis that the U.S. Department of Energy has presented in the application.

In developing this section of the Yucca Mountain Review Plan, there has been a specific effort to implement a risk-informed, performance-based philosophy based on current U.S. Nuclear

Regulatory Commission guidance. For example, "Emergency Planning," Section 2.5.7, assesses several items that represent the frequency and consequence components of risk. Each acceptance criterion in "Emergency Planning" has measurable and inspectable performance requirements. Information provided in the administrative and programmatic sections is based, to the extent possible, on prelicensing interactions. This is especially true for quality assurance. In most cases, however, the U.S. Department of Energy has not prepared specific administrative and programmatic procedures, and the level of detail in the Yucca Mountain Review Plan is minimal. U.S. Nuclear Regulatory Commission guidance is identified in the Yucca Mountain Review Plan, but selection of the compliance demonstration approach is left to the U.S. Department of Energy.

A1.4 Components of Each Review Section

Each Yucca Mountain Review Plan section provides the complete procedures and acceptance criteria for all areas of review pertinent to that section. Because the U.S. Nuclear Regulatory Commission is implementing a risk-informed, performance-based regulatory approach using risk insights, the staff reviewer may select and emphasize particular aspects from each Yucca Mountain Review Plan section, as appropriate. Consequently, in the review of the license application, the staff may not carry out in detail all the review steps listed in each Yucca Mountain Review Plan section. In some cases, the staff may rely on a more detailed evaluation made in the prelicensing consultative phase of the program. Thus, the staff may be able to use the technical understanding and basis for issue resolution developed during prelicensing to help focus its review on areas where a more detailed, prelicensing consultative review was not done, as appropriate. U.S. Nuclear Regulatory Commission staff are experienced in using this approach.

Each section of a U.S. Nuclear Regulatory Commission review plan typically contains areas of review, review methods, acceptance criteria, evaluation findings, and references.

Areas of Review Subsection

This subsection identifies the topical areas and defines the scope for the reviews. Having this scope in mind enables the reviewer to prepare for the review, including examining any technical or regulatory background material necessary to support the review.

Review Methods Subsection

The review methods provide the specific step-by-step procedures that the reviewer will use to assess compliance with regulatory requirements. The review methods are often technically specific, but their level of detail and complexity are determined by the particular regulatory requirements.

Acceptance Criteria Subsection

This subsection delineates criteria that can be applied by the reviewer to determine the acceptability of the applicant's compliance demonstration. The technical bases for these criteria have been derived from 10 CFR Part 63; the U.S. Nuclear Regulatory Commission regulatory guides; general design criteria; codes and standards; branch technical positions; standard testing methods (e.g., American Society for Testing and Materials standards); technical papers; and other similar sources. These sources typically include solutions and approaches previously determined to be acceptable by the staff for making compliance determinations for the specific area of review, or are based on the staff work from its first-of-a-kind reviews related to a high-level radioactive waste repository, such as the postclosure performance assessment.

The acceptance criteria have been defined so that staff reviewers can use consistent and well-documented approaches from prelicensing consultation to support the review of the license application, as appropriate. Flexibility is provided to enable the U.S. Department of Energy to implement the type of operations appropriate for the geologic repository operations area. The U.S. Department of Energy may take approaches, to demonstrating compliance, that are different from those presented in the Yucca Mountain Review Plan, as long as the U.S. Department of Energy can otherwise comply with the applicable regulations. However, the U.S. Department of Energy should recognize that, as is the case for all regulatory guidance, substantial staff time and effort have gone into the development of the review methods and acceptance criteria in the Yucca Mountain Review Plan as part of efforts to meet the 3-year statutory deadline for a Commission review on whether to issue a construction authorization. Thus, if the U.S. Department of Energy uses approaches other than those described in the Yucca Mountain Review Plan, it could result in longer review times and an increase in the number of U.S. Nuclear Regulatory Commission requests for additional information.

Evaluation Findings Subsection

This subsection presents general conclusions and findings of the staff that result from review of each area of the application as well as an identification of the applicable regulatory requirements. A conclusion is included in the safety evaluation report for each Yucca Mountain Review Plan section. The safety evaluation report contains a description of the review; the basis for the staff findings, including where the facility design or the applicant programs deviate from requirements; and the evaluation findings. An example of how the reviewer can document the evaluation findings is provided in each review section.

References Subsection

The references subsections of the review plan list any references used in the development of the Yucca Mountain Review Plan. Often, the U.S. Nuclear Regulatory Commission review plans reference more detailed information to support review methods, rather than reproducing detailed technical procedures or specifications within the review plan.

Yucca Mountain Review Plan Updates

The Yucca Mountain Review Plan will be revised and updated periodically, as the need arises, to clarify the content or correct errors and to incorporate modifications approved by the U.S. Nuclear Regulatory Commission. As noted above, such modifications could also result from revisions in the U.S. Nuclear Regulatory Commission regulations or requirements, following the normal public rulemaking process.

A1.5 References

National Research Council. "Technical Bases for Yucca Mountain Standards." Washington, DC: National Academy Press. 1995.

U.S. Nuclear Regulatory Commission. NUREG–1614, "FY2000–2005 Strategic Plan." Washington, DC: U.S. Nuclear Regulatory Commission. September 2000.

———. SECY–99–100. "Commission White Paper on Risk-Informed and Performance-Based Regulation." Washington, DC: U.S. Government Printing Office. March 11, 1999.

APPENDIX B

ACCEPTANCE REVIEW

B1 ACCEPTANCE REVIEW

B1.1 Description and Purpose of Acceptance Review

The staff will conduct an acceptance review of the license application to determine whether the license application is complete in accordance with 10 CFR 2.101(f). The reviewer will evaluate whether the information is sufficient to support a detailed review, and will assess the schedule for any later U.S. Nuclear Regulatory Commission milestones. The license application will be acceptable to docket if the information is complete in scope and detail. The docketing acceptance review does not determine the technical adequacy of the submitted information. That determination will be made during the follow-on detailed technical review that occurs once the license application is accepted and docketed.

If the license application passes the acceptance review, the application will be docketed, and the detailed technical review would begin. If the license application fails the acceptance review (for example, it is incomplete and there is insufficient information to support the detailed licensing review), the license application would be rejected and returned to the U.S. Department of Energy, or the deficiencies identified and additional information requested. Examples of ways an application may be found deficient in an acceptance review include omitted sections, illegible figures, missing analyses, or failure to submit required documents.

The acceptance review is an internal staff action that determines whether the application can be docketed. If the application is docketed, then a notice will be published in the *Federal Register* offering an opportunity for hearing and public participation in the licensing process. If deficiencies are limited, the staff could proceed with a detailed licensing review while awaiting additional specific information from the applicant. The U.S. Nuclear Regulatory Commission staff decision, at the acceptance review stage, to accept or reject an application would be made in writing 90 days of receipt of the application and would be based on consideration of the submitted information and the importance of the missing information for beginning the detailed technical review.

B1.2 Acceptance Review Checklist

The staff will conduct the acceptance review using a checklist, based on the structure of 10 CFR 63.21 ("Content of Application"). If the format of the U.S. Department of Energy license application structure differs from that of the Yucca Mountain Review Plan, the U.S. Department of Energy may provide a table that relates the sections of the license application to the regulatory requirements in 10 CFR Part 63.

To conduct the acceptance review, staff will apply extensive knowledge about the proposed Yucca Mountain repository developed during prelicensing and will specifically compare the contents of the license application with the requirements in 10 CFR 63.21 ("Content of Application"). The acceptance review will include an assessment of the legibility of drawings, the general adequacy of information, any proprietary information, and obvious technical inadequacies. Most license application sections incorporate multidisciplinary input. Therefore, the staff will conduct the acceptance review using teams of individuals from suitable disciplines. During the acceptance review, the staff will determine whether the U.S. Department of Energy has provided, in sufficient scope and detail, the following items that 10 CFR 63.21 requires in light of reasonably available information. The staff uses a simple scale of acceptability to help the reviewers document their results.

(1) A general description of the proposed geologic repository at the Yucca Mountain site. This description will identify the geologic repository operations area location, the general character of the proposed activities, and the basis for the U.S. Nuclear Regulatory Commission to exercise licensing authority.

- ☐ Accept for Review
- ☐ Accept, but Request for Additional Information Prepared
- ☐ Reject, Inadequate to Support Detailed Review

(2) Proposed schedules to build, receive waste, and emplace wastes at the geologic repository operations area.

- ☐ Accept for Review
- ☐ Accept, but Request for Additional Information Prepared
- ☐ Reject, Inadequate to Support Detailed Review

(3) A description of the detailed security measures for the physical protection of high-level radioactive waste. This plan must include the design for physical protection, the licensee's safeguards contingency plan, and the training and qualification plan for the security organization. The plan must list tests, inspections, audits, and other means to show compliance.

- ☐ Accept for Review
- ☐ Accept, but Request for Additional Information Prepared
- ☐ Reject, Inadequate to Support Detailed Review

(4) A description of the material control and accounting program.

- ☐ Accept for Review
- ☐ Accept, but Request for Additional Information Prepared
- ☐ Reject, Inadequate to Support Detailed Review

(5) A description of work conducted to characterize the Yucca Mountain site.

- ☐ Accept for Review
- ☐ Accept, but Request for Additional Information Prepared

☐ Reject, Inadequate to Support Detailed Review

(6) A description of the Yucca Mountain site, with appropriate attention to those features, events, and processes of the site that might affect the design of the geologic repository operations area and the performance of the geologic repository. The site description should include information about features, events, and processes outside the site to the extent the information is relevant and material to safety or performance of the geologic repository. The site description should include:

(a) Location of the geologic repository operations area with respect to the site boundary;

(b) Information about the geology, hydrology, and geochemistry of the site, including geomechanical properties and conditions of the host rock;

(c) Information about the surface-water hydrology, climatology, and meteorology of the site; and

(d) Information about the location of the reasonably maximally exposed individual and local human behaviors and characteristics, as needed, to select conceptual models and parameters used to define the reference biosphere and the reasonably maximally exposed individual.

☐ Accept for Review

☐ Accept, but Request for Additional Information Prepared

☐ Reject, Inadequate to Support Detailed Review

(7) Information relative to materials of construction of the geologic repository operations area (including geologic media, general arrangement, and approximate dimensions), and codes and standards that the U.S. Department of Energy proposes to apply to the design and construction of the geologic repository operations area.

☐ Accept for Review

☐ Accept, but Request for Additional Information Prepared

☐ Reject, Inadequate to Support Detailed Review

(8) A description and discussion of the design of the various components of the geologic repository operations area and the engineered barrier system, including:

(a) Dimensions, material properties, specifications, analytical and design methods used, and any applicable codes and standards;

(b) Design criteria and their relation to the preclosure and postclosure performance objectives; and

(c) Design bases and their relation to the design criteria.

- ☐ Accept for Review
- ☐ Accept, but Request for Additional Information Prepared
- ☐ Reject, Inadequate to Support Detailed Review

(9) A description of the kind, amount, and specifications of the radioactive material proposed for receipt and possession at the geologic repository operations area.

- ☐ Accept for Review
- ☐ Accept, but Request for Additional Information Prepared
- ☐ Reject, Inadequate to Support Detailed Review

(10) A preclosure safety analysis of the geologic repository operations area, for the period before permanent closure, that assumes that operations will be carried out at the maximum capacity and rate of receipt of radioactive waste stated in the license application.

- ☐ Accept for Review
- ☐ Accept, but Request for Additional Information Prepared
- ☐ Reject, Inadequate to Support Detailed Review

(11) A description of the program for control and monitoring of radioactive effluents and occupational radiological exposures to maintain such effluents and exposures in accordance with the preclosure performance objectives.

- ☐ Accept for Review
- ☐ Accept, but Request for Additional Information Prepared
- ☐ Reject, Inadequate to Support Detailed Review

(12) A description of plans for retrieval and alternate storage of the radioactive wastes.

- ☐ Accept for Review
- ☐ Accept, but Request for Additional Information Prepared
- ☐ Reject, Inadequate to Support Detailed Review

(13) A description of design considerations that are intended to facilitate permanent closure and decontamination or decontamination and dismantlement of surface facilities.

- ☐ Accept for Review
- ☐ Accept, but Request for Additional Information Prepared
- ☐ Reject, Inadequate to Support Detailed Review

(14) An assessment of the degree to which features, events and processes expected to materially affect compliance with the postclosure performance objectives have been characterized and the extent to which they affect waste isolation. Investigations must extend from the surface to a depth sufficient to determine principal pathways for radionuclide migration. Specific features events and processes must be investigated outside the site if they affect performance.

- ☐ Accept for Review
- ☐ Accept, but Request for Additional Information Prepared
- ☐ Reject, Inadequate to Support Detailed Review

(15) An assessment of the anticipated response of the geomechanical, hydrogeologic, and geochemical systems to the range of design thermal loadings, given the fracture patterns and other discontinuities and the heat transfer properties of the rock mass and water.

- ☐ Accept for Review
- ☐ Accept, but Request for Additional Information Prepared
- ☐ Reject, Inadequate to Support Detailed Review

(16) An assessment of the ability of the proposed repository to limit radiological exposures to the reasonably maximally exposed individual for the period after permanent closure.

- ☐ Accept for Review
- ☐ Accept, but Request for Additional Information Prepared
- ☐ Reject, Inadequate to Support Detailed Review

(17) An assessment of the ability of the proposed geologic repository to limit releases of radionuclides into the accessible environment.

- ☐ Accept for Review
- ☐ Accept, but Request for Additional Information Prepared

□ Reject, Inadequate to Support Detailed Review

(18) An assessment of the ability of the proposed geologic repository to limit radiological exposures to the reasonably maximally exposed individual for the period after permanent closure in the event of human intrusion into the engineered barrier system consistent with requirements at 10 CFR 63.321.

□ Accept for Review

□ Accept, but Request for Additional Information Prepared

□ Reject, Inadequate to Support Detailed Review

(19) An evaluation of the natural features of the geologic setting and design features of the engineered barrier system that are considered barriers important to waste isolation.

□ Accept for Review

□ Accept, but Request for Additional Information Prepared

□ Reject, Inadequate to Support Detailed Review

(20) An explanation of measures used to support models for performance assessments. These models should be supported using an appropriate combination of methods such as field tests *in situ* tests, laboratory tests representative of field conditions, monitoring data, and natural analog studies.

□ Accept for Review

□ Accept, but Request for Additional Information Prepared

□ Reject, Inadequate to Support Detailed Review

(21) An identification of those structures, systems, and components of the geologic repository, both surface and subsurface, that require research and development to confirm the adequacy of design. For structures, systems, and components important to safety and for the engineered and natural barriers important to waste isolation, the license application should provide a detailed description of the programs designed to resolve safety questions. This should include a schedule showing when the U.S. Department of Energy would resolve these questions.

□ Accept for Review

□ Accept, but Request for Additional Information Prepared

□ Reject, Inadequate to Support Detailed Review

(22) A description of the performance confirmation program.

- ☐ Accept for Review
- ☐ Accept, but Request for Additional Information Prepared
- ☐ Reject, Inadequate to Support Detailed Review

(23) An identification and justification for selecting those variables, conditions, or other items that are determined to be probable subjects of license specifications.

- ☐ Accept for Review
- ☐ Accept, but Request for Additional Information Prepared
- ☐ Reject, Inadequate to Support Detailed Review

(24) An explanation of how the U.S. Department of Energy used expert elicitation.

- ☐ Accept for Review
- ☐ Accept, but Request for Additional Information Prepared
- ☐ Reject, Inadequate to Support Detailed Review

(25) A description of the quality assurance program to be applied to the structures, systems, and components important to safety and to the engineered and natural barriers important to waste isolation, including a discussion of how the applicable requirements of 10 CFR 63.142 will be satisfied.

- ☐ Accept for Review
- ☐ Accept, but Request for Additional Information Prepared
- ☐ Reject, Inadequate to Support Detailed Review

(26) A description of the plan for responding to, and recovering from, radiological emergencies that may occur at any time before permanent closure and decontamination or decontamination and dismantlement of surface facilities.

- ☐ Accept for Review
- ☐ Accept, but Request for Additional Information Prepared
- ☐ Reject, Inadequate to Support Detailed Review

(27) The following information concerning activities at the geologic repository operations area, including:

(a) Organizational structure of the U.S. Department of Energy as it pertains to construction and operation of the geologic repository operations area, including a description of any delegations of authority and assignments of responsibilities, whether in the form of regulations, administrative directives, contract provisions, or otherwise;

(b) Identification of key positions that are assigned responsibility for safety at, and operation of, the geologic repository operations area;

(c) Personnel qualifications and training requirements;

(d) Plans for startup activities and startup testing;

(e) Plans for conduct of normal activities, including maintenance, surveillance, and periodic testing of structures, systems, and components of the geologic repository operations area;

(f) Plans for permanent closure and plans for the decontamination or decontamination and dismantlement of surface facilities; and

(g) Plans to use the geologic repository operations area for purposes other than disposal of radioactive wastes. The plans should include an analysis of the effects, if any, that such uses may have on the operation of structures, systems, and components important to safety and the engineered and natural barriers important to waste isolation.

☐ Accept for Review

☐ Accept, but Request for Additional Information Prepared

☐ Reject, Inadequate to Support Detailed Review

(28) A description of the program to be used to maintain records.

☐ Accept for Review

☐ Accept, but Request for Additional Information Prepared

☐ Reject, Inadequate to Support Detailed Review

(29) A description of the controls that the U.S. Department of Energy will apply to restrict access and to regulate land use at the Yucca Mountain site and adjacent areas. This

should include a conceptual design of monuments that would be used to identify the site after permanent closure.

- ☐ Accept for Review
- ☐ Accept, but Request for Additional Information Prepared
- ☐ Reject, Inadequate to Support Detailed Review

NRC FORM 335 (2-89) NRCM 1102, 3201, 3202	U.S. NUCLEAR REGULATORY COMMISSION **BIBLIOGRAPHIC DATA SHEET** *(See instructions on the reverse)*	1. REPORT NUMBER (Assigned by NRC, Add Vol., Supp., Rev., and Addendum Numbers, if any.) NUREG-1804, Revision 2

2. TITLE AND SUBTITLE

Yucca Mountain Review Plan
Final Report

3. DATE REPORT PUBLISHED

MONTH	YEAR
July	2003

4. FIN OR GRANT NUMBER

5. AUTHOR(S)

U.S. Nuclear Regulatory Commission/Center for Nuclear Waste Regulatory Analyses Staffs

6. TYPE OF REPORT

7. PERIOD COVERED *(Inclusive Dates)*

8. PERFORMING ORGANIZATION - NAME AND ADDRESS *(If NRC, provide Division, Office or Region, U.S. Nuclear Regulatory Commission, and mailing address; if contractor, provide name and mailing address.)*

Center for Nuclear Waste Regulatory Analyses
6220 Culebra Road
P.O. Drawer 28510
San Antonio, TX 78228-0510

9. SPONSORING ORGANIZATION - NAME AND ADDRESS *(If NRC, type "Same as above"; if contractor, provide NRC Division, Office or Region, U.S. Nuclear Regulatory Commission, and mailing address.)*

Division of Waste Management
Office of Nuclear Material Safety and Safeguards
U.S. Nuclear Regulatory Commission
Washington, D.C. 20555-0001

10. SUPPLEMENTARY NOTES

J. A. Ciocco, NRC Project Manager

11. ABSTRACT *(200 words or less)*

The Yucca Mountain Review Plan provides guidance for the U.S. Nuclear Regulatory Commission staff to evaluate a U.S. Department of Energy license application for a geologic repository. It is not a regulation and does not impose regulatory requirements. The licensing criteria are contained in the U.S.Code of Federal Regulations (CFR) Title 10, Part 63 (10 CFR Part 63), "Disposal of High-Level Radioactive Wastes in a Proposed Geologic Repository at Yucca Mountain, Nevada." The Secretary of Energy has recommended the Yucca Mountain site to the President for the development of a Yucca Mountain repository. The President has notified Congress that he considers the Yucca Mountain site qualified for application for a construction authorization for a repository. Nevada filed a notice of disapproval of the President's recommendation; however, Congress later approved the site recommendation. The U.S. Department of Energy may now submit a license application to the U.S. Nuclear Regulatory Commission. The principal purpose of the Yucca Mountain Review Plan is to ensure the quality, uniformity, and consistency of U.S. Nuclear Regulatory Commission staff reviews of the license application and any requested amendments. Each section of the Yucca Mountain Review Plan addresses determining compliance with specific regulatory requirements from 10 CFR Part 63. The regulations and the Yucca Mountain Review Plan are risk-informed, performance-based to the extent practical.

12. KEY WORDS/DESCRIPTORS *(List words or phrases that will assist researchers in locating the report.)*

10 CFR Part 63
biosphere
geologic repository
high-level waste
human intrusion
model abstraction
performance assessment
performance objective
postclosure safety
preclosure safety
preclosure safety analysis
representative volume
Yucca Mountain

13. AVAILABILITY STATEMENT

unlimited

14. SECURITY CLASSIFICATION

(This Page) unclassified

(This Report) unclassified

15. NUMBER OF PAGES

16. PRICE

NRC FORM 335 (2-89)

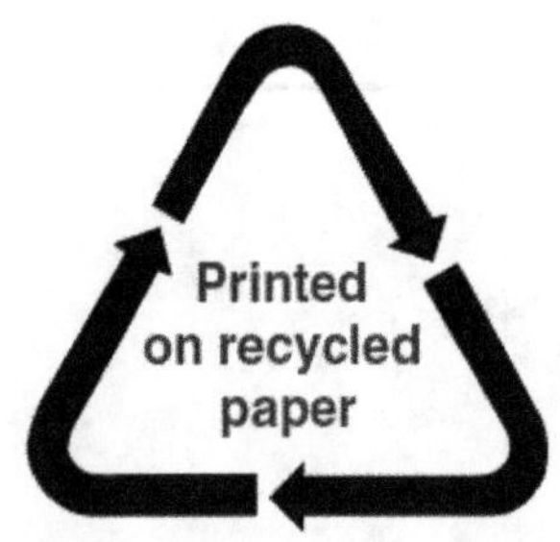
Printed
on recycled
paper

Federal Recycling Program

www.ingramcontent.com/pod-product-compliance
Lightning Source LLC
Chambersburg PA
CBHW080228180526
45167CB00006B/2244

* 9 7 8 1 5 0 0 1 1 3 1 9 3 *